Triumph TR7 Owners Workshop Manual

by J H Haynes
Member of the Guild of Motoring Writers
and Peter Ward

Models covered:
Triumph TR7 1998 cc (122 cu in) Fixed Head and Drophead

ISBN 0 85696 848 X

© Haynes Publishing Group 1977, 1980, 1981, 1983

All rights reserved. No part of this book may be reproduced or transmitted in any form or by any means, electronic or mechanical, including photocopying, recording or by any information storage or retrieval system, without permission in writing from the copyright holder.

Printed in England   (322 1J3)

HAYNES PUBLISHING GROUP
SPARKFORD YEOVIL SOMERSET BA22 7JJ ENGLAND
distributed in the USA by
HAYNES PUBLICATIONS INC
861 LAWRENCE DRIVE
NEWBURY PARK
CALFORNIA 91320
USA

Acknowledgements

Thanks are due to the Rover Triumph division of BL Cars UK Ltd, for the provision of technical information. Castrol Limited supplied the lubrication data and the Champion Sparking Plug Company supplied the illustrations showing the various spark plug conditions. The bodywork repair photographs used in this manual were provided by Holt Lloyd Limited who supply 'Turtle Wax', 'Dupli-color Holts' and other Holts range products.

Our project car was supplied by Wadham-Stringer Limited of Taunton.

Thanks are also due to all those people at Sparkford who helped in the production of this manual. Particularly Brian Horsfall who carried out the mechanical work, Les Brazier who took the photographs, Tim Parker who planned the layout of each page, John Fowler who wrote the Supplementary Chapter 13 and John Rose and David Neilson who edited the text.

About this manual

Its aim

The aim of this manual is to help you get the best value from your car. It can do so in several ways. It can help you decide what work must be done (even should you choose to get it done by a garage), provide information on routine maintenance and servicing, and give a logical course of action and diagnosis when random faults occur. However, it is hoped that you will use the manual by tackling the work yourself. On simpler jobs it may even be quicker than booking the car into a garage and going there twice to leave and collect it. Perhaps most important, a lot of money can be saved by avoiding the costs the garage must charge to cover its labour and overheads.

The manual has drawings and descriptions to show the function of the various components so that their layout can be understood. Then the tasks are described and photographed in a step-by-step sequence so that even a novice can do the work.

Its arrangement

The manual is divided into thirteen Chapters, each covering a logical sub-division of the vehicle. The Chapters are each divided into Sections, numbered with single figures, eg 5; and the Sections into paragraphs (or sub-sections), with decimal numbers following on from the Section they are in, eg 5.1, 5.2, 5.3 etc.

It is freely illustrated, especially in those parts where there is a detailed sequence of operations to be carried out. There are two forms of illustration: figures and photographs. The figures are numbered in sequence with decimal numbers, according to their position in the Chapter — eg Fig. 6.4 is the fourth drawing/illustration in Chapter 6. Photographs carry the same number (either individually or in related groups) as the Section or sub-section to which they relate.

There is an alphabetical index at the back of the manual as well as a contents list at the front. Each Chapter is also preceded by its own individual contents list.

References to the 'left' or 'right' of the vehicle are in the sense of a person in the driver's seat facing forwards.

Unless otherwise stated, nuts and bolts are removed by turning anti-clockwise, and tightened by turning clockwise.

Vehicle manufacturers continually make changes to specifications and recommendations, and these when notified are incorporated into our manuals at the earliest opportunity.

Whilst every care is taken to ensure that the information in this manual is correct, no liability can be accepted by the authors or publishers for loss, damage or injury caused by any errors in, or omissions from, the information given.

Introduction to the Triumph TR7

The TR7 is a fully equipped, closed or drop-head, 2-seater sports car. It uses a version of the well-proven Triumph slant-four, overhead camshaft engine. This engine, when used in the TR7 has a Dolomite Sprint-type block with a Dolomite-type cylinder head (the Dolomite is another Triumph model sold only in the UK).

When the TR7 was introduced to the USA market in 1975, a 4-speed manual gearbox only was available, but shortly after the introduction of the car to the UK market a 5-speed gearbox or the Borg-Warner model 65 automatic transmission were made available as options.

Externally, all models appear to be very similar, but there are many differences between UK and USA versions, mainly associated with the emission control systems. This results in the USA versions being considerably de-tuned, but the car has still proved very popular in a competitive market.

Contents

	Page
Acknowledgements	2
About this manual	2
Introduction to the Triumph TR7	2
Buying spare parts and vehicle identification numbers	5
Tools and working facilities	6
Safety first	8
Routine maintenance	9
Jacking and towing	11
Recommended lubricants and fluids	12
Chapter 1 Engine	13
Chapter 2 Cooling system	35
Chapter 3 Fuel and carburation, exhaust and emission control systems	40
Chapter 4 Ignition system	56
Chapter 5 Clutch	67
Chapter 6 Manual gearbox and automatic transmission	72
Chapter 7 Propeller shaft	93
Chapter 8 Rear axle	95
Chapter 9 Braking system	99
Chapter 10 Electrical system	107
Chapter 11 Suspension and steering	132
Chapter 12 Bodywork and fittings	141
Chapter 13 Supplement: Revisions and information on later models	155
General capacities, dimensions and weights	192
Conversion factors	193
Index	194

The Triumph TR7 Fixedhead Coupe used as the project car for this manual

The Triumph TR7 Drophead Coupe

Buying spare parts and vehicle identification numbers

Buying spare parts

Spare parts are available from many sources, for example: Triumph (Leyland) garages, other garages and accessory shops, and motor factors. Our advice regarding spare parts is as follows:

Officially appointed Triumph garages - This is the best source of parts which are peculiar to your car and otherwise not generally available (eg; complete cylinder heads, internal gearbox components, badges, interior trim etc). It is also the only place at which you should buy parts if your car is still under warranty; non-Triumph components may invalidate the warranty. To be sure of obtaining the correct parts it will always be necessary to give the storeman your car's engine and chassis number, and if possible to take the old part along for positive identification. Remember that many parts are available on a factory exchange scheme - any parts returned should always be clean! It obviously makes good sense to go to the specialists on your car for this type of part for they are best equipped to supply you.

Other garages and accessory shops - These are often very good places to buy material and components needed for the maintenance of your car (eg; oil filters, spark plugs, bulbs, fan belts, oils and grease, touch-up paint, filler paste etc). They also sell general accessories, usually have convenient opening hours, charge lower prices and can often be found not far from home.

Motor factors - Good factors will stock all of the more important components which wear out relatively quickly (eg; clutch components, pistons, valves, exhaust systems, brake cylinders/pipes/hoses/seals/shoes and pads etc). Motor factors will often provide new or reconditioned components on a part exchange basis - this can save a considerable amount of money.

Vehicle identification numbers

Modifications are a continuing and unpublicised process in vehicle manufacture quite apart from major model changes. Spare parts manuals and lists are compiled upon a numerical basis, the individual vehicle Commission Number being essential to correct identification of the component required.

The *Commission Number* is the identification number used for registration purposes; this appears on a plate on the rear edge of the driver's door on early models. It also carries the *paint and trim code* numbers (photo). On later models, the VIN is attached to the front right suspension tower.

The *engine number* is stamped on the cylinder block and can be seen by looking down, between the carburettors.

Engine number (the engine is out of the car in this photo)

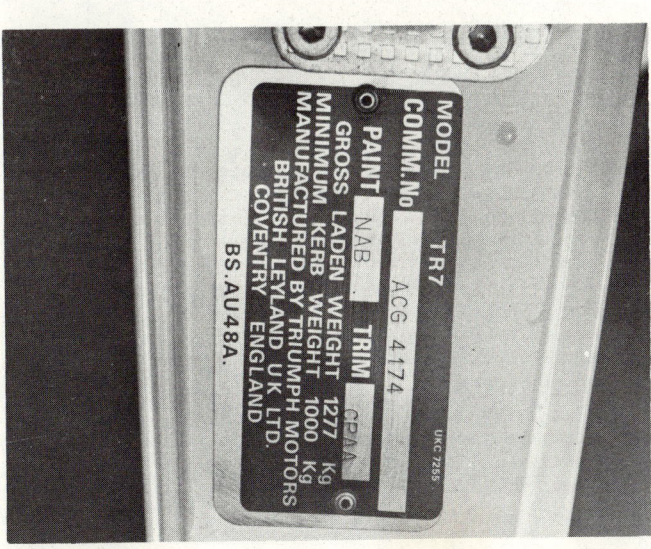

Commission number

Tools and working facilities

Introduction

A selection of good tools is a fundamental requirement for anyone contemplating the maintenance and repair of a motor vehicle. For the owner who does not possess any, their purchase will prove a considerable expense, offsetting some of the savings made by doing-it-yourself. However, provided that the tools purchased are of good quality, they will last for many years and prove an extremely worthwhile investment.

To help the average owner to decide which tools are needed to carry out the various tasks detailed in this manual, we have compiled three lists of tools under the following headings: Maintenance and minor repair, Repair and overhaul, and Special. The newcomer to practical mechanics should start off with the 'Maintenance and minor repair' tool kit and confine himself to the simpler jobs around the vehicle. Then, as his confidence and experience grows, he can undertake more difficult tasks, buying extra tools as, and when, they are needed. In this way, a 'Maintenance and minor repair' tool kit can be built-up into a 'Repair and overhaul' tool kit over a considerable period of time without any major cash outlays. The experienced do-it-yourselfer will have a tool kit good enough for most repair and overhaul procedures and will add tools from the 'Special' category when he feels the expense is justified by the amount of use these tools will be put to.

It is obviously not possible to cover the subject of tools fully here. For those who wish to learn more about tools and their use there is a book entitled 'How to Choose and Use Car Tools' available from the publishers of this manual.

Maintenance and minor repair tool kit

The tools given in this list should be considered as a minimum requirement if routine maintenance, servicing and minor repair operations are to be undertaken. We recommend the purchase of combination spanners (ring one end, open-ended the other); although more expensive than open-ended ones, they do give the advantages of both types of spanner.

Combination spanners - 7/16, ½, 9/16, 5/8, 11/16, ¾, 13/16, 15/16in
Combination spanners - 10, 13, 17, 19, 22mm
Adjustable spanner - 9 inch
Engine sump/gearbox/rear axle drain plug key (where applicable)
Spark plug spanner (with rubber insert)
Spark plug gap adjustment tool
Set of feeler gauges
Brake adjuster spanner (where applicable)
Brake bleed nipple spanner
Screwdriver - 4 in. long x ¼ in. dia. (plain)
Screwdriver - 4 in. long x ¼ in. dia. (crosshead)
Combination pliers - 6 inch
Hacksaw, junior
Tyre pump
Tyre pressure gauge
Grease gun (where applicable)
Oil can
Fine emery cloth (1 sheet)
Wire brush (small)
Funnel (medium size)

Repair and overhaul tool kit

These tools are virtually essential for anyone undertaking any major repairs to a motor vehicle, and are additional to those given in the Basic list. Included in this list is a comprehensive set of sockets. Although these are expensive they will be found invaluable as they are so versatile - particularly if various drives are included in the set. We recommend the ½ square-drive type, as this can be used with most proprietary torque wrenches. If you cannot afford a socket set, even bought piecemeal, then inexpensive tubular box spanners are a useful alternative.

The tools in this list will occasionally need to be supplemented by tools from the Special list.

Sockets (or box spanners) to cover range in previous list
Reversible ratchet drive (for use with sockets)
Extension piece, 10 inch (for use with sockets)
Universal joint (for use with sockets)
Torque wrench (for use with sockets)
'Mole' wrench - 8 inch
Ball pein hammer
Soft-faced hammer, plastic or rubber
Screwdriver - 6 in. long x 5/16 in. dia. (plain)
Screwdriver - 2 in. long x 5/16in. square (plain)
Screwdriver - 1½ in. long x ¼ in. dia. (crosshead)
Screwdriver - 3 in. long x 1/8 in. dia. (electricians)
Pliers - electricians side cutters
Pliers - needle nosed
Pliers - circlip (internal and external)
Cold chisel - ½ inch
Scriber (this can be made by grinding the end of a broken hacksaw blade)
Scraper (this can be made by flattening and sharpening one end of a piece of copper pipe)
Centre punch
Pin punch
Hacksaw
Valve grinding tool
Steel rule/straightedge
Allen keys
Selection of files
Wire brush (large)
Axle stands
Jack (strong scissor or hydraulic type)

Special tools

The tools in this list are those which are not used regularly, are expensive to buy, or which need to be used in accordance with their manufacturers instructions. Unless relatively difficult mechanical jobs are undertaken frequently, it will not be economic to buy many of these tools. Where this is the case, you could consider clubbing together with friends (or a motorists club) to make a joint purchase, or borrowing the tools against a deposit from a local garage or tool hire specialist.

The following list contains only those tools and instruments freely available to the public, and not those special tools produced by the vehicle manufacturer specifically for its dealer network. You will find occasional references to these manufacturers special tools in the text of this manual. Generally, an alternative method of doing the job without the vehicle manufacturers special tool is given. However, sometimes, there is no alternative to using them. Where this is the case and the relevant tool cannot be bought or borrowed you will have to entrust the work to a franchised garage.

Valve spring compressor
Piston ring compressor
Ball joint separator
Universal hub/bearing puller
Impact screwdriver
Micrometer and/or vernier gauge
Carburettor flow balancing device (where applicable)
Dial gauge
Stroboscopic timing light
Dwell angle meter/tachometer
Universal electrical multi-meter
Cylinder compression gauge
Lifting tackle
Trolley jack
Light with extension lead

Buying tools

For practically all tools, a tool factor is the best source since he will have a very comprehensive range compared with the average garage or accessory shop. Having said that, accessory shops often offer excellent quality tools at discount prices, so it pays to shop around.

Remember, you don't have to buy the most expensive items on the shelf, but it is always advisable to steer clear of the very cheap tools. There are plenty of good tools around, at reasonable prices, so ask the proprietor or manager of the shop for advice before making a purchase.

Tools and working facilities

Care and maintenance of tools

Having purchased a reasonable tool kit, it is necessary to keep the tools in a clean and serviceable condition. After use, always wipe off any dirt, grease and metal particles using a clean, dry cloth, before putting the tools away. Never leave them lying around after they have been used. A simple tool rack on the garage or workshop wall, for items such as screwdrivers and pliers is a good idea. Store all normal spanners and sockets in a metal box. Any measuring instruments, gauges, meters, etc., must be carefully stored where they cannot be damaged or become rusty.

Take a little care when the tools are used. Hammer heads inevitably become marked and screwdrivers lose the keen edge on their blades from time-to-time. A little timely attention with emery cloth or a file will soon restore items like this to a good serviceable finish.

Working facilities

Not to be forgotten when discussing tools, is the workshop itself. If anything more than routine maintenance is to be carried out, some form of suitable working area becomes essential.

It is appreciated that many an owner mechanic is forced by circumstance to remove an engine or similar item, without the benefit of a garage or workshop. Having done this, any repairs should always be done under the cover of a roof.

Wherever possible, any dismantling should be done on a clean flat workbench or table at a suitable working height.

Any workbench needs a vice: one with a jaw opening of 4 in. (100 mm) is suitable for most jobs. As mentioned previously, some clean dry storage space is also required for tools, as well as the lubricants, cleaning fluids, touch-up paints and so on which soon become necessary.

Another item which may be required, and which has a much more general usage, is an electric drill with a chuck capacity of at least 5/16 in. (8mm). This, together with a good range of twist drills, is virtually essential for fitting accessories such as wing mirrors and reversing lights.

Last, but not least, always keep a supply of old newspapers and clean, lint-free rags available, and try to keep any working area as clean as possible.

Spanner jaw gap comparison table

Jaw gap (in.)	Spanner size
0.250	¼ in. AF
0.275	7 mm AF
0.312	5/16 in. AF
0.315	8 mm AF
0.343	11/32 in. AF/1/8 in. Whitworth
0.354	9 mm AF
0.375	3/8 in. AF
0.393	10 mm AF
0.433	11 mm AF
0.437	7/16 in. AF
0.445	3/16 in. Whitworth/¼ in. BSF
0.472	12 mm AF
0.500	½ in. AF
0.512	13 mm AF
0.525	¼ in. Whitworth/5/16 in. BSF
0.551	14 mm AF
0.562	9/16 in. AF
0.590	15 mm AF
0.600	5/16 in. Whitworth/3/8 in. BSF
0.625	5/8 in. AF
0.629	16 mm AF
0.669	17 mm AF
0.687	11/16 in. AF
0.708	18 mm AF
0.710	3/8 in. Whitworth/7/16 in. BSF
0.748	19 mm AF
0.750	¾ in. AF
0.812	13/16 in. AF
0.820	7/16 in. Whitworth/½ in. BSF
0.866	22 mm AF
0.875	7/8 in. AF
0.920	½ in. Whitworth/9/16 in. BSF
0.937	15/16 in. AF
0.944	24 mm AF
1.000	1 in. AF
1.010	9/16 in. Whitworth/5/8 in. BSF
1.023	26 mm AF
1.062	1 1/16 in. AF/27 mm AF
1.100	5/8 in. Whitworth/11/16 in. BSF
1.125	1 1/8 in. AF
1.181	30 mm AF
1.200	11/16 in. Whitworth/¾ in. BSF
1.250	1 ¼ in. AF
1.259	32 mm AF
1.300	¾ in. Whitworth/7/8 in. BSF
1.312	1 5/16 in. AF
1.390	13/16 in. Whitworth/15/16 in. BSF
1.417	36 mm AF
1.437	1 7/16 in. AF
1.480	7/8 in. Whitworth/1 in. BSF
1.500	1 ½ in. AF
1.574	40 mm AF/15/16 in. Whitworth
1.614	41 mm AF
1.625	1 5/8 in. AF
1.670	1 in. Whitworth/1 1/8 in. BSF
1.687	1 11/16 in. AF
1.811	46 mm AF
1.812	1 13/16 in. AF
1.860	1 1/8 in. Whitworth/1 ¼ in. BSF
1.875	1 7/8 in. AF
1.968	50 mm AF
2.000	2 in. AF
2.050	1 ¼ in. Whitworth/1 3/8 in. BSF
2.165	55 mm AF
2.362	60 mm AF

A Haltrac hoist and gantry in use during a typical engine removal sequence

Safety first!

Professional motor mechanics are trained in safe working procedures. However enthusiastic you may be about getting on with the job in hand, do take the time to ensure that your safety is not put at risk. A moment's lack of attention can result in an accident, as can failure to observe certain elementary precautions.

There will always be new ways of having accidents, and the following points do not pretend to be a comprehensive list of all dangers; they are intended rather to make you aware of the risks and to encourage a safety-conscious approach to all work you carry out on your vehicle.

Essential DOs and DON'Ts

DON'T rely on a single jack when working underneath the vehicle. Always use reliable additional means of support, such as axle stands, securely placed under a part of the vehicle that you know will not give way.
DON'T attempt to loosen or tighten high-torque nuts (e.g. wheel hub nuts) while the vehicle is on a jack; it may be pulled off.
DON'T start the engine without first ascertaining that the transmission is in neutral (or 'Park' where applicable) and the parking brake applied.
DON'T suddenly remove the filler cap from a hot cooling system – cover it with a cloth and release the pressure gradually first, or you may get scalded by escaping coolant.
DON'T attempt to drain oil until you are sure it has cooled sufficiently to avoid scalding you.
DON'T grasp any part of the engine, exhaust or catalytic converter without first ascertaining that it is sufficiently cool to avoid burning you.
DON'T allow brake fluid or antifreeze to contact vehicle paintwork.
DON'T syphon toxic liquids such as fuel, brake fluid or antifreeze by mouth, or allow them to remain on your skin.
DON'T inhale dust – it may be injurious to health (see *Asbestos* below).
DON'T allow any spilt oil or grease to remain on the floor – wipe it up straight away, before someone slips on it.
DON'T use ill-fitting spanners or other tools which may slip and cause injury.
DON'T attempt to lift a heavy component which may be beyond your capability – get assistance.
DON'T rush to finish a job, or take unverified short cuts.
DON'T allow children or animals in or around an unattended vehicle.
DO wear eye protection when using power tools such as drill, sander, bench grinder etc, and when working under the vehicle.
DO use a barrier cream on your hands prior to undertaking dirty jobs – it will protect your skin from infection as well as making the dirt easier to remove afterwards; but make sure your hands aren't left slippery.
DO keep loose clothing (cuffs, tie etc) and long hair well out of the way of moving mechanical parts.
DO remove rings, wristwatch etc, before working on the vehicle – especially the electrical system.
DO ensure that any lifting tackle used has a safe working load rating adequate for the job.
DO keep your work area tidy – it is only too easy to fall over articles left lying around.
DO get someone to check periodically that all is well, when working alone on the vehicle.
DO carry out work in a logical sequence and check that everything is correctly assembled and tightened afterwards.
DO remember that your vehicle's safety affects that of yourself and others. If in doubt on any point, get specialist advice.
IF, in spite of following these precautions, you are unfortunate enough to injure yourself, seek medical attention as soon as possible.

Asbestos

Certain friction, insulating, sealing, and other products – such as brake linings, brake bands, clutch linings, torque converters, gaskets, etc – contain asbestos. *Extreme care must be taken to avoid inhalation of dust from such products since it is hazardous to health.* If in doubt, assume that they *do* contain asbestos.

Fire

Remember at all times that petrol (gasoline) is highly flammable. Never smoke, or have any kind of naked flame around, when working on the vehicle. But the risk does not end there – a spark caused by an electrical short-circuit, by two metal surfaces contacting each other, by careless use of tools, or even by static electricity built up in your body under certain conditions, can ignite petrol vapour, which in a confined space is highly explosive.

Always disconnect the battery earth (ground) terminal before working on any part of the fuel or electrical system, and never risk spilling fuel on to a hot engine or exhaust.

It is recommended that a fire extinguisher of a type suitable for fuel and electrical fires is kept handy in the garage or workplace at all times. Never try to extinguish a fuel or electrical fire with water.

Fumes

Certain fumes are highly toxic and can quickly cause unconsciousness and even death if inhaled to any extent. Petrol (gasoline) vapour comes into this category, as do the vapours from certain solvents such as trichloroethylene. Any draining or pouring of such volatile fluids should be done in a well ventilated area.

When using cleaning fluids and solvents, read the instructions carefully. Never use materials from unmarked containers – they may give off poisonous vapours.

Never run the engine of a motor vehicle in an enclosed space such as a garage. Exhaust fumes contain carbon monoxide which is extremely poisonous; if you need to run the engine, always do so in the open air or at least have the rear of the vehicle outside the workplace.

If you are fortunate enough to have the use of an inspection pit, never drain or pour petrol, and never run the engine, while the vehicle is standing over it; the fumes, being heavier than air, will concentrate in the pit with possibly lethal results.

The battery

Never cause a spark, or allow a naked light, near the vehicle's battery. It will normally be giving off a certain amount of hydrogen gas, which is highly explosive.

Always disconnect the battery earth (ground) terminal before working on the fuel or electrical systems.

If possible, loosen the filler plugs or cover when charging the battery from an external source. Do not charge at an excessive rate or the battery may burst.

Take care when topping up and when carrying the battery. The acid electrolyte, even when diluted, is very corrosive and should not be allowed to contact the eyes or skin.

If you ever need to prepare electrolyte yourself, always add the acid slowly to the water, and never the other way round. Protect against splashes by wearing rubber gloves and goggles.

When jump starting a car using a booster battery, for negative earth (ground) vehicles, connect the jump leads in the following sequence: First connect one jump lead between the positive (+) terminals of the two batteries. Then connect the other jump lead first to the negative (–) terminal of the booster battery, and then to a good earthing (ground) point on the vehicle to be started, at least 18 in (45 cm) from the battery if possible. Ensure that hands and jump leads are clear of any moving parts, and that the two vehicles do not touch. Disconnect the leads in the reverse order.

Mains electricity

When using an electric power tool, inspection light etc, which works from the mains, always ensure that the appliance is correctly connected to its plug and that, where necessary, it is properly earthed (grounded). Do not use such appliances in damp conditions and, again, beware of creating a spark or applying excessive heat in the vicinity of fuel or fuel vapour.

Ignition HT voltage

A severe electric shock can result from touching certain parts of the ignition system, such as the HT leads, when the engine is running or being cranked, particularly if components are damp or the insulation is defective. Where an electronic ignition system is fitted, the HT voltage is much higher and could prove fatal.

Routine maintenance

Maintenance is essential for ensuring safety and desirable for the purpose of getting the best in terms of performance and economy from the car. Over the years the need for periodic lubrication - oiling, greasing and so on - has been drastically reduced if not totally eliminated. This has unfortunately tended to lead some owners to think that because no such action is required the items either no longer exist or will last for ever. This is a serious delusion. It follows therefore that the largest initial element of maintenance is visual examination. This may lead to repairs or renewals.

Every 250 miles (400 km) or weekly

Check tyre pressures and inflate if necessary.
Check engine oil level and top-up if necessary (photos).
Check battery electrolyte level and top-up if necessary.
Check screen washer operation and fluid level. Top-up if necessary.
Check coolant level in reservoir and top-up if necessary (photo).
Check brake fluid level and top-up if necessary (photo).
Check clutch fluid level and top-up if necessary (photo).
Check operation of all lights, instruments and controls.

Every 3,000 miles (5,000 km)

Check for oil leaks from engine, transmission, steering box and rear axle.
Check condition of cooling system hoses.
Check condition and tension of all drivebelts.
Check condition of clutch system hydraulic pipes (as applicable).
Check shock absorbers for fluid leaks.
Check condition of steering/suspension joints and gaiters.
Check condition of brake system hydraulic pipes and hoses.
Check condition of brake discs and pads.
Check condition of brake servo hose.
Check headlamp alignment and adjust if necessary.
Check windscreen wiper blades and renew if necessary.
Check condition of pipes and hoses in fuel system.
Check exhaust system for leaks and security.
Check condition of tyre treads and sidewalls.
Check roadwheel nuts for tightness.
Check condition and security of seats, seatbelts and anchor points.
Check operation of seatbelt warning system.
Check cooling system for leaks.
Check footbrake operation.
Check operation of the brake system warning light (USA only).

Every 6,000 miles (10,000 km)

Renew engine oil filter (photo).
Renew engine oil (photo).
Top-up the carburettor piston damper oil level, using engine oil, to ½ in (13 mm) above the top of the hollow piston rod (except USA models).
Lubricate carburettor linkages and pivots with engine oil or light lubricating oil.
Check engine idle speed and mixture settings (except USA models).
Clean/adjust spark plugs (except USA models).
Clean, adjust or renew distributor contact breaker points (except USA models).
Lubricate distributor, as described in Chapter 4 (except USA models).
Check and adjust ignition timing (except USA models).
Check gearbox oil level and top-up if necessary (photo).
Check automatic transmission oil level and top-up if necessary.
Lubricate exposed linkage of automatic transmission selector linkage (as applicable).
Remove dust and mud from slots and screen on underside of torque converter housing (as applicable).

Check rear axle oil level and top-up if necessary (photo).
Fit a grease nipple in place of the centre plug of the steering rack damper plug and apply five strokes from a grease gun (photo).
Check wheel balance.
Lubricate the handbrake linkage, cable guides and pivots.
Check the operation of all door and boot locks and catches.
Check the output of the charging system (USA only).
Check front wheel alignment (toe-in).
Check handbrake operation and adjust if necessary.
Check condition of brake linings and drums.
Clean battery terminals and smear with petroleum jelly.
Lubricate all locks, hinges and pivots (except steering lock).
Lubricate the headlamp lift mechanism.

Every 12,000 miles (20,000 km) - UK models, or 12,500 miles - USA models

Renew carburettor air cleaner element.
Clean fuel pump filter and sediment bowl.
Check engine idle speed and mixture settings (USA models).
Unscrew the carburettor piston damper(s) and remove the air cleaner. Top-up the oil level using engine oil as described in Chapter 3, Section 10, paragraph 16 or Section 12, paragraph 8. (USA models).
Check/adjust carburettor deceleration bypass valve(s) (USA models).
Check security of EGR valve lines (USA models).
Check operation of EGR system (USA models).*
Check/adjust manual choke (USA models).
Check air intake temperature control system (some USA models).
Check crankcase breather and evaporative loss system hose for condition and security (USA models).
Lubricate distributor, as described in Chapter 4 (USA models).
Check/adjust ignition timing (USA models).
Check operation of distributor vacuum unit (USA models).
Renew spark plugs.
Check tightness of propeller shaft coupling bolts.
Lubricate clutch pedal pivots (as applicable).
Check security of suspension fixings.
Adjust front hub bearing endfloat.
Lubricate brake pedal pivots.
Check condition of fuel filler cap seal (USA models).
Check the operation of the window controls.
Check the air injection system pipes and hoses (USA only).
Check the air conditioning compressor fluid level (USA only).
Check the condition of the ignition wiring.
Clean and check the distributor cap.
Check the ignition coil on an oscilloscope (USA only).
This should be checked by your Triumph dealer, and the 12,500 mile EGR indicator reset using a special key.

Every 24,000 miles (40,000 km)

Drain, flush and refill cooling system with new antifreeze solution.*
At 2-yearly intervals if this occurs first.

Every 36,000 miles (60,000 km) or 3 years, whichever occurs first

Renew brake servo filter.
Renew all rubber seals and hoses in the braking system.
Renew the brake fluid.

Every 48,000 miles (80,000 km)

Renew absorption canister (USA models).
Renew the catalytic converter (USA only).

Engine oil filler cap

Engine oil dipstick tube on UK models. USA models have a short dipstick below the air cleaner and adjacent to the fuel pump

Coolant reservoir

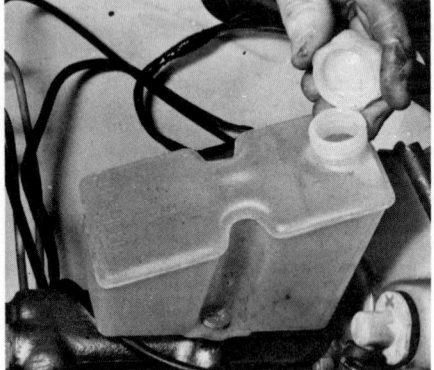
Brake fluid reservoir (right-hand drive shown)

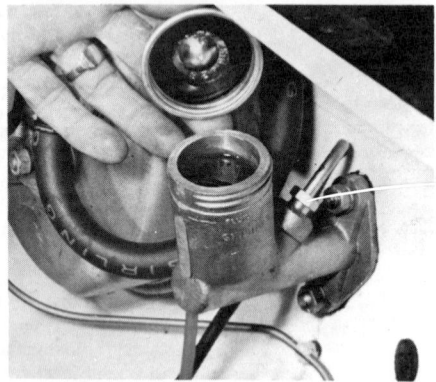
Clutch fluid reservoir (right-hand drive shown)

Engine oil filter being removed

Engine oil drain plug

Gearbox oil filler plug (5-speed - exhaust pipe removed for clarity)

Rear axle oil filler - some models have this on the rear of the differential housing

Steering rack showing centre plug which is removed for fitting a grease nipple (engine removed in this photograph)

Jacking point

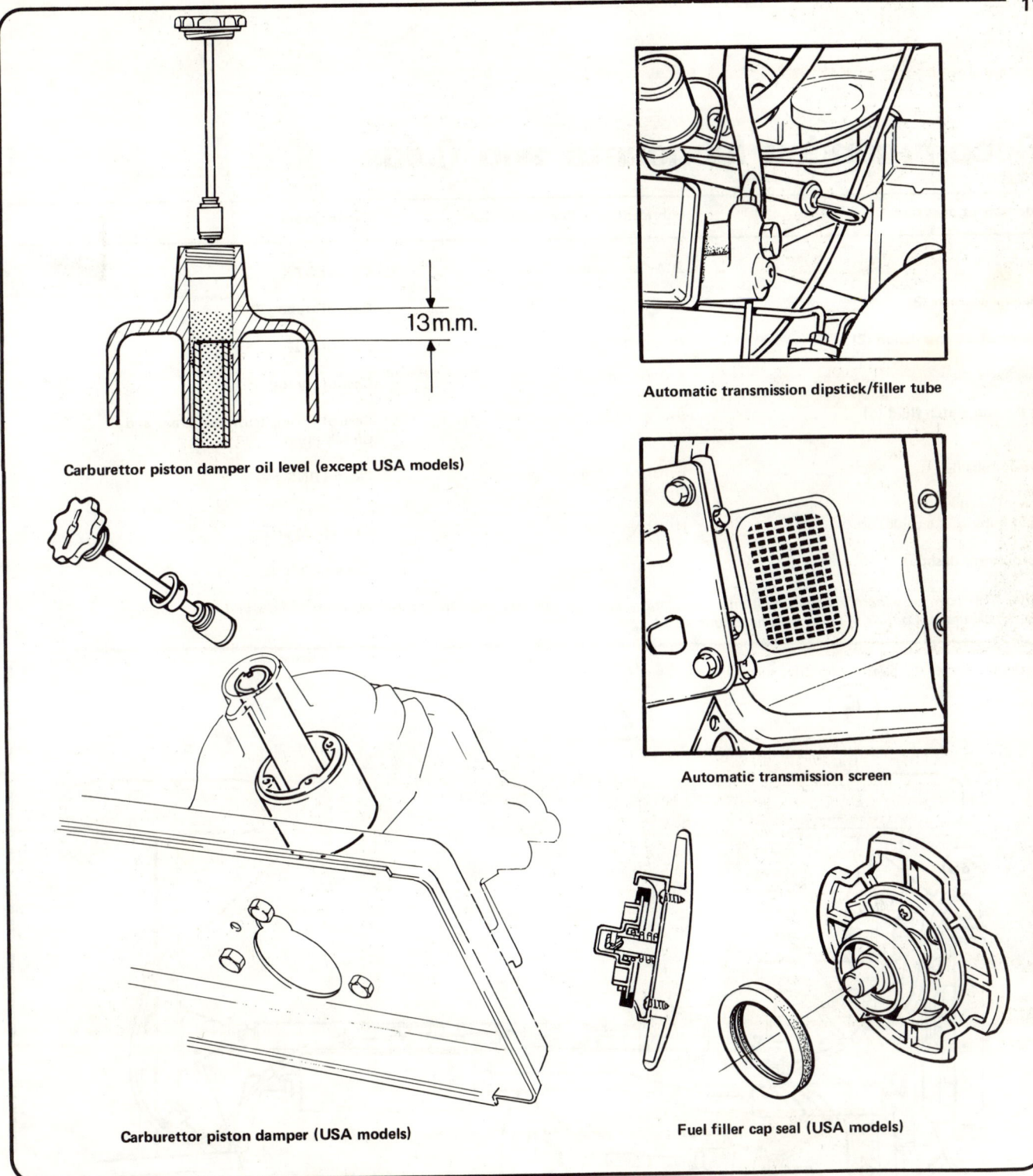

Carburettor piston damper oil level (except USA models)

Automatic transmission dipstick/filler tube

Automatic transmission screen

Carburettor piston damper (USA models)

Fuel filler cap seal (USA models)

Jacking and towing

The jack supplied with the car should be used to raise the car for changing a roadwheel. One jacking point is forward of the rear wheel and the other is to the rear of the front wheel; this applies to both sides of the car.

If the car is to be raised for maintenance/repair operations it is permissible to jack-up on the suspension subframe, rear axle or bodyframe sidemembers. Place a suitable wooden packing piece on the jack head to prevent damage to the car metalwork.

Towing eyes are provided at the front of the sidemembers, behind the shock absorbers. When on tow, ropes should be attached to either bracket, but not both. Note that vehicles with automatic transmission must not be towed for more than 20 miles (32 km), or at a speed in excess of 30 mph (48 km/hour), and this only if the transmission is undamaged. If any of these conditions are not complied with, either the propeller shaft must be removed or the car can be towed with the rear wheels lifted.

When towing a caravan or trailer, conform to the recommendations made by both the manufacturer of the caravan or trailer, and to the regulations existing in the territory concerned.

Recommended lubricants and fluids

Component or system	Lubricant type or specification	Castrol product
Engine (1)	20W/50 engine oil	Castrol GTX
Manual gearbox (2)	SAE 75 gear oil	Castrol Hypoy 75B
Automatic transmission (2)	Automatic transmission fluid	Castrol TQF
Rear axle (3)	SAE 90 EP gear oil	Castrol Hypoy
Brake and clutch fluid (4)	SAE J1703	Castrol Girling Universal Brake and Clutch Fluid
Front hubs (5)	NLGI No 2 high temperature grease	Castrol BNS grease
Rear hubs, brake cables etc	NLGI No 2 grease	Castrol LM grease
Carburettor dashpot	20W/50 engine oil	Castrol GTX

Note: *The above are general recommendations. Lubrication requirements vary from territory-to-territory and depend on the vehicle usage. Consult the operators handbook supplied with the car.*

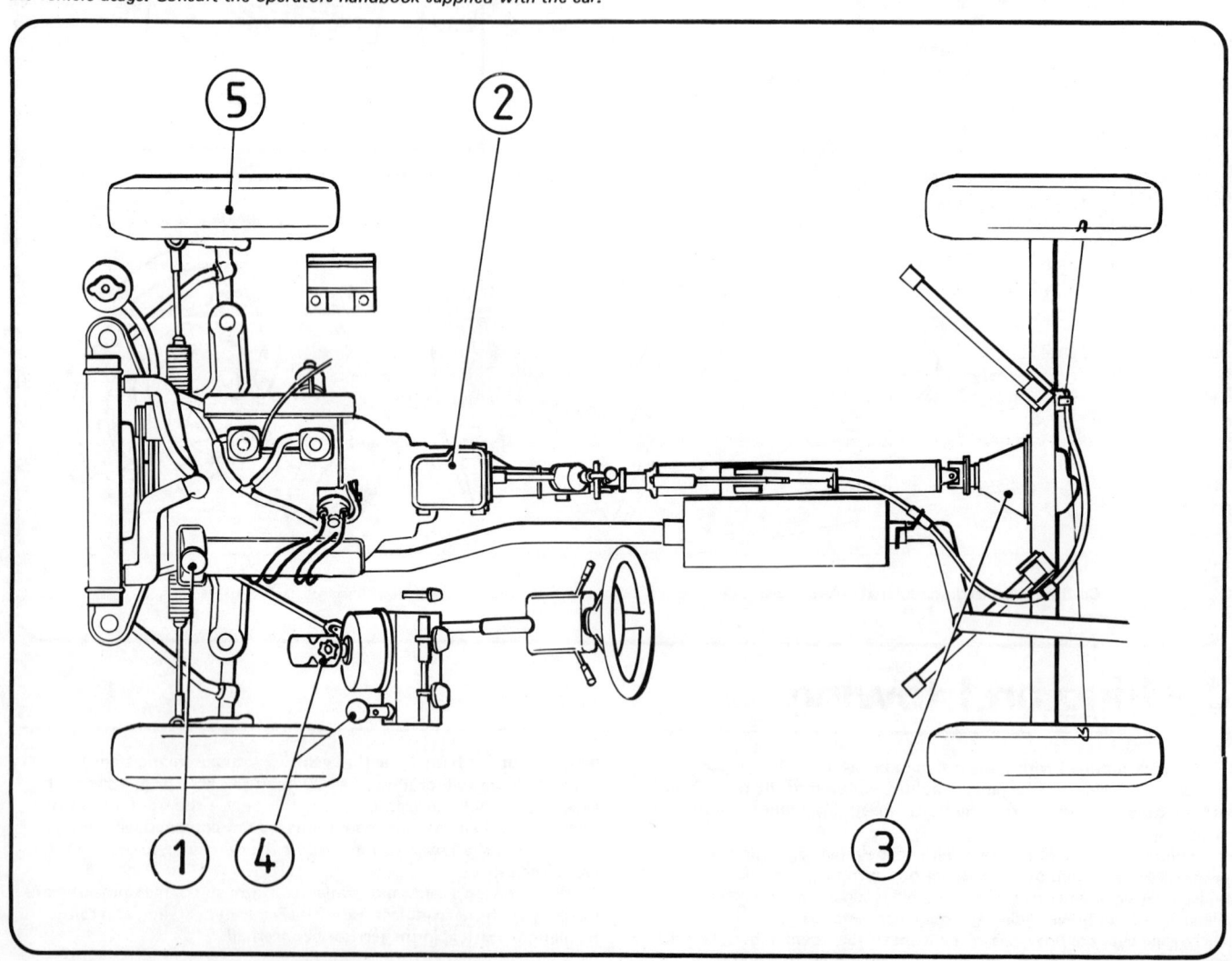

Chapter 1 Engine

For modifications, and information applicable to later models, see Supplement at end of manual

Contents

Big-end main bearing shells - examination and renovation	21
Camshaft removal	8
Components - cleaning, examination and storage - general note	19
Connecting rods to crankshaft - reassembly	39
Crankshaft - examination and renovation	20
Crankshaft, main bearings and rear oil seal - removal	14
Crankshaft - installation	35
Crankshaft pulley, clutch, flywheel and rear adaptor plate - removal	9
Crankshaft rear oil seal, adaptor plate, flywheel, oil strainer and sump - refitting	40
Cylinder bores - examination and renovation	22
Cylinder head and pistons - decarbonisation	32
Cylinder head - refitting	42
Cylinder head - reassembly (including camshaft fitment and valve clearance checking)	34
Cylinder head - removal (engine in the car)	7
Cylinder head - removal (engine out of the car)	6
Distributor and fuel pump - refitting	45
Engine (and gearbox) - removal	4
Engine - installation	48
Engine - final reassembly	47
Engine mountings and stabilizer - removal and installation	49
Engine reassembly - general note	33
Fault diagnosis - engine	50
Flywheel and starter ring - examination and renovation	29
General description	1
Gudgeon pin and small-end bearings - examination and renovation	24
Jackshaft, camshaft and camshaft bearings - examination and renovation	25
Major operations with the engine in the car	2
Major operations with the engine removed from the car	3
Miscellaneous components - examination and renovation	31
Oil pump - dismantling	18
Oil pump - examination and renovation	29
Oil transfer adaptor, oil filter and oil pump - removal	12
Oil transfer adaptor, oil pump and oil filter - reassembly and refitting	44
Pistons and connecting rods - dismantling	17
Pistons and connecting rods - reassembly	36
Pistons and connecting rods - refitting	38
Pistons and piston rings - examination and renovation	23
Piston rings - refitting	37
Separating the engine from the gearbox	5
Sump, pistons, connecting rods and big-end bearings - removal	13
Tappets and pallet shims - examination and renovation	28
Timing chain, sprockets and chain tensioner - examination and renovation	27
Timing cover, chain, sprockets, tensioner, guides and jackshaft - refitting	41
Timing cover, timing gear and jackshaft - removal with engine in place in the vehicle	11
Timing cover, timing gear and jackshaft - removal with engine out of the vehicle	10
Timing gear - reconnecting to camshaft	43
Valves, valve seats and guides - removal	16
Valves, valve seats and valve guides - examination and renovation	26
Water pump, inlet manifold and thermostat - installation	46
Water transfer housing - removal	15

Specifications

General

Type	Four cylinder in-line, overhead camshaft, cylinder inclined at 40 degrees to the vertical. (45° later models). Water cooled
Capacity	122 in^3 (1998 cm^3)
Bore	3.56 in (90.3 mm)
Stroke	3.07 in (78 mm)
Oil pump type	High capacity rotor
Oil light extinguishes at	3 to 5 lb/in^2 (0.21 to 0.35 kg/cm^2) pressure
Oil filter	Full flow type, replaceable element
Engine idle speed	See Chapter 3 Specifications
Fast idle speed	See Chapter 3 Specifications
Firing order	1 3 4 2 (number 1 cylinder at front)
Compression ratio	9.25 : 1 (UK models) 8.0 : 1 (USA models)
Valve clearance	
inlet	0.008 in (0.20 mm)
exhaust	0.018 in (0.457 mm)
Method of adjustment	Pallet shims between valve and tappet

Ignition timing, static	See Chapter 3 Specifications
Valve timing	
inlet	Opens at 16° B.T.D.C.
	Closes at 56° A.B.D.C.
exhaust	Opens at 56° B.B.D.C.
	Closes at 16° A.T.D.C.
Timing mark	
camshaft	Camshaft flange and bearing cap
crankshaft	Pulley and scale on timing cover

Jackshaft
Journal diameter	1.456 to 1.4565 in (36.982 to 36.995 mm)

Camshaft
Journal diameter	1.23 to 1.235 in (31.242 to 31.38 mm)

Connecting rods
Length between centres	5.123 to 5.127 in (130.12 to 130.23 mm)
Small end bush inside diameter	0.9377 to 0.9380 in (23.818 to 23.825 mm)

Gudgeon pins
Diameter	0.9374 to 0.9376 in (23.811 to 23.815 mm)
Clearance in connecting rod	0.0001 to 0.0006 in (0.002 to 0.015 mm)
Clearance in piston	Zero to 0.0004 in (Zero to 0.010 mm)

Pistons
Skirt clearance at right angles to gudgeon pin	0.0005 to 0.0015 in (0.013 to 0.039 mm)
Number of rings	3 (1 oil control, 2 compression)
Ring groove width, top	0.0705 to 0.0713 in (1.790 to 1.810 mm)
second	0.0700 to 0.0709 in (1.780 to 1.800 mm)
oil control	0.1579 to 0.1587 in (4.010 to 4.030 mm)

Piston rings (top ring has chrome periphery, second ring has TOP marked on upper face)
Compression	
Fitted gap (top and second)	0.015 to 0.025 in (0.40 to 0.65 mm)
Ring groove clearance, top	0.0019 to 0.0039 in (0.048 to 0.099 mm)
second	0.0015 to 0.0025 in (0.040 to 0.065 mm)
Oil control	
Gap in bore	0.015 to 0.055 in (0.40 to 1.40 mm)
Fitted gap	Parts to butt

Crankshaft
Journal diameter	2.126 to 2.1265 in (54 to 54.013 mm)
Minimum regrind diameter	2.086 to 2.0865 in (52.984 to 52.997 mm)
Crankpin diameter	1.75 to 1.7505 in (43.45 to 44.463 mm)
Minimum regrind diameter	1.71 to 1.7105 in (43.434 to 43.447 mm)
End thrust	Taken on thrust washers on centre main bearing
Crankshaft end float	0.003 to 0.011 in (0.07 to 0.28 mm)
Adjustment	Selective thrust washers

Main bearings
Number of bearings	5
Diametral clearance	0.0012 to 0.0022 in (0.030 to 0.058 mm)
Undersizes	0.010 in, 0.020 in, 0.030 in, 0.040 in
	(0.254 mm, 0.508 mm, 0.762 mm, 1.016 mm)

Big-end bearings
Diametral clearance	0.0008 to 0.0023 in (0.020 to 0.058 mm)
End-float on crankpin	0.006 to 0.013 in (0.15 to 0.33 mm)
Undersizes	0.010 in, 0.020 in, 0.030 in, 0.040 in
	(0.254 mm, 0.508 mm, 0.762 mm, 1.016 mm)

Valve seats
Outside diameter	
inlet	1.6695 to 1.6707 in (42.405 to 42.430 mm)
exhaust	1.3335 to 1.3345 in (33.870 to 33.896 mm)
Seat angle, inlet and exhaust	44½° (later models 42½°)

Valves
Seat angle, inlet and exhaust	45°
Head diameter	
inlet	1.560 in (39.62 mm)
exhaust	1.28 in (32.5 mm)
Stem diameter	
inlet	0.3107 to 0.3113 in (7.881 to 7.907 mm)
exhaust	0.31 to 0.3105 in (7.87 to 7.89 mm)

Stem/guide clearance
- inlet ... 0.0017 to 0.0023 in (0.043 to 0.058 mm)
- exhaust ... 0.0014 to 0.003 in (0.035 to 0.076 mm)

Valve springs
Free length ... 1.600 in (40.0 mm)

Oil pump
- Outer ring end float ... 0.004 in (0.1 mm)
- Inner ring end float ... 0.004 in (0.1 mm)
- Outer ring/body diametral clearance ... 0.008 in (0.2 mm)
- Rotor lobe clearance ... 0.01 in (0.25 mm)
- Spring, free length ... 1.700 in (43.18 mm)

Timing chain
- Type ... Single row
- Chain wheel alignment ... Alignment of crankshaft and jackshaft sprocket faces
- Adjustment ... Selective shims behind crankshaft sprocket
- Tensioner spring free length ... 2.75 in (69.8 mm)

Torque wrench settings

	lbf ft	Nm
Air pump link and brackets	20	27
Alternator link and brackets	20	27
Bearing caps to block	65	88
Camshaft chain wheel	10	14
Camshaft cover	5	7
Camshaft bearing cap	14	19
Clutch to flywheel	22	30
Clutch housing to rear engine plate (dowel bolt)	34	46
Clutch housing to rear engine plate (bolt)	20	27
Clutch housing and front engine support bracket	20	27
Connecting rod bolt	45	61
Crankshaft pulley attachment	120	160
Cylinder head to block (bolt and stud nut)	50	75
Drive plate to crank	45	61
Engine stay bracket	20	27
Engine stay attachment (jam nut)	34	46
Engine support to block (R.H)	20	27
Flywheel to crankshaft	45	61
Front lifting eye	34	46
Gearbox adaptor attachment	20	27
Hot air manifold to exhaust manifold	20	27
Inlet manifold to carburettor	20	27
Inlet manifold attachment	20	27
Jackshaft chainwheel attachment	38	51
Oil pump to cylinder block	20	27
Sump drain plug	25	34
Sump bolts	20	27
Oil transfer adaptor to block	34	46
Rear engine mounting to crossmember	21	28
Rear engine mounting to bracket	59	80
Rear engine mounting bracket to gearbox	21	28
Rear engine mounting crossmember to body	21	28
Rear engine plate to block	20	27
Right hand front engine mounting bracket to body	21	28
Right hand front engine mounting to engine bracket	21	28
Engine mounting attachments	37	50
Starter motor to clutch housing	34	46
Timing chain support bracket and guides	20	27
Timing chain cover bolts	20	27
Water pump impeller retaining bolt	14	19
Water pump cover	20	27
Water transfer housing	20	27

1 General description

The engine fitted to the Triumph TR7 is of the four cylinder overhead camshaft type with the cylinders inclined at 40° (45° on later models) from the vertical. This is an over-bored version of the engine previously used in the Triumph Dolomite.

The slant configuration allows for better accommodation under the bonnet and at the same time lends itself particularly well to a crossflow cylinder head. The carburettor(s) is/are centrally located above the block and the low mounting of the exhaust manifold ensures a good degree of cooling from the airflow of the moving car.

An unusual feature of the cylinder head is the ingenious manner in which it is secured to the block. The studs on the lower side of the head are parallel to the bore axes whilst the studs on the upper side are slightly angled. This means that they are accessible without removing the camshaft cover.

The overhead camshaft runs in five bearings, operating the valves via 'bucket' type tappets containing a shim for maintaining the correct clearance without constant adjustment. The crankshaft runs in five main bearings which, together with the big-end bearings, are of the

Chapter 1/Engine

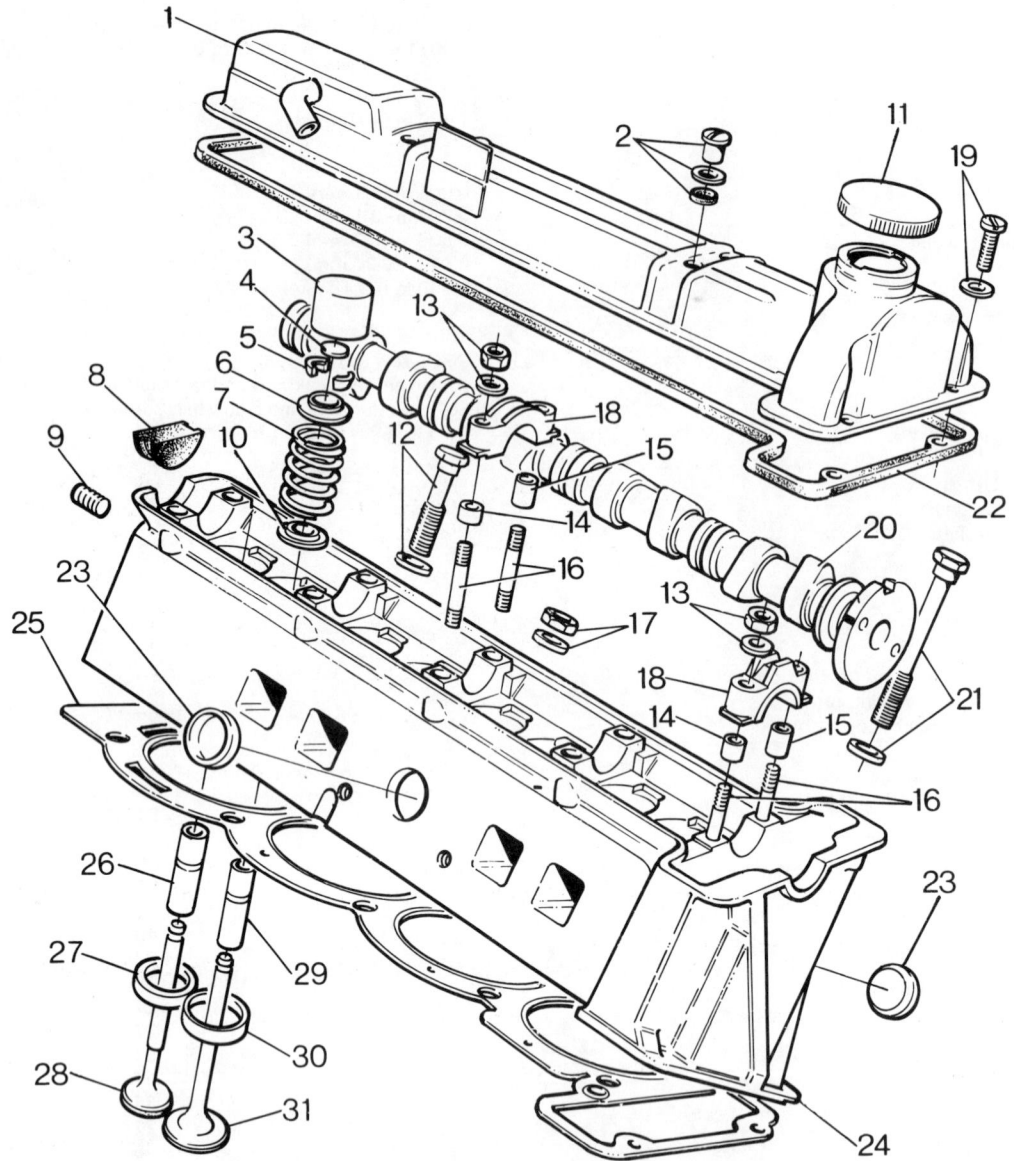

Fig. 1.1. Cylinder head - component parts (typical)
Note: This is basically a Dolomite * cylinder head, and there may be some detail differences

1	Camshaft cover	9	Oil gallery plug	17	Nut and washer	25	Gasket
2	Nut, washer and seal	10	Lower collar	18	Bearing cap	26	Valve guide
3	Tappet	11	Filler cap	19	Screw and washer	27	Valve seat
4	Pallet shim	12	Bolt and washer	20	Camshaft	28	Exhaust valve
5	Cotters	13	Nut and washer	21	Bolt and washer	29	Valve guide
6	Upper collar	14	Ring dowel	22	Camshaft cover gasket	30	Valve seat
7	Valve spring	15	Ring dowel	23	Core plug	31	Inlet valve
8	Rubber grommet	16	Studs	24	Cylinder head		

* A Triumph model not available in the USA.

renewable shell type. An intermediate jackshaft, running the full length of the engine drives the vane type oil pump, distributor and water pump. Like the camshaft, it is driven from the forward end of the crankshaft by a single hydraulically tensioned chain. The water pump is in an unusual position, vertically mounted on the block, above the front of the water heated inlet manifold.

Solid skirt pistons are used, each having two compression rings and an oil control ring. These, with the wedge shaped combustion chambers in the cylinder head and the separately ported aluminium inlet manifold are used to provide optimum operating efficiency.

It is important to note that, where air conditioning is fitted, the system must not be disconnected until discharged by properly qualified personnel. Failure to observe this rule can result in severe personal injury.

2 Major operations with the engine in the car

The following operations can be carried out with the engine installed in the car
1 Removal and installation of the cylinder head.
2 Removal and installation of the sump.
3 Removal and installation of the main and big-end bearings.
4 Removal and installation of the pistons and connecting rods.
5 Removal and installation of the timing chain and gears and the timing cover oil seal.

Chapter 1/Engine

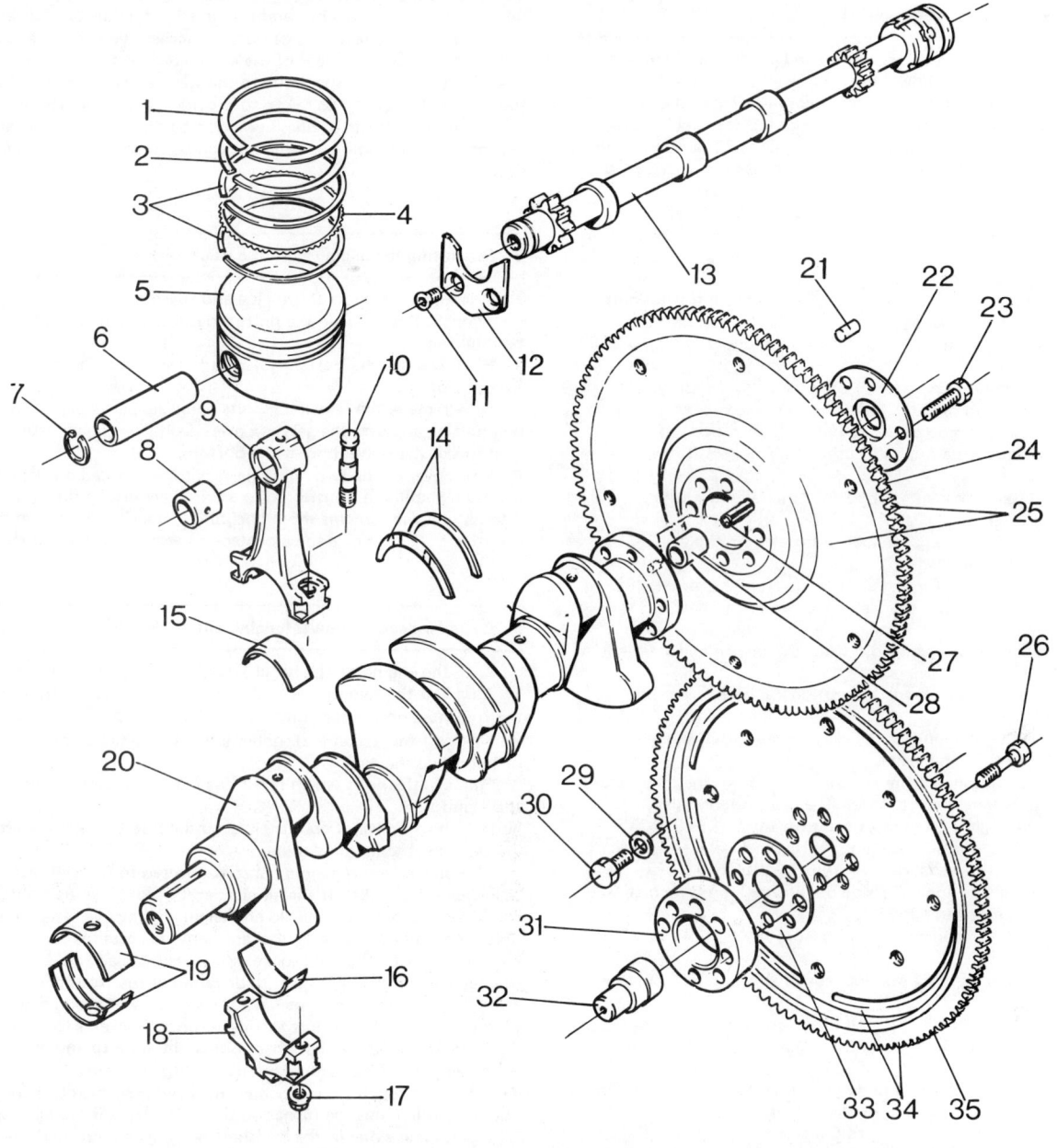

Fig. 1.2. Crankshaft, connecting rod, piston, jackshaft and flywheel (typical)
Note: These parts are basically as used on Dolomite models, and there may be some detail differences

1 Compression ring	10 Stud	19 Bearing shell	28 Bush
2 Scraper ring	11 Screw	20 Crankshaft	29 Washer
3 Oil control ring (chrome rail)	12 Thrust plate	21 Dowel	30 Bolt
4 Spacer or expander	13 Jackshaft	22 Bush	31 Distance piece
5 Piston	14 Thrust washer	23 Bolt	32 Bush
6 Gudgeon pin	15 Bearing shell	24 Flywheel starter ring	33 Washer
7 Circlip	16 Bearing shell	25 Flywheel (complete)	34 Flywheel, automatic (complete)
8 Small end bearing	17 Nut	26 Bolt	35 Flywheel starter ring, automatic
9 Connecting rod	18 Bearing cap	27 Dowel	

6 Removal and installation of the camshaft.
7 Removal and installation of the oil pump.
8 Removal and installation of the flywheel.
9 Removal and installation of the jackshaft.
10 Removal and installation of the engine/gearbox mountings.
11 Removal and installation of the crankshaft rear oil seal.

3 Major operations with engine removed from the car

The following operations can only be carried out after the engine has been removed from the car.
1 Removal and installation of the crankshaft.

4 Engine (and gearbox) - removal

Note: It is recommended that the engine and gearbox are removed together, or the gearbox removed first, followed by the engine. The engine cannot be removed with the gearbox remaining in the car except where the subframe has been removed; this is a much more complicated operation then removing the engine and gearbox together. Where air conditioning is installed, this must be de-pressurized by a suitably qualified person before commencement of work.

1 Initially, place large rags, paper or soft cardboard over the front wings, and tape them in position to prevent damage to the car's paintwork.
2 Disconnect the battery leads.
3 Mark the fitted position of the hinges, then remove the bonnet; refer to Chapter 12 if necessary.
4 Remove the fresh air duct as described in Chapter 12, Section 34, paragraph 3.
5 Remove the bonnet lock; refer to Chapter 12 if necessary.
6 Remove the bottom radiator hose. Unscrew and remove the drain plug from the left-hand side of the cylinder block in front of the starter motor. If the coolant is to be used again, collect it in a suitable container.
7 Remove the fixing brackets and lift out the radiator; refer to Chapter 2 if necessary.
8 Disconnect all the heater and coolant hoses from the engine. Do not forget those on the bulkhead.
9 Remove the air intake (or hot air) hose(s) from the air cleaner.
10 For convenience, remove the air cleaner; refer to Chapter 3 if necessary.
11 Disconnect the brake servo hose from the inlet manifold/plenum chamber.
12 Where applicable, detach the adsorption canister hoses; refer to Chapter 3 if necessary.
13 Where applicable, disconnect the vacuum hose from the inlet manifold to the anti-run-on valve.
14 As applicable, disconnect the electrical leads from the oil pressure switch, alternator, temperature transmitter, anti-run-on valve, compressor clutch, ignition coil, distributor, throttle jack, temperature sensor and thermotime switch. Tie on small labels to assist correct refitting of the leads. Remove the distributor cap.
15 Remove the pipe to the inlet side of the fuel pump or fuel rail. Plug the pipe to prevent fuel loss.
16 Detach the throttle and choke cables from the carburettor, as applicable.
17 Where applicable, pull off the float chamber spill pipe from the rear carburettor.
18 For convenience, remove the fan pulley assembly (and compressor, if applicable) as this provides more working room when lifting the engine out.
19 Remove the front exhaust pipe (see Chapter 3). On fuel injection vehicles, disconnect the Lambda (oxygen) sensor.
20 Detach the gear lever, referring to Chapter 6 or 13 as necessary.
21 Disconnect the propeller shaft at the gearbox end, after first marking the flanges to assist correct refitting.
22 Disconnect the reverse lamp/starter inhibitor wires, and the seat belt interlock wires (where applicable).
23 Remove the clamping plate and take out the speedometer cable from the drive pinion.
24 Where applicable, detach the clutch slave cylinder and move to one side. Remove the bellhousing bolts as necessary to release the clutch hydraulic pipe.
25 Remove the nut from the left-hand engine mounting.
26 Where fitted, remove the engine stabilizer. Refer to Section 49 for further information.
27 Where applicable, remove the starter motor shield.
28 Remove the leads from the starter motor solenoid.
29 Remove the main engine earth lead.
30 As applicable, remove the propeller shaft guard and lower the propeller shaft, and fold back the clips securing the cables around the bellhousing.
31 Support the gearbox weight with a trolley jack then detach the rear mounting from the underbody. Also remove the rear mounting from the gearbox (photo).
32 Attach chains to the engine lifting eyes, bearing in mind that a fairly steep angle of removal will be required, so that the weight is just taken up.
33 Remove the right-hand engine mounting nut.
34 The engine can now be carefully lifted out, while an assistant lowers the trolley jack beneath the gearbox as necessary. Take great care that cables and hoses at the rear of the engine do not foul the bulkhead, as there is little room to spare due to the steep angle of lifting. If necessary, the rear of the car can be raised to provide more clearance (photo).
35 Assuming that the engine is going to be dismantled, transfer the assembly to a suitable area where it can be cleaned with a water soluble solvent.

5 Separating the engine from the gearbox

1 Adequately support the engine and gearbox, either on the floor or on a bench, then slacken the bolts around the periphery of the bellhousing.
2 First, remove the starter motor and place it on one side. It is held by three bolts.
3 Now remove the remaining bolts from the periphery of the bellhousing and ease the gearbox away from the engine. Note the position of the dowel bolt at the bottom.
4 Whichever of the two major parts is to be worked on (ie; engine or gearbox) should be transferred to a suitable working surface. The assembly which remains should be put to one side and covered with a polythene sheet to give some protection against dust and dirt.

6 Cylinder head - removal (engine out of the car)

With the engine on the bench, remove the cylinder head as follows:
1 Take out the two screws at the front end of the camshaft cover and the four nuts on the top. Lift off the cover and put it on one side.
2 Take off the gasket and rubber end grommets before they get lost or fall inside the engine.
3 Remove the bolts which retain the exhaust manifold and lift away the manifold.
Note: This step is not absolutely essential but it makes the assembly lighter for subsequent handling.
4 Disconnect any remaining hose connected to the cylinder head or inlet manifold. Take off the air cleaner. Remove the distributor cap.
5 Remove the inlet manifold and engine lifting eye. There are three bolts at each inlet flange and all are fairly inaccessible so some degree of care with suitable flat and socket spanners is called for. As the manifold is lifted away it will also need to be raised at the front end to pull it away from the short pipe which connects the water pump cover to the inlet manifold (photo). For further information refer to Chapter 3.
6 On USA vehicles, disconnect either the pipe to the check valve,)n or the electrical lead to the Lambda (oxygen) sensor.
7 On the camshaft flange, behind the driving sprocket, there is a marking which must be turned so that initially, it is facing downwards.
8 Now remove one of the camshaft bearing cap nuts and use it to secure the camshaft sprocket spigot bolt to the support plate.
9 Remove the top bolt securing the sprocket to the camshaft.
10 Rotate the camshaft through 180° to align the marking on the camshaft flange with the marking on the end bearing cap (ie number 1 piston at top-dead-centre (TDC) on its firing stroke) then remove the remaining sprocket/camshaft bolt. Take great care that the tab washer does not fall into the engine. The sprocket is now retained solely by the support plate.
11 You can remove the camshaft at this stage if you wish. The full procedure is given in Section 8 of this Chapter.
12 The next step is to remove the cylinder head itself. There are two small nuts, bolts, and washers at the timing chain end of the head, five bolts and washers and five nuts which fit onto slotted studs. If the studs are difficult to remove, screw two nuts on and lock them against each other, then use a spanner on the underside nut to withdraw the stud. The studs and bolts should be loosened in the reverse order to that given in Section 42.
13 The head can now be removed; if necessary a heavy soft faced hammer can be used to break the gasket seal.
14 If the cylinder head is not going to be worked on immediately, it should be wrapped in polythene to protect it from dust and dirt. If the camshaft has been removed, keep the tappets and pallet shims in position at this stage to avoid the possibility of a mix-up.

4.31 Supporting the gearbox weight

4.34 Lifting out the engine - note the steep angle

6.5 The water pump/manifold connecting hose

6.10 The camshaft sprockets on the support plate

8.4 Removing a camshaft bearing cap

9.8a Flywheel locking device in position for removing the clutch and flywheel

7 Cylinder head removal (engine in the car)

The procedure is basically as described in the previous Section, but in addition carry out the procedures of paragraphs 1, 2, 3, 4, 5, 6, 11 and 19 of Section 4. It is preferable to leave the exhaust manifold attached until the cylinder head gasket seal has been broken, since the manifold can be used to apply some leverage to break the head-to-block seal.

8 Camshaft - removal

1 If the engine is still in the car, detach the battery earth lead, then remove the fresh air duct (see Chapter 12, Section 34, paragraph 3).
2 If the engine has been removed from car, initially carry out the procedure given in paragraphs 1 and 2; then paragraphs 7 to 10 inclusive, of Section 6 of this Chapter. If the engine is still in the car; first, remove the spark plug HT leads and their retainer on the camshaft cover. Take off the camshaft cover breather hose then carry out the procedure given in paragraph 1 and 2; then, paragraphs 7 to 10 inclusive of Section 6 of this Chapter.
3 Slacken the camshaft bearing caps by equal amounts and in pairs (except where one nut has already been removed) working along the length of the camshaft. This avoids undue strain on bearings, caps and camshaft.
4 Remove the bearing caps, note that they are numbered to avoid inadvertent mix-up. Then withdraw the camshaft (photo).

9 Crankshaft pulley, clutch, flywheel and rear adaptor plate - removal

Engine in the vehicle
1 Disconnect the battery.
2 Remove the fan blades (USA only).
3 Slacken the compressor adjustment and slip the drivebelt from the pulley (if applicable).
4 Slacken the alternator adjustment bolts and remove the drivebelt.
5 Place the vehicle in gear, apply the handbrake, and remove the crankshaft pulley bolt.
6 Withdraw the pulley/s, and retain the key.
7 Remove the clutch as described in Chapter 5, Section 7.
8 To remove the flywheel, mark the hub and crankshaft flange to assist refitting, remove the clutch. Lock the flywheel using a piece of bent mild steel. Remove the eight flywheel bolts, the retaining plate, and the spigot bush (photos). Unscrew the six rear adaptor plate securing bolts and remove the plate.

Engine on the bench
9 Lock the flywheel (see paragraph 8), remove the crankshaft pulley bolt, and withdraw the pulley (photo). Retain the key.
10 Reverse the locking bar, mark the relative positions of the clutch and flywheel, evenly remove the retaining bolts, and lift away the clutch.
11 Remove the flywheel as described in paragraph 8.
12 If necessary, take out the six bolts and remove the rear adaptor plate.

10 Timing cover, timing gear and jackshaft - removal with engine out of vehicle

Timing cover (all models)
1 Remove the crankshaft pulley/s (see Section 9).
2 Remove the alternator, alternator mounting bracket, and adjusting link.

Timing cover (UK and Europe)
3 Remove four of the sump bolts, two at the front and one at each side of the timing cover.
4 Remove the fan and viscous coupling (see Chapter 2).
5 Remove the two timing cover to cylinder head bolts and nuts, and the remaining cover bolts including the centre bolt.
6 Take off the cover and gasket.
7 Cover the opening to the sump.

Timing cover (USA and Canada)
8 Remove the air pump (see Chapter 3), and the diverter relief valve and bracket.
9 Remove the air pump and compressor brackets.
10 Remove the two timing cover to cylinder head bolts and nuts.

11 Slacken the four air compressor bracket bolts.
12 Remove the three air compressor adjusting bolts.
13 Remove the air compressor adjustment bracket.
14 Proceed as described in paragraph 3.
15 Looking from the front, remove the centre and lower left bolts from the timing cover.
16 Remove the fan and Torquatrol unit (see Chapter 2).
17 Proceed as described in paragraphs 6 and 7.

Timing chain, sprockets, tensioner and guides
18 Remove the timing chain cover.
19 Remove the camshaft cover (Section 6, paragraphs 1 and 2).
20 Proceed as described in Section 6, paragraphs 7 to 10.
21 Remove the timing chain tensioner bolts, followed by the tensioner and backplate (photo).
22 Remove the two bolts retaining the adjustable chain guide, and remove the guide.
23 Hold the camshaft sprocket, remove the remaining securing bolt from the fixed guide and camshaft support bracket, and remove the fixed guide.
24 Release the chain from the sprockets and withdraw the chain, sprocket and sprocket bracket upwards (photo).
25 Remove the bolt and lock washer, and take off the jackshaft sprocket (photo).
26 Remove the oil thrower and the crankshaft sprocket.
27 Remove the crankshaft key and alignment shims (photo).

Jackshaft
28 Remove the timing cover.
29 Remove the inlet manifold complete with carburettors or fuel rail (see Chapter 3).
30 Remove the water pump (see Chapter 2).
31 Remove the fuel pump (see Chapter 3).
32 Remove the camshaft cover (Section 6, paragraphs 1 and 2) and position the timing mark on the camshaft flange in line with the groove on the front bearing cap.
33 Remove the distributor (see Chapter 4).
34 Proceed as in paragraphs 21 and 22.
35 Remove the two Allen screws, and then the jackshaft keeper plate (photo).
36 Lift the chain and withdraw the jackshaft (photo).

11 Timing cover, timing gear and jackshaft - removal with engine in place in the vehicle

Timing cover, chain, sprockets, tensioner and guides
1 Disconnect the battery.
2 Remove the bonnet.
3 Remove the radiator (see Chapter 2).
4 Proceed as described in Section 10.

Jackshaft
5 Disconnect the battery and remove the radiator.
6 Remove the fresh air duct.
7 Remove the air conditioning condenser, if fitted.
8 Proceed as in Section 10, paragraphs 28 to 36.

12 Oil transfer adaptor, oil filter and oil pump - removal

1 Disconnect the battery.
2 If the engine is still in the car, initially remove the electrical lead(s) to the oil pressure switch.
3 Remove the centre retaining bolt and pull off the adaptor (photo).
4 To remove the oil filter, unscrew the centre retaining bolt then withdraw the bowl. Remove the seal from the cylinder block annular groove (photo).
5 If the engine has been removed from the car, remove the four bolts and spring washers which retain the oil pump. Withdraw the pump, O-ring and hexagon driveshaft (photo).
6 If the engine is still in the vehicle, remove the two bolts securing the clutch slave cylinder.
7 Carefully withdraw the slave cylinder and move it to one side.
8 Remove the bolts and spring washers securing the pump (additionally, it will be necessary to remove the nut and bolt securing the bellhousing at this point).
9 Withdraw the pump, O-ring and hexagon driveshaft.

13 Sump, pistons, connecting rods and big-end bearings - removal

1 The sump, pistons and connecting rods can be removed with the engine either in the car or on the bench. In either case remove the cylinder head (Sections 6 or 7) if the pistons and connecting rods are to be removed.

Engine in the vehicle
2 Disconnect the battery, and drain the sump.
3 Take the weight of the engine on a hoist.
4 Remove the fresh air duct.
5 Remove the fan guard.
6 Remove the bolts from the sump coupling plate.
7 Remove the engine stabiliser.
8 Remove the two right-hand engine mounting bolts.
9 Remove the left-hand engine mounting nut.
10 Remove the sump nuts and bolts.
11 Raise the engine, and withdraw the sump with the left-hand mounting attached.
12 Remove the oil strainer by taking out the two bolts and spring washers (photo).
13 Remove the nuts securing the big-end bearing caps. Take the caps off one at a time. Note that the caps are numbered; always make sure that the bearing shells are kept with their correct caps. This also applies to the connecting rods and pistons which can now be removed one at a time from the top of the cylinder block (photo).
14 Place all the connecting rods, bearings, pistons, etc, safely on one side for attention later.

Engine out of the vehicle
15 Proceed as above, omitting paragraphs 1 to 9, and 11.

14 Crankshaft main bearings and rear oil seal - removal

1 The main bearings may be removed with the engine in or out of the car. The crankshaft and rear oil seal can only be removed with the engine out.

Engine out of the vehicle
2 Remove the clutch, flywheel and adaptor plate (Section 9).
3 Remove the sump (Section 13).
4 Remove six bolts, and remove the rear oil seal housing (photo).
5 Remove the timing cover, chain, and crankshaft sprocket (Section 10).
6 Remove the connecting rod bearing caps (Section 13).
7 Remove the main bearing bolts, caps and bearings. Note the numbering, and keep related items together (photo).
8 Lift out the crankshaft, followed by the upper bearing shells and thrust washers (photo).

Engine in the vehicle
9 Disconnect the battery. Drain and remove the sump (Section 13).
10 Proceed as in paragraphs 7 and 8.
11 Slide the upper shells out, tag end first.

15 Water transfer housing - removal

1 If the engine is in the vehicle, disconnect the battery and drain the cooling system.
2 Remove the fresh air duct.
3 Remove the heater hose from the housing.
4 Remove the five bolts and washers, followed by the housing.

16 Valves, valve seats and guides - removal

1 Keep the springs, tappets, etc with the valve to which they belong, in a suitable container with sections numbered 1 to 8.
2 Remove each pallet shim, and place it in the appropriate section of the box.
3 Remove the split cotters using a valve spring compressor, followed by the upper collar, lower collar, and valve (photos).
4 Examine the valve guides and seats (see Section 26).
5 Renewal of the guides and seats should be carried out by specialist engineers.

9.8b Clutch pilot bearing (spigot bush)

9.9 Flywheel locking device in position for removing the pulley

10.6 Removing the timing cover

10.21 Removing the chain tensioner

10.25 Removing the jackshaft sprocket

10.27 The crankshaft pulley shims (behind the Woodruff key)

10.24 Removing the camshaft sprocket support

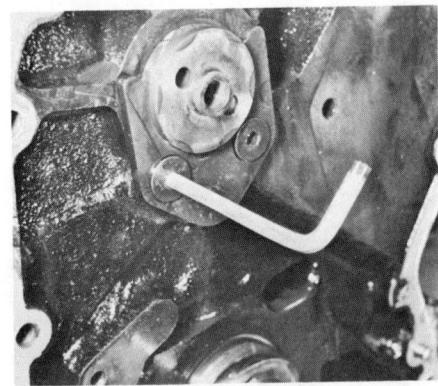

10.35 Removing the thrust plate ...

10.36 ... and jackshaft

12.3 Removing the oil transfer housing

12.4 Removing the oil filter

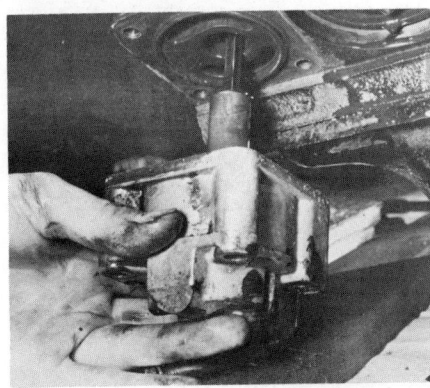

12.5 Removing the oil pump and driveshaft

13.12 Removing the oil strainer

13.13 Removing a big-end bearing cap and shell

14.4 Removing the crankshaft rear oil seal housing

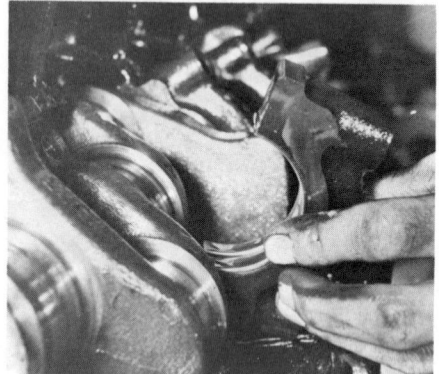
14.7 Removing a main bearing cap

14.8 The crankshaft, ready for removal

16.3a Using a spring compressor ...

16.3b ... when removing ...

16.3c ... a valve spring ...

16.3d ... and valve

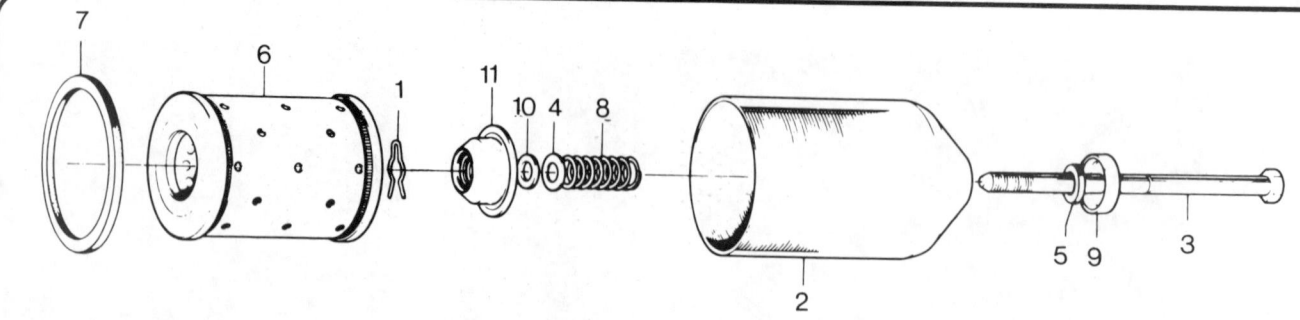

Fig. 1.3. The oil filter - exploded view

1	Spring clip	4	Retainer	7	Sealing ring	10	Spacer
2	Bowl	5	Seal	8	Spring	11	Seat
3	Centre bolt	6	Element	9	Cap		

17 Pistons and connecting rods - dismantling

It is extremely important that all parts associated with each piston are kept together so that they can eventually be refitted (if suitable) in the same position.

1 To remove the gudgeon pin and free the piston from the connecting rod, remove one of the circlips at either end of the pin with a pair of circlips pliers.
2 Press out the pin from the rod and piston with your fingers.
3 If the pin shows reluctance to move, then on no account force it out, as this could damage the piston. Immerse the piston in a pan of boiling water for three minutes. On removal the expansion of the aluminium should allow the gudgeon pin to slide out easily.
4 Make sure the pins are kept with the same piston.
5 To remove the piston rings, slide them carefully over the top of the piston, taking care not to scratch the aluminium alloy. Never slide them off the bottom of the piston skirt. It is very easy to break the iron piston rings if they are pulled off roughly, so this operation should be done with extreme caution. It is helpful to make use of an old hacksaw blade with the teeth ground off, or better still, an old 0.020 in (0.5 mm) feeler gauge.
6 Lift one end of the piston ring to be removed, out of its groove, and insert the end of the feeler gauge under it.
7 Turn the feeler gauge slowly round the piston and, as the ring comes out of its groove, apply slight upward pressure so that it rests on the land above. It can then be eased off the piston with the feeler gauge stopping it from slipping into any empty groove, if it is any but the top piston ring that is being removed.

18 Oil pump - dismantling

1 Take out the hexagonal driveshaft then remove the screws which retain the pump cover.
2 Remove the rotors.
3 Take out the O-ring from the pump body.
4 To remove the relief valve, withdraw the split-pin in the pump body.
5 Take out the plug, spring and relief valve.
Note: If the plug is reluctant to come out, due to the O-ring sticking, it can be carefully drifted out from inside the pump body.
6 Remove the O-ring from the plug.

19 Components: cleaning, examination and storage - general note

1 With the engine stripped down and all parts dismantled they must now be cleaned in petrol or proprietary solvents. Remove any traces of gasket material remaining using a blunt scraper but take care not to score any of the sealing surfaces.
2 Once the parts are cleaned, both before and after inspection, make sure that they stay clean by covering or wrapping them in polythene.

20 Crankshaft - examination and renovation

Examine the crankpins and main journal surfaces for signs of scoring or scratches. Check the ovality of the crankpins at different positions with a micrometer. If more than 0.001 inch (0.0254 mm) out of round, the crankpins will have to be reground. They will also have to be reground if there are any scores or scratches present. Also check the journals in the same fashion. If it is necessary to regrind the crankshaft and fit new bearings your local Triumph garage or engineering works will be able to decide how much metal to grind off and therefore, the correct undersize shells to fit.

21 Big-end and main bearing shells - examination and renovation

1 Big-end bearing failure is often accompanied by a noisy knocking from the crankcase, and a slight drop in oil pressure. Main bearing failure is accompanied by vibration, which can be quite severe as the engine speed rises and falls, and a drop in oil pressure.
2 Bearings which have not broken up, but are badly worn will give rise to low oil pressure and some vibration. Inspect the big-ends, main bearings and thrust washers for signs of general wear, scoring, pitting and scratches. The bearings should be matt grey in colour. With lead-indium bearings, should a trace of copper colour be noticed, the bearings are badly worn as the lead bearing material has worn away to expose the indium underlay. Renew the bearings if they are in this condition, or if there is any sign of scoring or pitting.
3 The undersizes available are designed to correspond with the regrind sizes, ie - 0.010 in (0.2540 mm) bearings are correct for a crankshaft reground - 0.010 in (0.2540 mm) undersize. The bearings are in fact, slightly more than the stated undersize, as running clearances have been allowed for, during their manufacture.
4 Very long engine life can be achieved by changing big-end bearings at intervals of 3,000 miles and main bearings at intervals of 50,000 miles, irrespective of bearing wear. Normally, crankshaft wear is infinitesimal and a change of bearings will ensure mileages of between 100,000 to 120,000 miles before crankshaft regrinding becomes necessary. Crankshafts normally have to be reground because of scoring due to bearing failure.

22 Cylinder bores - examination and renovation

1 The cylinder bores must be examined for taper, scoring and scratches. Start by carefully examining the top of the cylinder bores. If they are at all worn, a very slight ridge will be found on the thrust side. This marks the top of the piston ring travel. The owner will have a good indication of the bore wear prior to dismantling the engine, or removing the cylinder head. Excessive oil consumption, accompanied by blue smoke from the exhaust, is a sure sign of worn cylinder bores and piston rings.

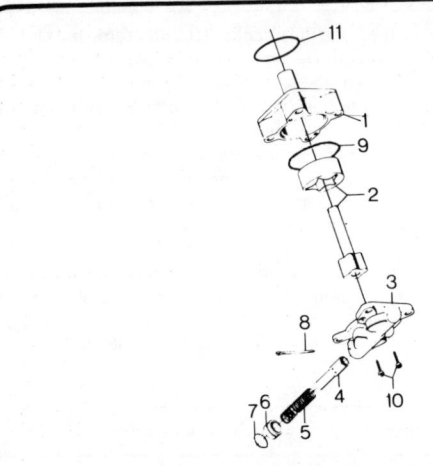

Fig. 1.4. The oil pump - exploded view

1 Pump body
2 Rotors
3 Cover
4 Piston
5 Spring
6 Stop
7 Seal
8 Split pin
9 Seal
10 Screws
11 Seal

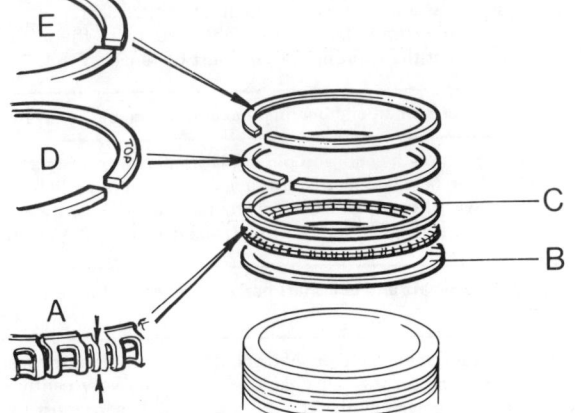

Fig. 1.5. Piston rings

A Spacer (expander)
B Bottom oil control ring
C Top oil control ring
D Second compression (scraper) ring
E Top compression ring

2 Measure the bore diameter just under the ridge with a micrometer, and compare it with the diameter at the bottom of the bore, which is not subject to wear. If the difference between the two measurements is more than 0.006 in (0.1524 mm) then it will be necessary to fit special pistons and rings or to have the cylinders rebored and to fit oversize pistons. If no micrometer is available, remove the rings from a piston and place the piston in each bore in turn about ¾ in (19 mm) below the top of the bore. If an 0.010 in (0.2540 mm) feeler gauge can be slid between the piston and the cylinder wall on the thrust side of the bore, then remedial action must be taken. Your Triumph dealer or an automobile engineering specialist will be able to advise you on rebore and oversize piston sizes.
3 If the bores are slightly worn but not so badly worn as to justify reboring them, then special oil control rings and pistons can be fitted which will restore compression and stop the engine burning oil. Several different types are available, and the manufacturer's instructions concerning their fitting must be followed closely.
4 If new pistons are being fitted and the bores have not been reground, it is essential to slightly roughen the hard glaze on the sides of the bores with fine glass paper so the new piston rings will have a chance to bed in properly.

23 Pistons and piston rings - examination and renovation

Note: For decarbonisation of the pistons, refer to Section 32.
1 If new rings are to be fitted to the old pistons, then the top ring should be stepped so as to clear the ridge left above the previous top ring. If a normal but oversize new ring is fitted, it will hit the ridge and break, because the new ring will not have worn in the same way as the old which will have worn in unison with the ridge.
2 Before fitting the rings on the pistons, each should be inserted approximately 3 in (76 mm) down the cylinder bore and the gap between the two ends of the ring measured with a feeler gauge. The correct value is given in the Specifications Section. The gap must be measured at the bottom of the ring travel, as if it is measured at the top of a worn bore it could easily seize at the bottom. If the ring gap is too small, rub down the ends of the ring with a very fine file until the gap, when fitted, is correct. Keep the rings square in the bore for measurement by inserting an old piston in the bore upside down and using it to push the ring down about 3 in. Remove the piston and measure the ring gap.
3 When fitting new pistons and rings to a rebored engine, the ring gap can be measured at the top of the bore as the bore will not taper. It is not necessary to measure the side clearance in the ring grooves with the rings fitted, as the dimensions are accurately machined during manufacture. When fitting new oil control rings to an old piston, it may be necessary to have the grooves widened by machining to accept the new wider rings. Where this is the case follow the ring manufacturer's instructions.
4 Fit the rings to the piston, and measure the gap between the ring and the wall of the groove, and check against the Specifications.
5 A 0.001 in (0.025 mm) oversize piston is available for fitting to a standard bore, and where these are fitted the bores must first be honed by a specialist engineer to provide the specified clearance.
6 The gudgeon pin should be a finger tight press fit in the piston, and wear means that the appropriate part must be renewed.

24 Gudgeon pin and small-end bearings - examination and renovation

1 Examine the fit of the gudgeon pin in the small-end bearing. If any wear is detectable, new bearings and/or a new gudgeon pin will be required. Renewal of bearings is best left to a specialist since they require to be pressed in then reamed to suit the gudgeon pin size.

25 Jackshaft, camshaft and camshaft bearings - examination and renovation

1 The bearing surfaces of either shaft should show no signs of wear, very shallow scoring on the cams can be removed by very gently rubbing with fine emery cloth. The greatest possible care must be taken to retain the original profile.
2 Examine the camshaft bearings for any signs of scoring or cracking. If they appear to be in first class condition, (and provided that no attention is required to the camshaft journals) they should be satisfactory for further use.

Note: The bearings cannot be renewed. If their condition warrants replacement it will mean purchasing a new cylinder head complete.
3 Examine the jackshaft gears for wear or chipping of the teeth. Such defects will adversely affect the ignition timing, and if necessary the shaft must be renewed.
Note: Check the mating gears on the distributor and water pump.

26 Valves, valve seats and valve guides - examination and valve grinding

1 Examine the heads of the valves for pitting and burning, especially the heads of the exhaust valves. The valve seatings should be examined at the same time. If the pitting on valve and seat is very slight, the marks can be removed by grinding the seats and valves together with coarse, and then fine, grinding paste. Where bad pitting has occurred to the valve seats, it will be necessary to recut them and fit new valves. If the valve seats are so worn that they cannot be recut, then it will be necessary to fit new valve seat inserts. These latter two jobs should be entrusted to the local Triumph garage or an engineering works. In practice, it is seldom necessary. Normally, it is the exhaust valve that is too worn for re-use. The owner can purchase a new set of valves and match them to the seats by grinding. Before commencing work, decide which valves are going to be used in the reassembly, then check them in the valve guides. If the valves are a loose fit in the guides and there is the slightest suspicion of lateral rocking using a new valve, new guides will have to be fitted (see Section 16).
2 Whether new valves and seats are used or the original ones replaced, valve grinding should be carried out as follows:
Smear a trace of coarse carborundum paste on the seat face, and apply a suction grinder tool to the valve head. With a semi-rotary motion, grind the valve head to its seat, lifting the valve occasionally to redistribute the grinding paste. When a dull matt even surface finish is produced on both the seat and valve, wipe off the paste and repeat the process with fine carborundum paste, lifting and turning the valve to redistribute the paste as before. A light spring placed under the valve head will greatly ease this operation. When a smooth unbroken ring of light grey matt finish is produced on both faces, the grinding operation is completed.
3 Scrape away all carbon from the valve head and the valve stem. Carefully clean away every trace of grinding compound, taking care to leave none in the ports or guides. Clean the valves and seat with a paraffin soaked rag, and then with a clean rag. If an air line is available, blow the valve guides and valve ports clean.
4 Check the valve springs for cracks and obvious defects. Check the free length against the specification and discard the springs as necessary.

27 Timing chain, sprockets and chain tensioner - examination and renovation

1 Examine the teeth on the camshaft, jackshaft and crankshaft sprockets for wear. Each tooth forms an inverted 'V' with the sprocket periphery, and if worn, the side of each tooth, ie, one side of the inverted 'V' will be concave when compared with the other. If any sign of wear is present the sprockets must be renewed.
2 Examine the links of the chain for side slackness, and renew the chain if any slackness is noticeable when compared with a new chain. It is a sensible precaution to renew the chain if the engine is stripped down for a major overhaul. The actual rollers on a very badly worn chain may be slightly grooved.
3 Examine the chain tensioner slipper and the chain guides for wear, and if any present these items should be renewed also. It is almost certain that if wear in the chain is bad enough to require fitment of a new chain, the running surfaces will also need renewal (photo).

28 Tappets and pallet shims - examination and renovation

1 Even after a very high mileage the tappets and pallet shims should still be in satisfactory condition for re-use. Examine them for any signs of indentation, wear, scoring or cracking.
2 Using a micrometer, measure the thickness of the shims and keep a note of the dimensions of each one. When the valve clearances

27.3 Timing chain tensioner in released position

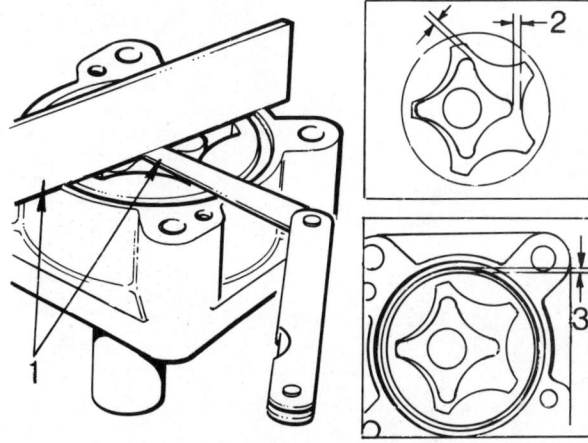

Fig. 1.6. Checking wear on oil pump

1 Endfloat check
2 Rotor lobe clearance check
3 Outer ring/pump body diametral clearance

are being set you will undoubtedly need to purchase some new ones, but some of the existing ones may be suitable for re-use.
3 In the unlikely event of scoring of the surfaces of the tappets and pallet shims, it is permissible to lap the surface a little if the proper facilites are at hand. The important thing to remember is that the surfaces must be smooth, flat and parallel when they are assembled.

29 Flywheel and starter ring - examination and renovation

1 If the flywheel clutch face is deeply scored, a new flywheel should be obtained or the surface skimmed using a lathe. The maximum allowable flywheel face run-out is 0.002 in (0.05 mm) at a 4 in (102 mm) radius.
2 If the teeth on the flywheel starter ring are badly worn, or if some are missing, then it will be necessary to remove the ring. This is achieved by splitting the ring with a cold chisel. The greatest care should be taken not to damage the flywheel during this process. It is sometimes advantageous to drill a ¼ inch (6 mm) hole at the intersection point of two teeth, and to strike this point with the cold chisel.
3 To fit a new ring, heat it gently and evenly to a temperature of 170 to 175°C (338 to 347°F) in an oil bath or circulatory oven; do not exceed this temperature. With the ring at this temperature, fit it to the flywheel with the front teeth facing the flywheel register. The ring should be tapped gently down onto its register and left to cool naturally when the shrinkage of the metal on cooling will ensure that it is a secure and permanent fit. The maximum permissible gap between the flywheel and starter ring on one length of 6 in (15 cm) is 0.025 in (0.6350 mm).

30 Oil pump - examination and renovation

Thoroughly clean all the component parts in petrol and then check the rotor endfloat and lobe clearances in the following manner:
1 Position the rotors in the pump and place the straight edge of a steel ruler across the joint face of the pump. (Ensure that the chamfered edge of the outer rotor is at the driving end of the rotor pocket). Measure the gap between the bottom of the straight edge and the top of the rotors with a feeler gauge. If the measurement exceeds 0.004 in (0.1016 mm) then check the lobe clearances as described in the following paragraphs. If the lobe clearances are correct, then lap the joint face on a sheet of plate glass.
2 Measure with a feeler gauge the gap between the inner and outer rotors. It should not be more than 0.010 in (0.254 mm).
3 Then measure the gap between the outer rotor and the side of the pump body which should not exceed 0.008 in (0.2032 mm). It is essential to renew the pump if the measurements are outside these figures. It can be safely assumed that at any major reconditioning the pump will need renewal.
4 Measure the free length of the relief valve spring which should be 1.700 in (43.18 mm). If this length is not obtained, renew the spring.
5 Check the relief valve and its bore for scoring and wear, renewing the parts if any are present.

6 Check the condition of the driveshaft. If wear is noted on the driving faces it must either be renewed or, if the services of a specialist welder are available, weld may be applied and the profile regained by suitable grinding.

31 Miscellaneous components - examination and renovation

Sump
1 Thoroughly wash out the sump and wipe dry using a clean lint-free cloth.
2 Inspect the casing for signs of cracking which could be caused by incorrect positioning of a jack or hitting a high object on the road surface. If a crack is evident, it may be repaired by welding using the services of a firm of specialists, otherwise obtain a new item.
3 Check that the sealing faces are flat with no signs of scoring.

Water pump/thermostat housing connecting tube:
4 If the tube has separate O-rings fitted, they should be renewed.
5 If any dents, cracks or distortion is present on the tube itself, it should be renewed.
6 Where the tube has a bonded rubber covering, check the sealing ridges for nicks and burrs. If any are present the tubes must be renewed.
7 Check that the tube is a good firm fit in the water pump and thermostat housing connecting bores. If not, the tube must be renewed or very carefully swaged out a little; for example, by using a vice to very slowly press a suitable profiled socket spanner into the end.
8 Check that the water pump and thermostat housing connecting bores are in good condition. If any corrosion or scale is present it should be scraped out with a screwdriver or penknife.

Clutch pilot bearing (spigot bush)
9 Check the clutch pilot bearing for wear; if evident obtain a replacement.

Studs, nuts and bolts
10 Examine all studs, nuts and bolts for wear and damage. Where self-locking nuts are used, these should be renewed.

32 Cylinder head and pistons - decarbonisation

1 Remove the spark plugs.
2 With the cylinder head off, carefully remove with a wire brush and blunt scraper, all traces of carbon deposits from the combustion spaces and the ports. The valve stems and valve guides should also be freed from any carbon deposits. Wash the combustion spaces and ports down with petrol and scrape the cylinder head surface free of any foreign

matter with the side of a steel rule, or similar article.
3 Clean the pistons and top of the cylinder bores. If the pistons are still in the block, then it is essential that great care is taken to ensure that no carbon gets into the cylinder bores as this could scratch the cylinder walls or cause damage to the piston rings. To ensure this does not happen, first turn the crankshaft so that two of the pistons are at the top of their bores. Stuff rag into the other two bores, or seal them off with paper and masking tape. The waterways should also be covered with small pieces of masking tape to prevent particles of carbon entering the cooling system and damaging the water pump.
4 There are two schools of thought as to how much carbon should be removed from the piston crown. One school recommends that a ring of carbon should be left around the edge of the piston and on the cylinder bore wall as an aid to low oil consumption. Although this is probably true for early engines with worn bores, on later engines the thought of the second school can be applied, which is that for effective decarbonisation, all traces of carbon should be removed.
5 If all traces of carbon are to be removed, press a little grease into the gap between the cylinder walls and the two pistons which are to be worked on. With a blunt scraper, carefully scrape away the carbon from the piston crown, taking great care not to scratch the aluminium. Also scrape away the carbon from the surrounding lip of the cylinder wall. When all carbon has been removed, scrape away all the grease which will now be contaminated with carbon particles, taking care not to press any into the bores. To assist prevention of carbon build-up, the piston crown can be polished with a metal polish. Remove the rags or masking tape from the other two cylinders, and turn the crankshaft so that the two pistons which were at the bottom are now at the top. Place rag or masking tape in the cylinders which have been decarbonised and proceed as just described.
6 If a ring of carbon is going to be left round the piston, this can be helped by inserting an old piston ring into the top of the bore to rest on the piston and ensure that carbon is not accidentally removed. Check that there are no particles of carbon in the cylinder bores. Decarbonising is now complete.
7 As soon as the spark plugs have been cleaned or renewed they can be refitted.

33 Engine reassembly - general note

1 To ensure maximum life with minimum trouble from a rebuilt engine, not only must everything be correctly assembled, but all the parts must be spotlessly clean, all the oilways must be clear, locking washers and spring washers must always be fitted where indicated, and all bearing and other working surfaces must be thoroughly lubricated during assembly. Before assembly begins, renew any bolts or studs if the threads of which are in any way damaged, and whenever possible use new spring washers.
2 Check the core plugs for signs of weeping and always renew the plug at the front of the engine as it is normally covered by the engine end plate.
3 Drive a punch through the centre of the core plug.
4 Using the punch as a lever, lift out the old core plug.
5 Thoroughly clean the core plug orifice and using a small diameter headed hammer as an expander, firmly tap a new core plug in place, convex side facing out.
6 Apart from normal tools, a supply of clean rags, an oil can filled with engine oil (an empty plastic detergent bottle thoroughly cleaned and washed out, will invariably do just as well), a new supply of assorted spring washers, a set of new gaskets, a set of new oil seals and O-rings, a set of new valve springs and a torque wrench, should be collected together. You will also need new pallet shims as determined during the assembly stage of the cylinder head, and new big-end bearing cap nuts.

34 Cylinder head - reassembly (including camshaft refitment and valve clearance checking)

1 After a major overhaul, it is recommended that rebuilding should commence with the cylinder head. Then, if it is necessary to obtain pallet shims, other work can proceed in the meantime.

Valves - refitting
2 Fit each valve in turn, wiping down and lubricating each stem before inserting it into the guide from which it was removed.
3 On the camshaft side of the cylinder head fit the lower collar, spring, upper collar and split cotters, using a spring compressor. Repeat for each valve.

Camshaft refitting (cylinder head on the bench)
4 Fit the tappets and pallet shims in their original positions (photos).
5 Oil each camshaft bearing and cap.
6 Place the camshaft into position, position each cap correctly in accordance with the number on it, and refit the retaining nuts. Tighten to the specified torque evenly and in pairs, working along the length of the shaft (photo).
7 Rotate the shaft, using a spanner on the hexagonal end, and check the gap as described in paragraphs 13 to 16.

Camshaft refitting (engine in the car)
8 Ensure that the crankshaft position remains as it was in Section 6, paragraph 10. If it has been disturbed, proceed first as described in paragraphs 9 and 10.
9 Turn the crankshaft until the timing mark on the crankshaft pulley coincides with the zero mark on the timing cover scale, whilst at the same time the rotor arm on the distributor is pointing to the rear bolt securing the inlet manifold to the cylinder head. This positions the crankshaft at TDC on No 1 cylinder.
10 Turn the crankshaft through 90°, thus taking the pistons half-way down the cylinders. This permits the camshaft to be turned by a spanner on the hexagon, without interference between the pistons and valves.
11 Refit the camshaft and associated items as described in paragraphs 4 to 7.
12 With the peak of each cam vertical in turn, check the gap against the Specifications, using feeler gauges.
13 If the measured gap is too large by, say, 0.008 in (0.203 mm) it means that the pallet shim fitted at the moment needs a replacement which is 0.008 in (0.203 mm) thicker. Measure the existing pallet shim thickness, add the hypothetical 0.008 in (0.203 mm) to it and this is the thickness of pallet shim required.
14 If the measured gap is too small by, say, 0.003 in (0.076 mm), it means that the pallet shim fitted at the moment needs to be 0.003 in (0.076 mm) thinner. Measure the existing pallet shim thickness, subtract the hypothetical 0.003 in (0.076 mm) from it and this is the thickness of pallet shim required.
15 When all the gaps have been checked, note the errors, since it may be found that by interchanging some of the existing pallet shims that you will reduce the number which you have to purchase.
16 Remove the camshaft, fit the correct pallet shims, replace the camshaft and recheck the gaps. Exercise care when removing and refitting the camshaft and lubricate the bearings.
17 If the engine is in the car, turn the camshaft until the timing mark on the flange is in line with the mark on the front bearing cap. Then turn the crankshaft through 90° to TDC, in reverse of the direction described in paragraph 10. Do not disturb again until the timing gear is fitted, or damage may be caused to the pistons and valves.

35 Crankshaft - installation

Ensure that the crankcase is thoroughly clean and that all oilways are clear. A thin twist drill or a nylon pipe cleaner is useful for cleaning them out. If possible, blow them out with compressed air.

Treat the crankshaft in the same manner and then inject engine oil into the crankshaft oilways.

Commence work on rebuilding the engine by installing the crankshaft and main bearings:
1 If the old main bearing shells are to be renewed (it is false economy not to do so unless they are virtually new), fit the upper halves of the main bearing shells to their locations in the crankcase, after wiping the locations clean (photo).
2 Note that at the back of each bearing is a tab which engages in the locating grooves in either the crankcase or the main bearing cap housings.
3 If new bearings are being fitted, carefully clean away all traces of the protective grease with which they are coated.
4 With the upper bearing shells securely in place, wipe the lower

Chapter 1/Engine

34.4a Fitting a pallet shim ...

34.4b ... and tappet

34.6 Fitting the bearing caps

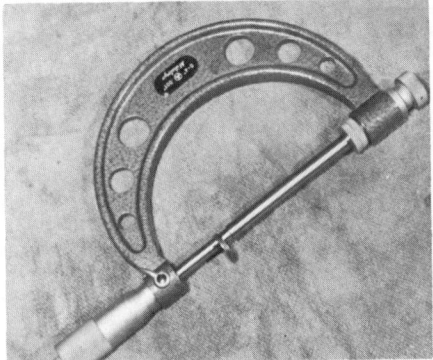
34.13 Measuring the thickness of a pallet shim

35.1 Replacing a main bearing shell in the block

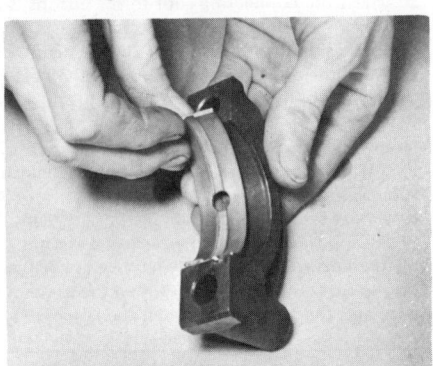
35.4 Replacing a main bearing shell in its cap

35.5 The crankshaft thrust washer being fitted

35.8 Fitting the crankshaft

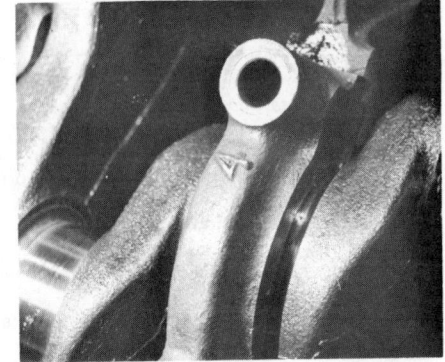
35.9 Fitting the main bearing caps (note the numbers)

35.10 Torque tightening the main bearing caps

bearing cap housings and fit the lower shell bearings to their caps, ensuring that the right shell goes into the right cap if the old bearings are refitted (photo).

5 Wipe the recesses either side of the centre main bearing which locate the thrust washers. Smear some grease onto the thrust washers and place the upper halves in position (photo).

6 Note the milled faces of the thrust washers face outwards as shown in photo.

7 Generously lubricate the crankshaft journals and the upper and lower main bearing shells.

8 Carefully lower the crankshaft into position (photo).

9 Fit the main bearing caps in position ensuring they locate correctly. The mating surfaces must be spotlessly clean or the caps will not seat correctly. As the bearing caps were assembled to the cylinder block and then line bored during manufacture, it is essential that they are returned to the same positions from which they were removed (photo).

10 Refit the main bearing cap bolts and washers and tighten the bolts to the specified torque (photo).

11 Test the crankshaft for freedom of rotation. Should it be stiff to turn or possess high spots, a most careful inspection must be made, preferably by a qualified mechanic, with a micrometer to get to the root of the trouble. It is very seldom that any trouble of this nature will be experienced when fitting the crankshaft.

12 Check the crankshaft endfloat with a feeler gauge measuring the longitudinal movement between the crankshaft and the centre main bearing cap. Endfloat should be 0.003 in to 0.011 in (0.07 mm to 0.28 mm). If the endfloat is incorrect, selective use of thrust washers must be made.

36 Pistons and connecting rods - reassembly

1 If the old pistons are being re-used then they must be mated to the same connecting rod with the same gudgeon pin. If new pistons are being fitted, it does not matter which connecting rod they are used with, but the gudgeon pins should be fitted on the basis of selective assembly.
2 Because aluminium alloy, when hot, expands more than steel, the gudgeon pin may be a very tight fit in the piston when cold. To avoid damage to the piston, it is best to heat it in boiling water when the pin will slide in easily.
3 The same rod and piston must go back into the same bore. If new pistons are being used, ensure that the right connecting rod is placed in each bore.
4 Fit a gudgeon pin circlip in position at one end of the piston.
5 Refit the connecting rods to the pistons, so that the identification numbers and shell bearing keeper recesses on the rods are on the opposite side to the raised part of the piston crown.
6 Slide the gudgeon pin in through the hole in the piston and through the connecting rod little end, until it rests against the previously fitted circlip.
Note: The pin should be a push fit.
7 Fit the second circlip in position. Repeat this procedure for all four pistons and connecting rods.
8 Where special oil control pistons are being fitted, should the position of the top ring be the same as the position of the top ring on the old piston, ensure that a groove has been machined on the top of the new ring so no fouling occurs between the unworn portion at the top of the bore and the piston ring, when the latter is at the top of its stroke.

37 Piston rings - refitting

1 Check that the piston ring grooves and oilways are thoroughly clean and unblocked. Piston rings must always be fitted over the head of the piston and never from the bottom.
2 The easiest method to use when fitting rings, is to wrap a 0.020 in (0.5 mm) feeler gauge round the top of the piston and place the rings on one at a time, starting with the bottom oil control ring over the feeler gauge.
3 The feeler gauge, complete with ring, can then be slid down the piston over the other piston ring grooves until the correct groove is reached. The piston ring is then slid gently off the feeler gauge into the groove.
4 An alternative method is to fit the rings by holding them slightly open with the thumbs and both index fingers. This method requires a steady hand and great care as it is easy to open or twist the ring too much and break it.
5 Fit the oil control expander rail with the ends abutting, but not overlapping. Fit the lower, and then the upper, chrome rail.
6 Fit the scraper ring to the piston centre groove, with the word TOP uppermost.
7 Fit the chromed ring to the top groove.

38 Pistons and connecting rods - refitting

1 To fit the pistons, initially wipe the cylinder bores with a clean rag.
2 Position the ring gaps at approximately 90° to each other and away from the thrust side of the piston.
3 Lightly oil the piston and bores then fit a piston ring clamp. Slide the pistons into their bores from the top of the block, align with the crankpin, and push home with a hammer handle.
4 Ensure that the raised flat part of the piston crown is to the right of the engine as from the driving seat. Arrows which are stamped on the skirt adjacent to the gudgeon pin bores should also face to the right-hand side. Arrows which are stamped on the crown must point to the front.

39 Connecting rods to crankshaft - reassembly

1 Wipe clean the connecting rod half of the big-end bearing cap and the underside of the shell bearing and fit the shell bearing in position with its locating tongue engaged with the corresponding rod (photo).
2 If the bearings are nearly new and are being refitted, then ensure they are replaced in the correct rods.
3 Generously lubricate the crankpin journals with engine oil, and turn the crankshaft so that the crankpin is in the most advantageous position for the connecting rod to be drawn onto it.
4 Wipe clean the connecting rod bearing cap and back of the shell bearing and fit the shell bearing in position, ensuring that the locating tongue at the back of the bearing engages with the locating groove in the connecting rod cap.
5 Generously lubricate the shell bearing and offer up the connecting rod bearing cap to the connecting rod (photo).
6 Fit new connecting rod nuts and tighten to the specified torque.
7 When all the connecting rods have been fitted, rotate the crankshaft to check that everything is free and that there are no high spots causing binding.

40 Crankshaft rear oil seal, adaptor plate, flywheel, oil strainer and sump - refitting

1 Fit the crankshaft rear oil seal housing using a new gasket and oil seal. Check that there is an arrow on the outer face of the seal which indicates the direction of rotation of the crankshaft. Take care not to damage the seal when it is being fitted (photo).
2 The next step is to fit the gearbox adaptor plate (photo).
Note: Spring washers are not used with the bolts.
3 Fit the flywheel with the marks on the hub and crankshaft flange aligned. Lubricate the clutch pilot bearing with grease and press home.
4 Install the retainer plate and progressively tighten the retaining bolts to the specified torque (photo).
5 Fit the oil strainer, and secure with the two bolts and washers (photo).
6 Fit a new sump gasket, and refit the sump in reverse of the procedure given in Section 13 (photos).

41 Timing cover, chain, sprockets, tensioner, guides and jackshaft - refitting

Jackshaft
1 Fit the sprocket to the jackshaft, using a new tab washer. Tighten the bolt to the specified torque.
2 Oil the shaft, and fit it to the cylinder block (photo).
3 Fit the keeper plate, and the two Allen screws (photo).
4 Check that the camshaft flange mark is aligned with that on the front bearing.
5 Check that the crankshaft is at TDC on the firing stroke (temporarily refit the timing cover and pulley to check, if necessary).
6 Turn the jackshaft until the scribed line is as depicted in Fig.13.2. Fit the chain over the sprocket.
7 Fit the adjustable chain guide, leaving the bolt slack.
8 Fit the chain tensioner (see paragraphs 27 to 30) (photo).
9 Fit the fuel pump (see Chapter 3) and the water pump (see Chapter 2).
10 Fit the distributor so that the rotor points to the last manifold to cylinder head bolt hole.
11 Refit the timing cover (see Chapter 13).
12 Refit the camshaft cover (see Section 43).
13 Refit the air conditioning condenser, if applicable.
14 Refit the radiator (see Chapter 2).
15 Refit the inlet manifold with carburettors or fuel rail.
16 Refit the fresh air duct.
17 Fill the cooling system and reconnect the battery.

Timing chain, sprockets, tensioners and guides
18 Proceed as described in paragraph 1.
19 Fit the crankshaft sprocket and shims, and check across to the jackshaft with a straight edge. Correct any misalignment by altering the shims behind the sprocket (photo).
20 Remove the crankshaft sprocket.
21 Proceed in the reverse of Section 10, paragraphs 27 to 22 (excluding paragraph 25). Leave the guide bolts loose (photos).
22 Fit a slave bolt to the centre hole of the timing cover.
23 Secure the camshaft sprocket to the camshaft with one bolt, using a new tab washer, and ensuring that the camshaft flange mark is aligned with that on the front bearing.
24 Turn the jackshaft sprocket to the position indicated in Fig.13.2.

38.3 Installing the piston and connecting rod

39.1 Fitting a big-end bearing shell in its cap

39.5 Fitting the bearing cap to the connecting rod

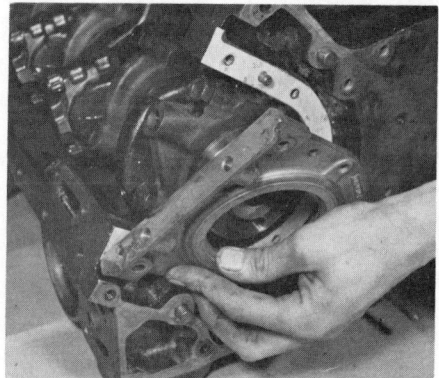
40.1 Fitting the crankshaft rear oil seal housing

40.2 Fitting the gearbox adaptor plate

40.3 Note the flywheel alignment markings

40.4 Installing the retainer plate

40.5 Fitting the oil strainer

40.6a Fitting the sump

40.6b Note the engine crossmember

41.2 Sliding the jackshaft into the block

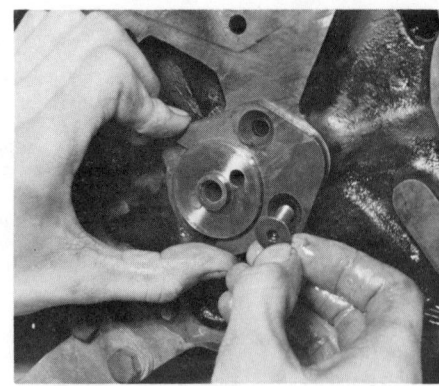
41.3 Fitting the jackshaft thrust plate

41.8 The chain tensioner in position

41.19 Fitting the crankshaft sprocket

41.21a Fitting the timing chain guides ...

41.21b ... and camshaft sprocket support

41.21c The oil slinger in position

Fig. 1.7. Crankshaft/jackshaft alignment

1 Straight edge

25 Check that the distributor rotor arm points to the head of the rear bolt securing the inlet manifold.
26 Proceed as in paragraph 5.
27 Fit the timing chain tensioner ratchet in the bore, and turn clockwise with an Allen key to lock. Insert a temporary spacer between the body and slipper.
28 Fit the tensioner assembly and backplate, secure with two bolts and spring washers, and remove the spacer.
29 Insert a 0.10 in (2.54 mm) feeler gauge between the slipper and the body.
30 Press down on the adjustable chain guide until the feeler gauge is just held, and tighten first the adjustable guide bolt followed by the two remaining bolts. Remove the feeler gauge.
31 Ensure that the camshaft spigot does not foul the support bracket.
32 Revolve the engine to permit the remaining bolt securing the camshaft sprocket to be tightened and locked.
33 Refit the camshaft cover (see Section 43).
34 Refit the distributor cap.
35 Fit the timing cover (see Chapter 13).

Timing cover
36 To refit the timing cover, see Chapter 13.

42 Cylinder head - refitting

1 Ensure that the mating surfaces are clean.
2 Align the marks on the camshaft spigot and bearing cap (photo).
3 Rotate the crankshaft to put the number one piston at TDC, as indicated by the flywheel or crankshaft pulley marks (photo).
4 Position the head gasket on the block, and fit the head (photo).
5 Fit the five studs, and replace the nuts and washers finger tight, together with the two plug lead brackets.
6 Fit the five bolts and washers, finger tight, and the nuts, bolts and washers at the front of the head (photo).
7 Tighten the nuts and bolts in the correct sequence (see Fig.1.8), to the specified torque figure.
Note: Do *not* rotate the camshaft or crankshaft again until the procedure in Section 43 has been carried out.
8 If the engine is in the vehicle, reverse the remaining operations described in Section 7.
9 After the engine has been fully assembled and run for approximately 20 minutes, allow it to cool and then re-tighten all cylinder head nuts and bolts between 55 and 60 lbf ft (77.5 and 80 Nm) in the proper sequence.

43 Timing gear - reconnecting to camshaft

1 Ensure that the conditions described in Section 42, paragraphs 2 and 3, exist.
2 Proceed as in Section 41, paragraphs 23 and 32.
3 Remove the nut holding the camshaft sprocket to the support plate.
4 Refit the semi-circular grommet to the cylinder head.
5 Refit the two cylinder head to timing cover nuts and bolts.
6 Grease the camshaft cover gasket and fit, together with the cover.
7 Fit the four sleeve nuts with new washers and seals.
8 Fit the two front screws and spring washers.

44 Oil transfer adaptor, oil pump and oil filter - reassembly and refitting

1 Refit the oil transfer adaptor using the single bolt and washer. Use new O-rings if necessary (photo).
2 Refit the oil pressure relief valve to the pump, inserting the closed coiled end of the spring first. Use a new O-ring. Reverse the procedure given in Section 18 to reassemble the oil pump.
3 Refit the oil pump and drive, using a new O-ring. This is the reverse of the procedure given in Section 12.
4 Refit the oil filter assembly using a new element and O-ring, reversing the procedure given in Section 12. Run the engine, check for leaks, and if necessary top up the oil level.

45 Distributor and fuel pump - refitting

1 The valve timing should be correct, as described in Chapter 13, Section 3.
2 Refit the distributor as described in Chapter 4, Sections 5 or 7.
3 Refit the fuel pump, as described in Chapter 3, Section 5.

42.2 Alignment marks for the camshaft

42.3 Set number 1 piston at TDC

42.4 Installing the cylinder head

42.6 Don't forget the bolts at the front of the head

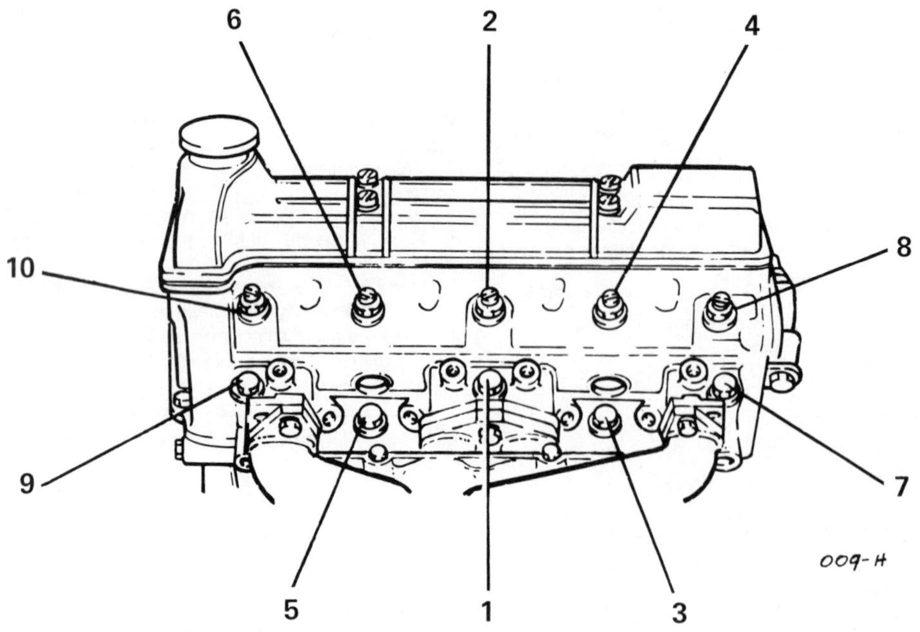

Fig. 1.8. Cylinder head tightening sequence

41.32 Fitting the camshaft sprocket

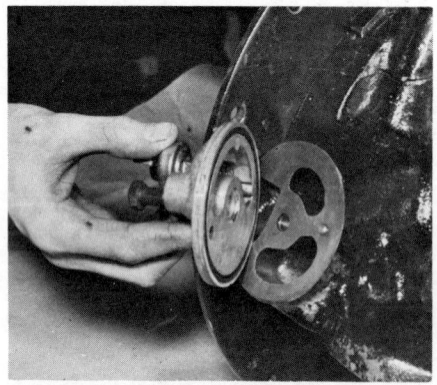

44.1 Fitting the oil transfer housing

44.3 Fitting the oil pump

44.4a Fitting the O-ring ...

44.4b ... and oil filter assembly

48.2 Installing the bellhousing bolts - note the cable clips

46 Water pump, inlet manifold and thermostat - installation

Refer to Chapters 2 and 3 for these procedures.

47 Engine - final reassembly

1 Refit the exhaust manifold as described in Chapter 3.
2 Refit the clutch; refer to Chapter 5 if necessary.
3 Refit the cylinder block and sump drain plugs if not already fitted.
4 Refit the right-hand engine mounting.

48 Engine - installation

1 If the gearbox was removed with the engine it will need to be recoupled. Carefully draw the two together, ensuring that no strain is put onto the gearbox input shaft.
2 Fit all the gearbox/engine bellhousing bolts and at the same time fit the starter motor. Note the position of the cable retaining clips around the bellhousing. The bolt above the starter motor is used for the engine earth strap and the lowest one is used for the exhaust pipe bracket; these can be left out at this stage, if wished. Note that the bellhousing dowel bolt is the lowest one (photo).
3 Using a hoist, lift the combined assembly and swing it into position above the engine compartment of the car. Note the sort of angle that will be required to ensure that the engine is fitted with the least amount of difficulty.
4 Ease the engine into position and allow it to rest on the mountings. As with removal, it will be necessary to support the clutch area with a trolley jack while the engine is being installed. If necessary, the rear of the car can be raised to ease this job.
5 The remainder of the installation procedure is the reverse of that used when removing. Where applicable, adjust the engine stabilizer as described in Section 49. When installing the propeller shaft, ensure that the marks made during removal are aligned. Adjust the choke and throttle cables to remove all free play. Top up the cooling system as described in Chapter 2. Top up the engine and gearbox with the recommended grades of oil.
6 Start the engine and run it up to normal operating temperature whilst checking for coolant and oil leaks. Add coolant if necessary as it circulates through the engine. As soon as possible, check the carburettor and ignition timing settings, and road test the car.

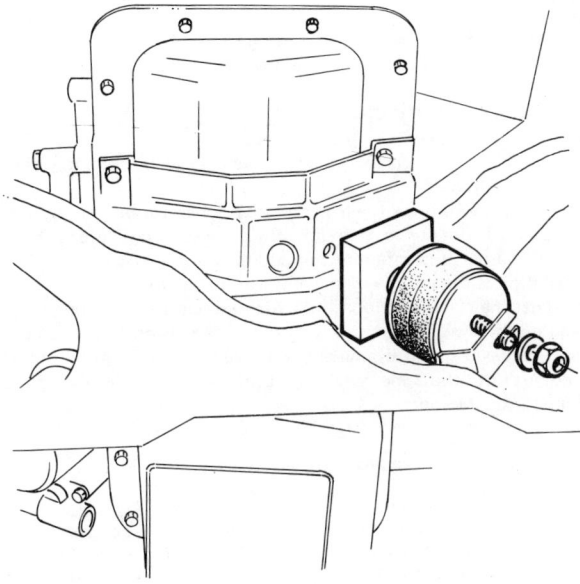

Fig. 1.9. Left-hand engine mounting

49 Engine mountings and stabilizer - removal and installation

Front left hand mounting
1 Detach the battery earth lead and remove the fan guard.
2 Remove the fresh air duct as described in Chapter 12, Section 34, paragraph 3.
3 Raise the car for access to the mounting.
4 Where applicable, remove the stabilizer as described in paragraphs 20 to 22.
5 Remove the nut and washer securing the mounting to the subframe.
6 Support the engine with a jack and wooden packer beneath the sump coupling plate.
7 Remove the nut securing the right-hand mounting to the engine.
8 Raise the jack sufficiently for the left-hand mounting to be removed. Note that on some models a packing piece is also used between the mounting and the sump crossmember.
9 Installation is the reverse of the removal procedure. Where applicable, adjust the stabilizer as described in paragraph 24.

Front right-hand mounting
10 Detach the battery earth lead.
11 Support the engine with a jack and wooden packer beneath the sump coupling plate.
12 Remove the nut securing the mounting to the body bracket, and the two bolts securing the rubber mounting to the engine bracket.
13 Installation is the reverse of the removal procedure.

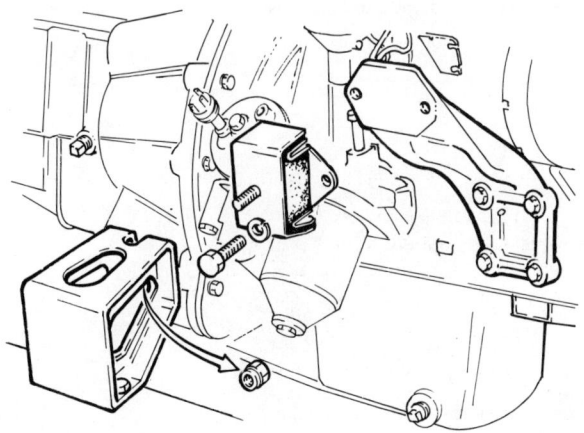

Fig. 1.10. Right-hand engine mounting

Rear centre mounting

14 Disconnect the battery earth lead, then raise the car for access to the rear mounting.
15 Support the gearbox with a jack and wooden packer.
16 Detach the rear mounting from the engine, noting where the various bolts, distance pieces and rubbers are. This is important because several different types of mounting are in service.
17 Remove the four bolts and washers securing the crossmember to the underframe.
18 Where applicable, detach the mounting rubber from the crossmember.
19 Installation is the reverse of the removal procedure.

Engine stabilizer (early models only)

20 Raise the car for access to the stabilizer.
21 Remove the nylon nuts securing the stabilizer to the cylinder block bracket and subframe.
22 From below, withdraw the lower washer, lower rubber and nylon distance piece. From above, withdraw the stabilizer complete with top rubber, top washer, top retaining nut and lock nut.
23 Installation is the reverse of the removal procedure.
24 To adjust the stabilizer, loosen the locknut and top retaining nut, then tighten the lower nylon nut to obtain a dimension of 0.040 in (1.0 mm) between the subframe and the lower rubber. Tighten the top retaining nut until the upper washer just nips the nylon distance piece, then tighten the locknut.

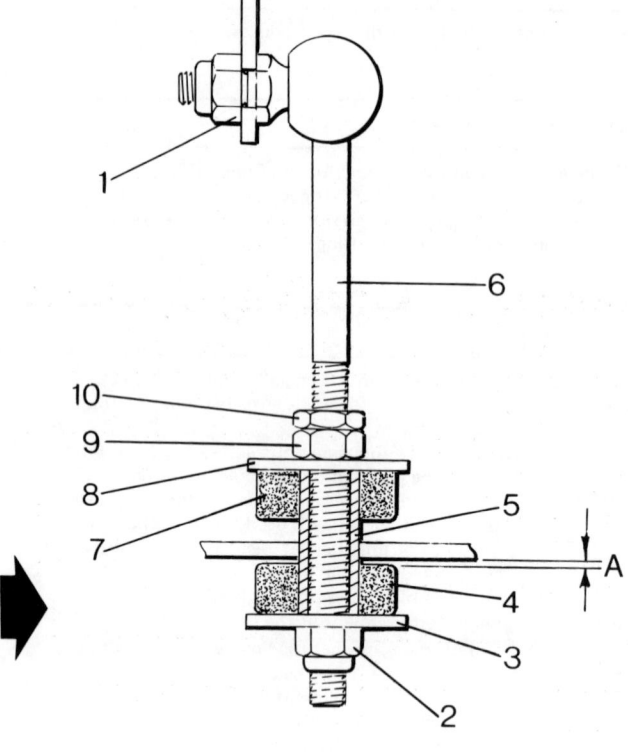

Fig. 1.11. Stabilizer (fitted to early cars)

1 Nylon nut
2 Nylon nut
3 Lower washer
4 Lower rubber
5 Nylon distance piece
6 Stabilizer
7 Top rubber
8 Top washer
9 Top retaining nut
10 Locknut

A = 0.040 in (1.0 mm)

50 Fault diagnosis - engine

Symptom	Reason/s
Engine will not turn over when starter switch is operated	Flat battery. Bad battery connections. Bad connections at solenoid switch and/or starter motor. Defective starter motor.
Engine turns over normally but fails to start	No spark at plugs. No fuel reaching engine. Too much fuel reaching the engine (flooding).
Engine starts but runs unevenly and misfires	Ignition and/or fuel system faults. Incorrect valve clearances. Burnt out valves. Worn out piston rings.
Lack of power	Ignition and/or fuel system faults. Incorrect valve clearances. Burnt out valves. Worn out piston rings.
Excessive oil consumption	Oil leaks from crankshaft oil seals, timing cover gasket, camshaft cover gasket, oil filter gasket, sump gasket, sump plug. Worn piston rings or cylinder bores resulting in oil being burnt by engine. Worn valve guides and/or defective inlet valve stem seals.
Excessive mechanical noise from engine	Wrong valve clearances. Worn crankshaft bearings. Worn cylinders (piston slap). Slack or worn timing chain and sprockets.

Note: When investigating starting and uneven running faults do not be tempted into snap diagnosis. Start from the beginning of the check procedure and follow it through. It will take less time in the long run. Poor performance from the engine in terms of power and economy is not normally diagnosed quickly. In any event the ignition and fuel systems must be checked first before assuming any further investigation needs to be made.

Chapter 2 Cooling system

For modifications, and information applicable to later models, see Supplement at end of manual

Contents

Anti-freeze ... 5	Fan pulley and viscous coupling - removal and installation ... 6
Cooling system - draining ... 2	Fault diagnosis - cooling system ... 12
Cooling system - flushing ... 3	General description ... 1
Cooling system - filling ... 4	Radiator - removal and installation ... 9
Fan/alternator drive belt - removal and installation ... 8	Thermostat - removal, testing and installation ... 10
Fan blades - removal and installation ... 7	Water pump - removal, overhaul and installation ... 11

Specifications

System type ... Pressurized spill-return with thermostatic control; pump and viscous coupling fan assisted.

Pump ... Centrifugal

Thermostat opening temperature
Up to engine No CL1468 ... 82°C (180°F)
From engine No CL1469 ... 88°C (190°F)

Pressure cap setting ... 15 lb f/in^2 (1.05 kg f/cm^2)

Cooling system capacity ... Approx. 13.5 Imp pt (16.0 US pt, 7.7 litres)

Anti-freeze type ... Ethylene glycol with inhibitors for mixed-metal engines (Specification SAE J1034, BS 3151 or BS 3152)

Torque wrench settings

	lb f ft	Nm
Fan to coupling bolts	9	12.0
Fan pulley assembly bolts	14	19.0
Water pump cover bolts	20	27.0
Water pump impeller bolt (left hand thread)	14	19.0
Water outlet elbow to manifold	20	27.0
Water transfer housing to cylinder head	20	27.0

1 General description

The engine cooling water is circulated by a water pump, and the coolant is pressurised. This is primarily to prevent premature boiling in adverse conditions and also to allow the engine to operate at its most efficient running temperature. Two cooling system arrangements have been fitted, with the change having taken place during the 1979 year manufacturing programme. The earlier system was fitted with an expansion tank, and the later with a header tank, both systems being of the "no loss" type, allowing for coolant expansion and contraction.

The cooling system comprises the radiator, water pump, thermostat, interconnecting hoses and the waterways in the cylinder block. The water pump is gear-driven from the engine jackshaft.

The cooling fan is connected to a Holset viscous coupling which limits the fan speed at high engine revolutions. This works in a similar manner to a torque converter, and provides a 'slipping clutch' effect; its aim is to reduce noise and engine loading. This type of fan and coupling is called a Torquatrol unit.

2 Cooling system - draining

1 Place the car on level ground.
2 Disconnect the battery.
3 If the engine is cold, remove both the radiator filler plug and the expansion/header tank cap by turning them anti-clockwise. If the engine is hot, turn the cap only slightly, to relieve pressure in the system, but use a rag over the cap for protection against any escaping steam. If the engine is very hot, the sudden release of the cap (causing a drop in pressure) can result in the water boiling. After relieving the system pressure, remove the filler plug and the filler cap.
4 If antifreeze is used in the cooling system, and has been in use for less than two years, drain the coolant into a container of suitable capacity. This is done by simply removing the bottom radiator hose. There is an additional drain plug on the left-hand side of the engine just forward of the starter motor (the lower one of the two plugs); also remove this.

3 Cooling system - flushing

1 With the passing of time, the cooling system will gradually lose its efficiency as the radiator becomes choked with rust, scale deposits from water and other sediment. To clear the system out, initially drain the system as described previously, then allow water from a hose to run through from the top hose connection and out of the bottom hose connection for several minutes. Install the cylinder block drain plug while this is being done.
2 Refit the hoses and top-up the system with fresh, 'soft' water, (use rain water if possible) and a proprietary flushing compound, following the manufacturer's instructions carefully. (Refer to the next Section).
3 If the cooling system is very dirty, reverse flush it, by forcing water from a hose up through the bottom hose connection for about five to ten minutes, then complete the operation as described in paragraph 2, of this Section.

4 Cooling system - filling

Expansion tank system

1 Refit the bottom radiator hose and cylinder block drain plug.
2 Half fill the expansion tank with the coolant. If antifreeze is to be added, refer to the following Section for the correct proportions, and half fill the expansion tank with antifreeze only.
3 Now move the heater control to "Hot" and remove the filler plug from the thermostat housing dome.
4 Fill the system with coolant through the filler plug orifice then refit the filler plug. Do not refit the expansion tank pressure cap.
5 Run the engine at moderate speed for about three minutes, then switch off.
6 Refit the expansion tank pressure cap.
7 Remove the thermostat housing filler plug.
8 Squeeze the large top hose to expel air.
9 Top up the coolant via the filler plug.
10 Recheck the expansion tank level, and top up as required.
11 Subsequent topping-up of the cooling system should be via the expansion tank, which should be kept half-full.

5 Antifreeze solution

1 Prior to the onset of cold winter weather, it is essential that antifreeze is added to the engine coolant if it has not been previously used.
2 Any antifreeze which conforms with specification 'SAE J1034', 'BS 3151' or 'BS 3152' can be used. Never use an antifreeze with an alcohol base as evaporation is too high.
3 Antifreeze can be left in the cooling system for up to two years, but after six months it is advisable to have the specific gravity of the coolant checked at your local garage, and thereafter, every three months.
4 Listed below are the amounts of antifreeze which should be added to ensure adequate protection down to the temperature given:

Percentage antifreeze	Protects down to	
25%	−12°C	10°F
30%	−16°C	3°F
35%	−20°C	−4°F
50%	−36°C	−33°F

6 Fan pulley and viscous coupling - removal and installation

1 Disconnect the battery earth lead.
2 On USA models, remove the bonnet and air pump (refer to Chapters 3 and 12 as necessary).
3 Loosen the alternator adjustment bolts (refer to Chapter 10 if necessary), and remove the fan belt.
4 Where applicable, remove the five air compressor-to-steady bracket bolts. Then remove the two remaining fan pulley assembly bolts.
5 Where no air conditioning is installed, remove the four bolts retaining the pulley assembly (photo).
6 If there is sufficient room, withdraw the pulley assembly from the left-hand side of the engine. Alternatively, remove the fan blade guard (two bolts), then loosen the four radiator top location bolts to enable the radiator to be moved forwards. Withdraw the fan assembly.
7 If necessary, remove the fan blades (four nuts and bolts) (photo).
8 To remove the viscous coupling, support the Torquatrol unit as close to the centre as possible, and press the pulley assembly through the unit.
9 To reassemble, reverse the dismantling procedure. Use a new tolerance ring, place it on the shaft, and compress it with the fingers whilst locating the coupling. Make sure that the coupling is the right way round.

7 Fan blades - removal and installation

1 Disconnect the battery earth lead.
2 Remove the fan blade guard (two bolts).
3 Loosen the four radiator top location bolts and move the radiator away from the engine.
4 Remove the four nuts and bolts, and detach the fan blades from the viscous coupling.
5 Installation is the reverse of the removal procedures.

8 Fan/alternator drive belt - removal and installation

1 Disconnect the battery earth lead.
2 Where applicable, loosen the air conditioning compressor belt adjustment bolts, then remove the fan blade guard (two bolts). Also loosen the four fan blade attachment bolts to provide clearance when removing the belt.
3 Where applicable, remove the air pump drive belt (refer to Chapter 3 if necessary).
4 Loosen the alternator adjustment bolts (refer to Chapter 10 if necessary), and remove the drive belt. Feed it over the fan blades, air pump belt and/or compressor belt, as applicable.
5 Installation is the reverse of the removal procedure. Adjust the belt tension as described in Chapter 10, Section 8.

9 Radiator - removal and installation

1 Disconnect the battery earth lead.
2 Drain the cooling system as described in Section 2.
3 Detach the two top radiator hoses.
4 Where applicable detach the temperature sensor leads from the top left-hand side of the radiator.
5 Remove the four nuts and washers securing the two radiator top attachment brackets (photo).
6 Remove the brackets, and withdraw the radiator vertically.
7 Installation is the reverse of the removal procedure, but ensure that the top mounting rubbers are in the brackets, and the two bottom ones are in position on the radiator. Refer to Section 4 when fitting the cooling system (photo).

10 Thermostat - removal, testing and installation

1 Disconnect the battery earth lead then partially drain the cooling system to bring the coolant level to below the thermostat height.
2 Unscrew the two set bolts and spring washers from the thermostat housing, and lift the housing and gasket away. Take out the thermostat (photo).
3 Test the thermostat for correct functioning by placing it in a container of warm water together with a thermometer.
4 Heat the water and note when the thermostat begins to open. This temperature is stamped on the flange of the thermostat and is also given in the Specifications Section.
5 Discard the thermostat if it opens too early. Continue heating the water until the thermostat is fully open, then let it cool down naturally. If the thermostat will not open fully in boiling water, or does not close down as the water cools, then it must be renewed.
6 If the thermostat is stuck open when cold, this will be apparent when removing it from the housing.
7 Installation of the thermostat is the reverse of the removal procedure. Remember to use a new gasket between the thermostat housing and the thermostat.

4.4 Engine filler plug

6.5 Pulley assembly bolts

6.7 Fan and viscous coupling

9.5 Radiator top attachment

9.7 Radiator mounting rubber

10.2 Removing the thermostat

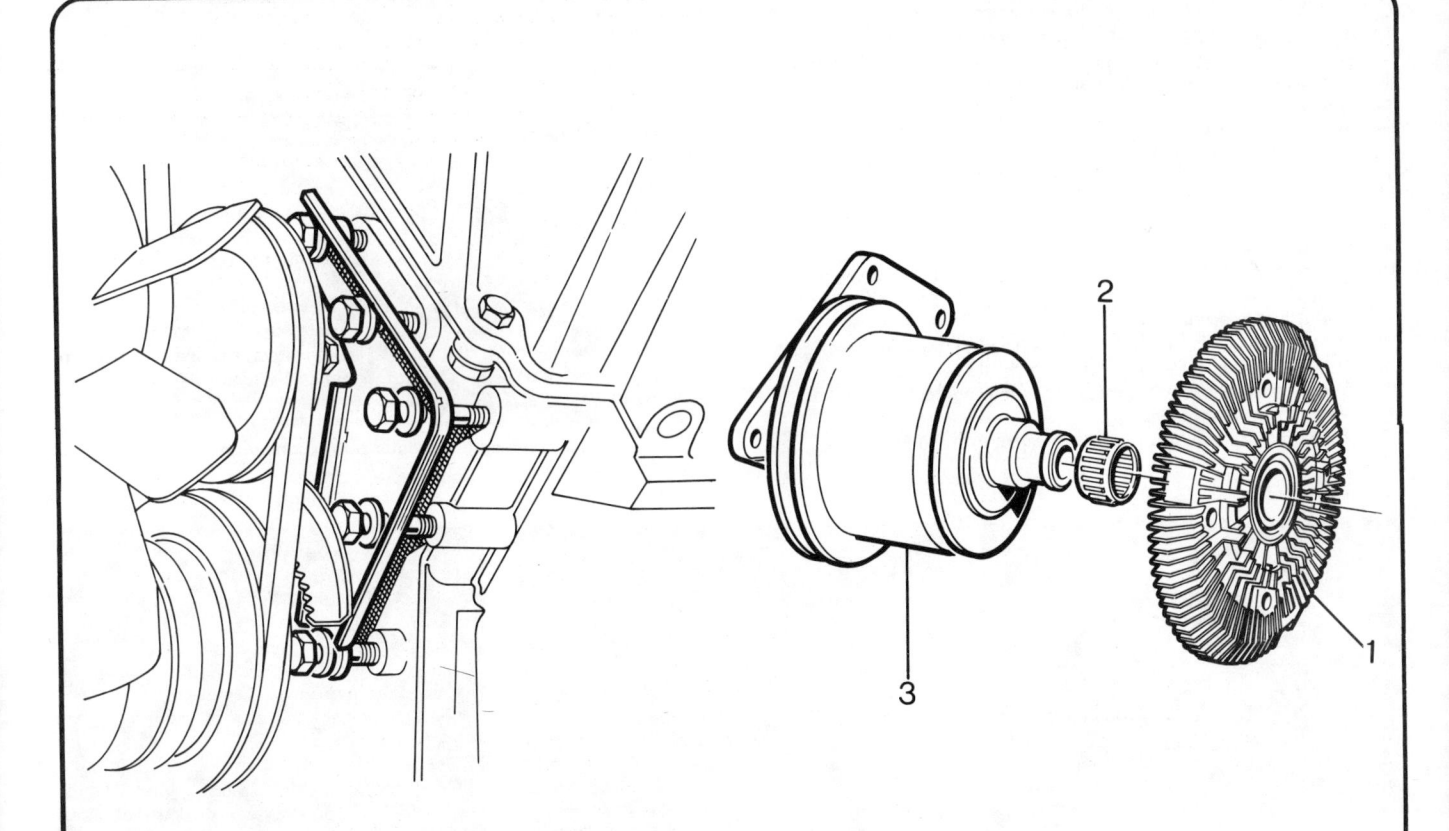

Fig. 2.1. Air compressor-to-steady bracket bolts
(The two top bolts also retain the pulley assembly)

Fig. 2.2. Torquatrol unit

1 Viscous coupling
2 Tolerance ring
3 Pulley assembly

11 Water pump - removal, overhaul and installation

Note: Before attempting to remove the water pump from the block it must be appreciated that there is a likelihood that special tools will be required. These are identified as Triumph part numbers 'S4235A-10' and '4235A'.

1 Detach the battery earth lead.
2 Remove the inlet manifold complete with carburettor(s) as described in Chapter 3.
3 Remove the connecting tube and water hose connected to the pump cover (photo).
4 Remove the three bolts and flat washers securing the pump cover. Take the cover off.
5 Remove the impeller retaining bolt (photo) by turning clockwise (ie, left-hand threaded bolt) until:
a) either the water pump is released from the jackshaft and can be lifted out or
b) the centre bolt alone comes out.
6 If (b) applies, fit the impact tool 'S4235A-10' and adaptor '4235A' and drive out the pump.

Note: If it is only necessary to remove the water pump drive to enable the jackshaft to be withdrawn and (b) applies, it is possible to remove the impeller by fitting a 'Mole' grip to the top spigot and levering off (photo). If a large washer is then fitted beneath the head of the bolt, the driveshaft can be levered out (photo).

7 With the water pump removed from the block, drive the shaft out of the impeller. Make sure that the back of the impeller is adequately supported.
8 Drive out the driveshaft using a soft faced hammer.
9 Take off the O-ring, graphite seal, water thrower, oil seal and circlip in that order.
10 Carefully drift the bearing off the shaft.
11 Take off the oil thrower, and remove the O-ring from the housing.
12 Discard the bearing, seals and O-ring. It is false economy not to do this even if they are in an apparently serviceable condition.
13 Carefully examine the housing for cracks and corrosion. If cracked it must be renewed, but corrosion can be carefully dressed out with a fine file and emery paper; unless it is so severe that it will not permit the O-rings to seal properly. Make sure that no swarf remains after any local dressing.
14 Examine the shaft for wear. If this is present, or if there is any wear evident on the gears, the shaft should be renewed. It is permissible to carefully dress out any light scoring.
15 Check the impeller for corrosion and cracks. Light corrosion can be carefully dressed out but if any other defect is present, it too, must be renewed.
16 To reassemble, fit the oil thrower, concave side towards the gear.
17 Fit the bearing to the shaft, then the circlip. The original one may be reused if undamaged.
18 Now carefully press the shaft into the housing, then fit the oil seal inside the housing with the flat face towards the bearing.
19 Carefully press the water thrower onto the shoulder of the shaft, concave side towards the bearing.
20 Fit the graphite seal, flat face downwards, into the housing.

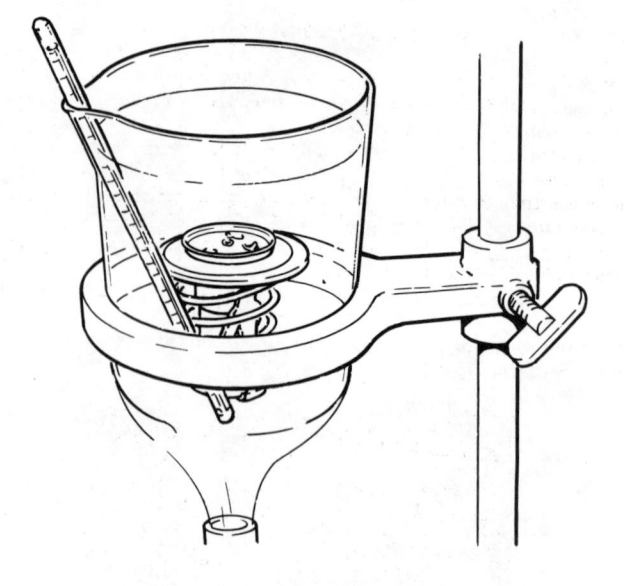

Fig. 2.3. Testing a thermostat

11.3 Removing the water pump connecting tube

11.5 The impeller and left-hand threaded bolt

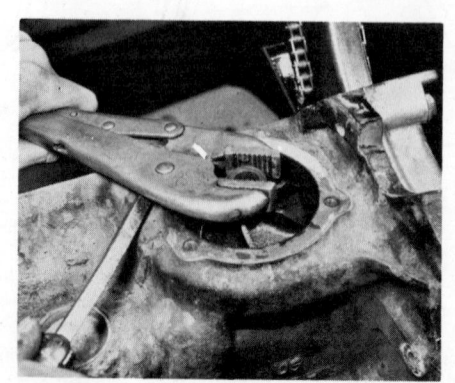

11.6a Using a Mole grip to remove the impeller

11.6b Levering out the water pump driveshaft

Chapter 2/Cooling system

21 Fit the shaft O-ring followed by the impeller. Tighten the left-hand threaded bolt to the specified torque.
22 Fit two new O-rings to the grooves in the housing, the smaller one near the gear.
23 Smear a little rubber grease on the surface of the housing O-rings then fit the pump to the cylinder block. Turn the impeller anticlockwise to engage the gears but do not use force either for turning, or pressing in. Don't forget to make sure that the cavity walls of the cylinder block are clean before fitting the pump.
24 Clean the mating faces of the pump cover and block, and refit the pump cover without the gasket. Fit the cover bolts finger tight only, and check the gap with a feeler gauge, adjusting the bolts a little if necessary until the gap is equal all round.
25 Select gaskets of a thickness equal to the gap between the mating surfaces plus 0.010 to 0.025 in (0.25 to 0.5 mm).
26 Fit the gaskets and cover evenly, and tighten to the specified torque (photo).
27 Fit the connecting tube, then install the manifold (see Chapter 3 for further information).

11.26 Water pump cover in position

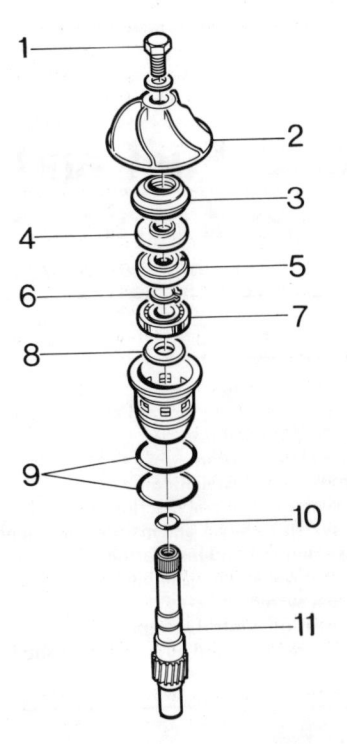

Fig. 2.4. The component parts of the water pump

1 Centre bolt and washer
2 Impeller
3 Graphite seal
4 Water thrower
5 Oil seal
6 Circlip
7 Bearing
8 Oil thrower
9 Housing O-rings
10 Shaft O-ring
11 Shaft

12 Fault diagnosis - cooling system

Symptom	Reason/s
Heat generated in engine not being successfully disposed of by radiator	Insufficient water in cooling system. Drive belt slipping (accompanied by a shrieking noise on rapid engine acceleration). Radiator core blocked or radiator grille restricted. Water hose collapsed, impeding flow. Thermostat not opening properly. Ignition advance and retard incorrectly set (accompanied by loss of power and perhaps misfiring). Carburettor incorrectly adjusted (mixture too weak). Exhaust system partially blocked. Oil level in sump too low. Blown cylinder head gasket (water/steam being forced down the radiator overflow pipe under pressure). Engine not yet run-in. Brakes binding. Viscous coupling faulty.
Too much heat being dispersed by radiator	Thermostat jammed open. Incorrect grade of thermostat fitted allowing premature opening of valve. Thermostat missing. Viscous coupling faulty.
Leaks in system	Loose clips on water hoses. Water hoses perished and leaking. Radiator core leaking. Thermostat gasket leaking. Pressure cap spring worn or seal ineffective. Blown cylinder head gasket (pressure in system forcing water/steam down expansion tank pipe). Cylinder wall or head cracked.
Temperature gauge reading incorrect	Poor connection of earth cable (either at the battery negative post or on the body)

Chapter 3 Fuel and carburation, exhaust and emission control systems

For modifications, and information applicable to later models, see Supplement at end of manual

Contents

Air cleaner - general description ... 2	Fuel pump - removal, filter cleaning and installation ... 5
Air cleaner (Twin carburettors) - removal and installation ... 3	Fuel tank - removal and installation ... 17
Air cleaner element (Twin carburettors) - renewal ... 4	General description ... 1
Carburettors - general description ... 6	Inlet manifold - removal and installation ... 18
Carburettors - removal and installation ... 7	Mixture (choke) control cable - removal and installation ... 15
Emission control systems - general description ... 21	Single Zenith-Stromberg carburettor - repair operations ... 13
Emission control systems - repair and maintenance operations ... 22	Single Zenith-Stromberg carburettor - timing and adjusting ... 12
Exhaust manifold - removal and installation ... 19	Throttle pedal assembly - removal and installation ... 14
Exhaust system - removal and installation ... 20	Twin SU carburettors - dismantling, overhaul and reassembly ... 9
Fault diagnosis - fuel system ... 23	Twin SU carburettors - tuning and adjusting ... 8
Fault diagnosis - emission control system ... 24	Twin Zenith-Stromberg carburettors - repair operations ... 11
Fuel filler cap, filler assembly and hose - removal and installation ... 16	Twin Zenith-Stromberg carburettors - tuning and adjusting ... 10

Specifications

Air cleaner
Dry paper element. USA models have inlet air temperature control for twin carburettor models

Carburettors: *UK and general market models*
Type	Twin side-draught SU HS6
Needle size	BDM
Jet size	100

Carburettors: *USA*
Type
1975/76 models, Federal specification	Twin Zenith-Stromberg 175CD-SEVX
1975/76 models, California specification	Single Zenith-Stromberg 175CD 4TV
1977 models	Twin Zenith-Stromberg 175CD FEVX
1978/79 models	Twin Zenith-Stromberg 175CD FVX
Choke size	1.625 in (41.275 mm)

Needle
1975/76 models	45C
1977 to 79 models, Federal specification	B1 DH
1980 models, Federal specification	B1 EP
1977 to 79 models, California specification	B1 EP
Orifice	0.100
Piston return spring	Red

Carburettors (general)
Engine idle speed:
UK and general market models	650/850 rpm
USA models	700/900 rpm
Fast idle speed	1500/1700 rpm

CO emission at idle speed:
UK and general market models	2.5 to 4.5%
USA models	Refer to emission control decal in engine compartment

Fuel pump
Type	Mechanical, diaphragm type
Operating pressure	2.5 to 3.5 lb f/in^2 (0.17 to 0.24 kg f/m^2)

Fuel tank capacity
12 Imp gal (14.4 US gal/54 litre)

Fuel octane requirement (research method rating):
UK and general market models	97 octane (4-star rating)
USA, except catalytic converter models	91 octane (regular fuel), leaded or unleaded
USA catalytic converter models	91 octane unleaded only

Fuel injection
See Chapter 13

Chapter 3/Fuel and carburation, exhaust and emission control systems

Torque wrench settings

	lb f ft	Nm
Air pump bolts ...	20	27
Carburettor and flexible mounting bolts	14	19
Fuel pump bolts ...	14	19
Hot air manifold-to-exhaust manifold bolts	20	27
Inlet manifold bolts ...	20	27
Exhaust manifold bolts ...	34	46
Exhaust pipe-to-manifold bolts	37	50
Exhaust pipe bracket bolts ...	21	28

1 General description

The fuel system comprises a fuel tank at the rear of the car, from which the mechanically operated pump delivers fuel to the carburettor(s). A combined air cleaner and silencer incorporates a paper element which must be renewed at the specified intervals. USA models also have an inlet air temperature control on the air cleaner.

A description of the components is given in the relevant Section in the Chapter.

Note: Where a fuel injection system is fitted, see Chapter 13 regarding fuel system depressurisation.

2 Air cleaner - general description

1 The air cleaner used is a metal container housing a paper filter element.
2 Models with an inlet air temperature control system have either, in the case of the single carburettor model, a bi-metal strip which opens or closes a flap valve under the influence of temperature, or, in the case of the twin carburettor model, a flap valve operated by a separate temperature sensor unit. In either case, the design purpose is to maintain a carburettor inlet air temperature of 100°F (38°C).

3 Air cleaner (twin carburettors) - removal and installation

1 Disconnect the air inlet pipe(s). Where applicable, also detach the vacuum line (photo).
2 Where applicable, detach the engine dipstick support from the air cleaner; also loosen the nut securing the throttle linkage bracket to the air cleaner.
3 Release the cover retaining clips; remove the cover and air cleaner element.
4 Where applicable, remove the two fuel traps, disconnect the vacuum pipe from the sensor unit and remove the single retaining nut from behind the backplate.
5 Remove the six bolts and spring washers, and withdraw the backplate. Remove the gaskets (photo).
6 Installation is the reverse of the removal procedure, new gaskets being used for each carburettor.

4 Air cleaner element (twin carburettors) - renewal

1 Proceed as described in Paragraphs 1, 2 and 3 of previous Section, but do not loosen the throttle linkage bracket nut.
2 Clean the inside of the cover and air filter body, then install a new element. Note that the metal frame or rubber seal (as applicable) should be towards the carburettors (photo).
3 The remainder of the installation is the reverse of the removal procedure.

5 Fuel pump - removal, filter cleaning and installation

Filter cleaning
1 Remove the fresh air duct as described in Chapter 12, Section 34, Paragraph 3.
2 Remove the single screw and take off the pump cover.
3 Lift out the filter, and remove sediment from it and the cover.
4 Install the filter and cover, using a new rubber sealing ring and/or fibre washer as necessary.
5 Install the fresh air duct.

Removal and installation
6 Remove the fresh air duct (see Paragraph 1).
7 Remove the pump inlet and outlet hoses, plugging each one to prevent loss of fuel.
8 Remove the retaining bolts and flat washers, and lift away the pump and gasket.
9 Installation is the reverse of the removal procedure. Use a new gasket if necessary, and ensure that the pump actuating lever sits on top of the jackshaft cam. The carburettor connection is the one on the pump body.

6 Carburettors - general description

The carburettors are of the variable choke type. The fuel, which is drawn into the air passage through a jet orifice, is metered by a tapered needle which moves in and out of the jet, thus varying the effective size of the orifice. This needle is attached to, and moves with, the air valve piston which controls the variable choke opening.

At rest, the air valve piston is right down, choking off the air supply and the tapered needle is fully home with the jet virtually cutting off the fuel outlet from the jet. For starting, the choke control is used. On the SU carburettor the jet is pulled downwards away from the needle, thus increasing the amount of fuel which is mixed with the incoming air and providing a richer mixture. The Zenith-Stromberg carburettor incorporates a disc valve for cold starting which allows additional fuel to flow into the mixture stream. The disc valve itself incorporates several orifices which are progressively uncovered as the disc is moved when the choke control is pulled. On both types of carburettor the throttle butterfly is also opened a small amount. Also included in the Zenith-Stromberg carburettor is a temperature controlled valve which weakens the mixture under light load and idling conditions when the engine is hot. The choke on single carburettor models is operated automatically by a bi-metallic spring arrangement which is heated by water from the engine cooling system.

As soon as the engine fires, the suction from the engine (or manifold depression) is partially diverted to the upper side of the chamber in which the diaphragm attached to the air valve piston is positioned. This causes the valve to rise and provides sufficient air flow to enable the engine to run. As the throttle is opened further, manifold depression is reduced and now it is the speed of air through the venturi which causes the depression in the upper chamber, thus causing the piston to rise further. If the throttle is opened suddenly, the natural tendency of the air valve piston to rise - causing a weak mixture when it is least required (ie during acceleration) - is prevented by a hydraulic damper which delays the piston in its upward travel. The air intake is thus restricted and a proportionately larger quantity of fuel to air is drawn through.

7 Carburettors - removal and installation

Note: *At the time of compilation of this Manual, no details were available for removal of the single carburettor used on catalytic converter models.*

SU carburettors
1 Detach the battery earth lead.

3.1 Air cleaner air inlet pipes (typical)

3.5 Air cleaner backplate (typical)

4.2 Installing the air cleaner element

6.1a General views ...

6.1b ... of twin SU ...

6.1c ... carburettor linkage

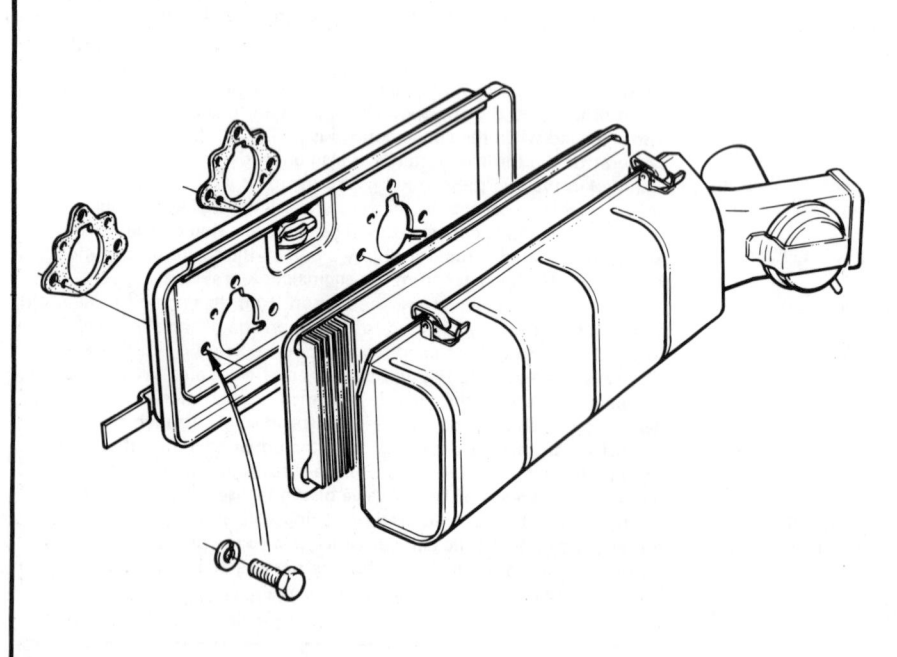

Fig. 3.1. Air cleaner - USA twin carburettor model shown

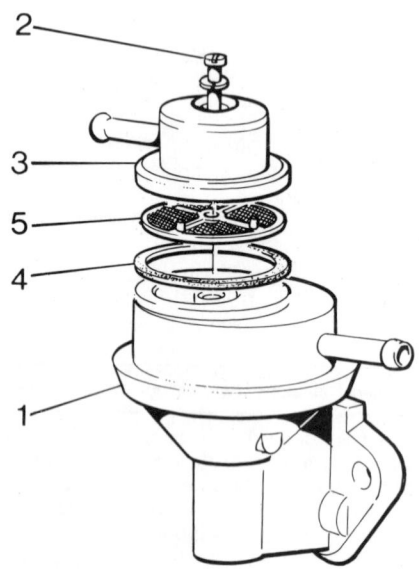

Fig. 3.2. Fuel pump - exploded view

1 Pump body
2 Screw and washer
3 Pump cover
4 Sealing ring
5 Filter

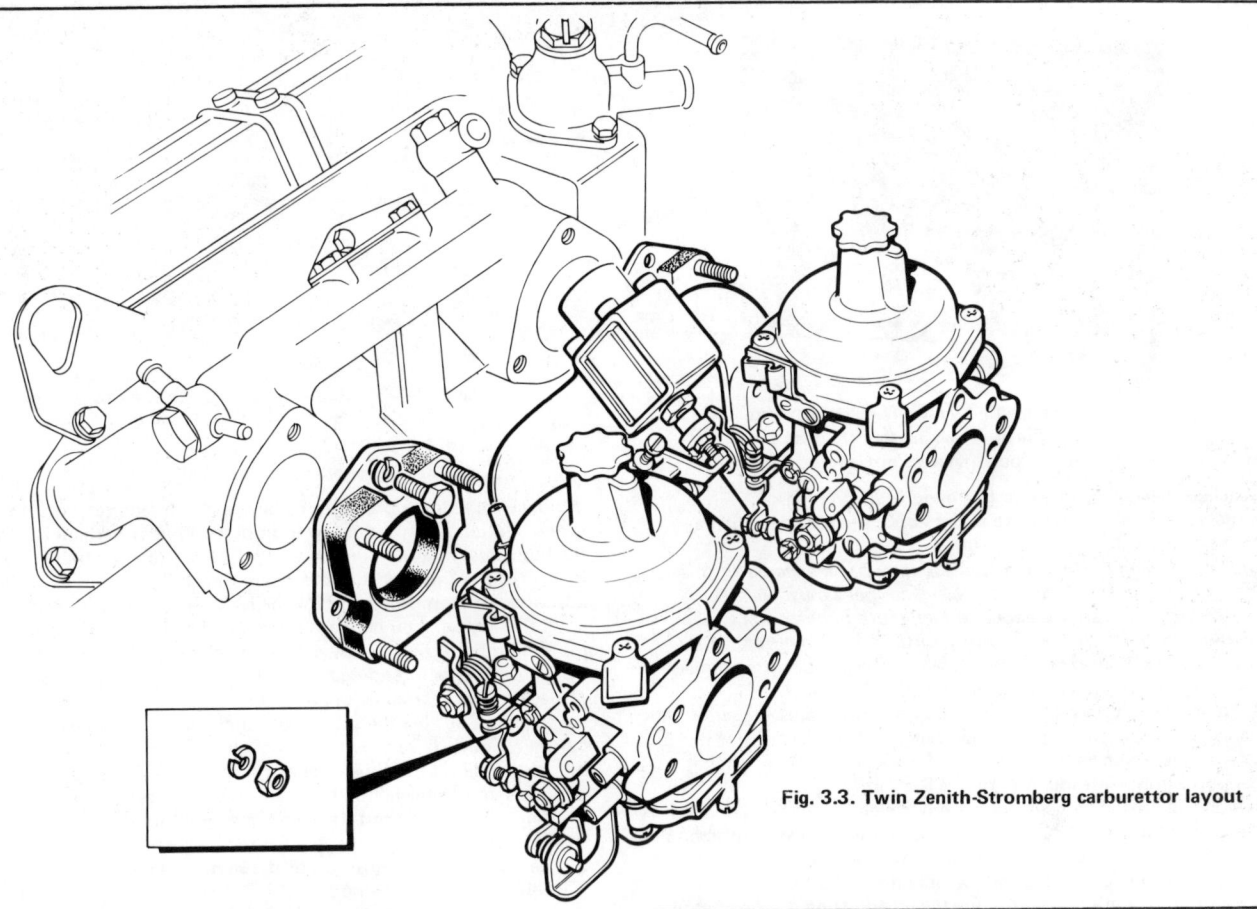

Chapter 3/Fuel and carburation, exhaust and emission control systems

Fig. 3.3. Twin Zenith-Stromberg carburettor layout

2 Remove the air cleaner as described in Section 3.
3 Remove the fresh air duct as described in Chapter 12, Section 34, Paragraph 3.
4 Pull off the fuel overflow pipes from each float chamber.
5 Pull off the vacuum advance pipe from the rear carburettor.
6 Disconnect the engine breather pipes and fuel feed pipes from the carburettors.
7 Disconnect the vertical link from the throttle interconnection shaft, and the three throttle shaft return springs.
8 Remove the four nuts, flat washers and spring washers securing each carburettor to the inlet manifold.
9 Withdraw the two carburettors and remove the flange gaskets.
10 If necessary, the flexible mountings can be removed by undoing the three bolts on each one.
11 Installation of the carburettors is the reverse of the removal procedure. Ensure that the mating faces are clean, and use new flange gaskets. Check that the mixture control pins are located in their respective forks. Tune and adjust the carburettors before the air cleaner is installed (see Section 8).

Zenith-Stromberg carburettors
12 Detach the battery earth lead.
13 Remove the air cleaner as described in Section 3.
14 Remove the fresh air duct as described in Chapter 12, Section 34, Paragraph 3.
15 Disconnect the emission control, vacuum and fuel pipes from the carburettors.
16 Disconnect the throttle return spring and the throttle link.
17 Remove the four nuts, flat washers and spring washers securing each carburettor to the inlet manifold.
18 Withdraw the two carburettors complete with gasket.
19 Detach the throttle jack leads and remove the jack mounting bracket, where relevant.
20 If necessary, the flexible mountings can be removed by undoing the three bolts on each one.
21 Installation of the carburettors is the reverse of the removal procedure. Ensure that the mating faces are clean and use new gaskets. Tune and adjust the carburettors before the air cleaner is installed (see Section 10).

8 Twin SU carburettors - tuning and adjusting

Before attempting to tune and adjust the carburettors, unless you have already stripped and overhauled them yourself, it is important that the reader is aware of the delicate nature of this task. In this respect it may well be considered worthwhile to entrust the job to a properly equipped Triumph garage. For those motorists intending to do the job themselves it is useful to have an airflow balancing meter, although, listening carefully at the end of a rubber or plastic tube of about ¼ in (6 mm) diameter will usually be sufficient. It is essential that before any attempt is made to adjust the carburettors, ignition timing, tappet clearances, spark plugs and the distributor contact breaker are checked and adjusted as necessary.

1 Remove the air cleaner (if not already removed) - see Section 3.
2 Top-up the carburettor piston dampers with engine oil until the level is ½ in (13 mm) above the top of the hollow piston rod.
3 Check that when the throttle pedal is depressed, the linkage moves and that it returns when the throttle pedal pressure is released.
4 Check that a small clearance exists between the ends of the fast idle screws and their cams.
5 Disconnect the mixture (choke) control cable at the trunnion.
6 Turn the throttle adjusting screws until they just clear the throttle levers when the throttles are closed; now turn the screws 1½ turns clockwise (photo).
7 Raise the piston of each carburettor either by means of the piston lifting pin (where fitted) or by using a small screwdriver; check that the piston falls freely when released. **Note:** If the piston does not fall freely, mark the fitted position of the suction chamber and remove it. Clean off any fuel deposits with fuel or methylated spirits, wipe the parts dry and refit the cover. If the piston will still not fall freely, remove the carburettor from the car and carry out the procedures given in Section 9, Paragraph 35.
8 Lift and support each piston clear of the carburettor bridge so that the jet can be seen. Turn each jet adjusting nut to bring the jet flush with the carburettor bridge.
9 Check that the needle shank is flush with the underside of the piston, then turn the jet adjusting nuts down two turns.
10 Run the engine at a fast idle until the normal running temperature

44 Chapter 3/Fuel and carburation, exhaust and emission control systems

8.6 Throttle adjusting screw. The arrangement on some carburettors may be slightly different, but the relative position does not alter

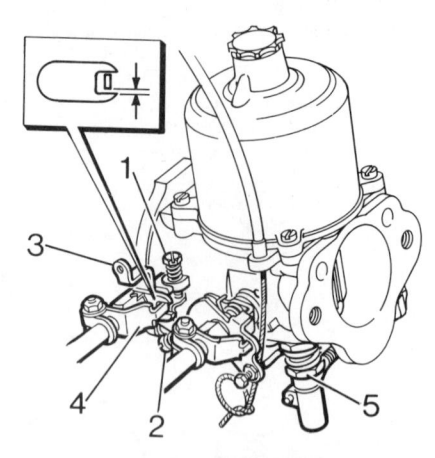

Fig. 3.4. SU carburettor adjustment points

1 Throttle adjusting screw *
2 Fast idle adjusting screw
3 Throttle lever stop
4 Throttle lever pin
5 Jet adjusting nut

* Position varies on some carburettors - see photo 8.6

is reached then allow a further five minutes of running.
11 Increase the engine speed to about 2500 rpm for about half a minute.
Note: *The tuning procedure is now about to commence but if this takes longer than three minutes, the engine speed must again be increased to 2500 rpm for about half a minute to purge the system. This must then be repeated for every further three minutes of tuning.*
12 Slacken both nuts and bolts on the throttle spindle interconnections, and both nuts and bolts on the jet control connections.
13 Adjust the idle speed by means of the throttle adjusting screws, then balance the carburettor inlet airflow using an airflow meter. If this is not available, listen to the airflow 'hiss' using the small bore tube referred to in the introduction to this Section.
14 If the idle speed and balance are not correct, turn the idle speed screws to obtain the correct airflow balance then adjust both screws by equal amounts to obtain the correct idle speed.
15 If a smooth idle speed cannot be obtained, adjust the mixture on each carburettor in turn by rotating the jet adjusting nut one flat at a time in the same direction until the fastest speed is obtained.
16 Now turn each adjusting screw up one flat at a time until the engine speed just commences to fall.
17 Now turn the adjusting nuts down by the minimum amount until the fastest speed is regained.
18 If necessary, readjust the idle speed by turning each throttle adjusting screw by the same amount.
19 At this stage, the exhaust gas should be analysed if possible to check its CO content. It may be found necessary to make a further minor adjustment, but only turn each jet adjusting nut by the minimum amount necessary.
20 With the throttle lever held against its stop, rotate the lever pins on the throttle interconnection shaft until a gap of 0.010in (0.25mm) exists between the pins and the lower arms of the forks (see Fig. 3.4).
21 Ensure that the link rod is at the top of the elongated hole in the throttle shaft actuating lever, then tighten the lever pin clamping nuts and bolts.
22 Take up any slack by adjusting the throttle cable.
23 If necessary adjust the throttle pedal stop to provide full opening of the throttle plate.
24 Recheck the balance by running the engine at 1500 rpm (see Paragraph 13).
25 With the fast idle cams against their stops, tighten the jet control interconnection clamps so that both cams start to move simultaneously.
26 Connect the mixture (choke) control cable to the trunnion ensuring that 1/16 in (1.5mm) of free movement exists before the cams move.
27 Pull out the mixture control knob until the linkage is about to move the jet, then recheck the balance while turning the adjusting screws to obtain the specified fast idle speed.
28 Where applicable, adjust the automatic transmission kickdown cable as described in Chapter 13.
29 On completion, install the air cleaner as described in Section 3.

9 Twin SU carburettors - dismantling, overhaul and reassembly

Before commencing to completely dismantle a carburettor, whether it is for overhaul or examination, it must be appreciated that you are dealing with a precision metering device which requires a great deal of care and attention for it to work properly. Where overhaul is concerned, the individual may well decide that a service exchange unit is preferable in view of the time and trouble saved. Generally speaking, it is not advisable, or even necessary, for the carburettor to be dismantled for any other reason, and here again the individual may well consider the job should be carried out by a Triumph dealer or carburettor specialist. For those people intending to do the job themselves, there are no major obstacles to be overcome provided that care is taken. No special tools are required, but for the subsequent tuning and adjustment, an airflow balancing meter is very useful, although listening carefully at the end of a rubber or plastic tube of about ¼ in (6mm) diameter will usually be sufficient. To dismantle the carburettors proceed as follows:
1 Remove the carburettors from the engine and thoroughly clean their exteriors with petrol.
2 Mark the installed position of the suction chamber to ensure that it is refitted to the same body.
3 Unscrew the piston damper, then remove the sealing washer.
4 Remove the suction chamber (three screws) and take out the spring.
5 Carefully lift out the piston assembly and empty any residual oil from the reservoir.
6 Take out the guide locking screw and remove the needle assembly. Separate the needle, guide and spring, taking care that they are not damaged in any way.
7 Where fitted, push the piston lifting pin upwards, then remove the circlip and separate the assembly.
8 Unclip the pick-up lever return spring from its retaining lug on the body.
9 Protect the plastic moulded base of the jet then remove the screw retaining the jet pick-up link and link bracket.
10 Remove the float chamber/jet base tube. Note the gland, washer and ferrule.
11 Take off the jet adjusting nut and spring. Unscrew the jet locking nut then withdraw the bearing from the nut.
12 Unscrew the lever pivot bolt and spacer then detach the lever assembly and return springs. Note the pivot bolt tubes and the location of the cam and pick-up lever springs.
13 Take out the float chamber/body securing bolt and remove the float chamber and spacer.
14 Mark the installed position of the float chamber cover then remove the cover (three screws and washers) and joint washer. Retain the tag bearing the part number.
15 Take out the float hinge pin from its serrated end. Do not allow the float to drop.
16 Withdraw the float needle and unscrew the seating from the cover.
17 Close the throttle butterfly and mark the relative installed position of the butterfly plate with respect to the flange face.
18 Press the split ends of the butterfly retaining screws together, remove the screws and, with the butterfly spindle open, remove the disc from the spindle slot.
19 Fold back the tabs on the spindle washer then remove the nut, fork lever, lever arm, washer and throttle spindle. Note the location of the lever arm with respect to the spindle and body.
20 Wash all the parts in clean fuel then allow them to drain. Compressed air may be used provided that it is clean and dry. if ultrasonic

Chapter 3/Fuel and carburation, exhaust and emission control systems

cleaning facilities are available, all the internal parts should be cleaned using this method also.

21 When all the parts have been cleaned, lay them on a clean surface for inspection. Discard all the old seals and gaskets.

22 Examine the needle valve and seating, and if any ridging and grooving of the needle sealing face is detected, renew both items. Examine the metering needle using a watchmaker's eyeglass. Look for scoring and any signs of wear or bending. If the needle is defective in any way it must be renewed. Check the body, float chamber and float chamber cover for distortion and cracks, particularly distortion of the body/manifold mounting flange. Check the float, butterfly and spindle for wear. Renew any defective or suspect parts.

23 If there is going to be some delay before reassembly is commenced, eg, if new parts have to be purchased, store the remaining parts carefully in a polythene bag.

24 When reassembling, first refit the spindle to the body, countersunk holes facing outwards.

25 Assemble the spacing washer, lever, fork lever, tab washer and securing nut. Ensure that the idle stop in the lever is against the idle screw abutment on the body in the throttle closed position.

26 Tighten the spindle nut and fold over the tab washer.

27 Insert the butterfly plate into the spindle slot noting the identification markings.

28 Manoeuvre the disc in the spindle and centralize it in the bore by opening and closing the throttle several times.

29 Fit the butterfly retaining screws, but only pinch them up very lightly, until it is ascertained that the butterfly has not moved. Spread the split ends of the screws once they are fully tightened.

30 Screw the needle valve housing into the float chamber core; take care not to overtighten.

31 Insert the needle, cone end first, then fit the float and hinge pin to the cover.

32 Re-align the float chamber and cover markings, then refit the cover using a new gasket.

33 Refit the float chamber and spacer to the carburettor body and tighten the bolt.

34 Refit the piston lifting pin and spring from beneath, then the seal, washer and circlip from above (where applicable).

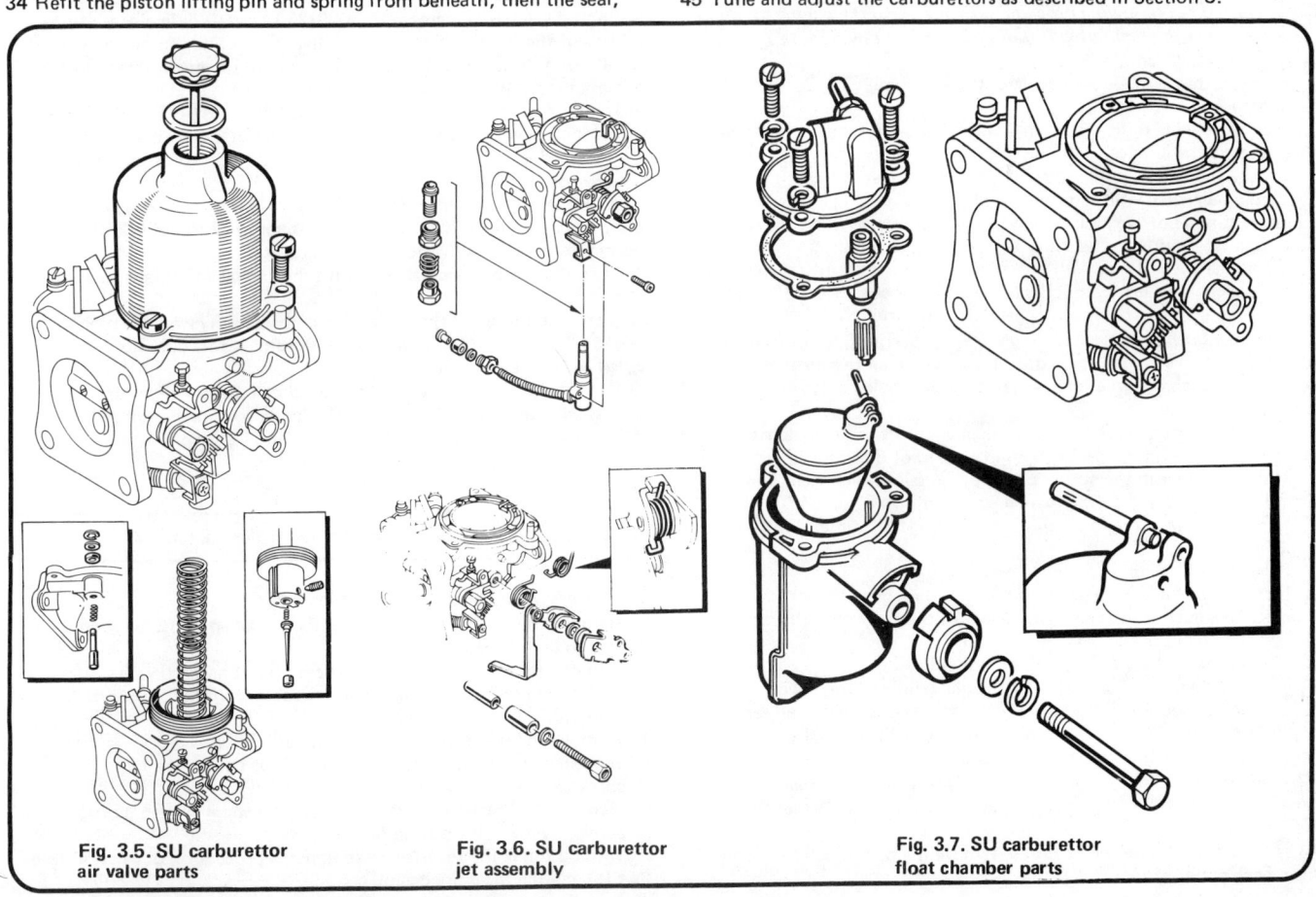

35 Clean any deposits from the suction chamber and piston using fuel, methylated spirit or isopropyl alcohol (never use an abrasive) then carry out the following serviceability check:

a) *Fit the piston damper and washer to the suction chamber then temporarily plug the air transfer holes in the piston with plasticine or adhesive tape.*

b) *Fit the piston, then screw a bolt through one of the fixing holes in the suction chamber and fit a large washer and nut.*

c) *Invert the assembly (ie, damper cap lowermost), hold the piston up and then check that the time taken to fall is between five and seven seconds. If this cannot be achieved the assembly must be renewed. Dismantle the assembly on completion of this check.*

36 Fit the spring and guide to the needle then insert them into the piston. The shoulder of the needle must be flush with the lower face of the piston and the slot in the guide positioned adjacent to the locking screw. When aligned, fit and tighten a new locking screw.

37 Check that the piston key in the body is secure then fit the jet bearing and tighten the locking nut.

38 Fit the spring and jet adjustment nut, screwing the nut up as far as possible.

39 Insert the jet then fit the sleeve nut, washer and gland to the end of the flexible tube. Make sure that the tube projects at least 3/16in (4.8mm) beyond the gland. Tighten the sleeve nut sufficiently to compress the gland to effect an adequate seal.

40 Refit the piston, spring and suction chamber and tighten the securing nuts evenly. Don't forget to align the marks made during dismantling.

41 Reassemble the lever assembly and return springs, pivot bolt tubes, cam and pick-up lever springs. Next fit the lever pivot bolts and spacer.

42 Hold the choke lever up, to relieve the pressure of the jet pick-up link, then refit the link bracket. Support the end of the jet and tighten the securing screw.

43 Screw the jet adjusting nut down two complete turns to provide an initial datum setting.

44 Refit the carburettors to the engine as described in Section 7.

45 Tune and adjust the carburettors as described in Section 8.

Fig. 3.5. SU carburettor air valve parts

Fig. 3.6. SU carburettor jet assembly

Fig. 3.7. SU carburettor float chamber parts

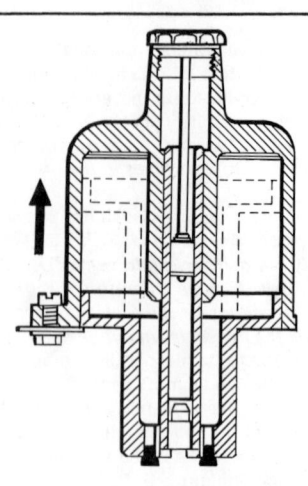

Fig. 3.8. SU carburettor suction chamber and piston checks

10 Twin Zenith-Stromberg carburettors - tuning and adjusting

Before attempting to tune and adjust the carburettors, it is important to realise that certain items of equipment will be required. If these are not available, the job should not be attempted, but should be carried out by a suitably equipped Triumph dealer. For those motorists intending to do the job themselves, it is essential that service tool number 'S 353' or a similar item is purchased beforehand to enable the jet needle to be repositioned. An airflow balancing meter is also a very useful device for tuning and balancing although listening carefully at the end of a rubber or plastic tube of about ¼ in (6mm) diameter will usually be sufficient. It is essential also that a non-dispersive infra-red exhaust gas analyser is used to check the CO content of the exhaust gas. Before any attempt is made to adjust the carburettors, ignition timing, tappet clearances, spark plugs and distributor pick-up air gap should be checked and adjusted as necessary.

Note: *During the following procedure do not allow the engine to idle for longer than three minutes without purging for one minute's duration at 2000 rpm.*

1 Remove the fresh air duct as described in Chapter 12, Section 34, Paragraph 3.
2 Run the engine until the normal operating temperature is attained, then remove the air cleaner (refer to Section 3 if necessary).
3 Ensure that the fast idle screw is clear of the fast idle cam then check the balance of the carburettor inlet air flow using an airflow meter. If this is not available, listen to the airflow hiss using the small bore tube referred to in the introduction to this Section.
4 To adjust the airflow, hold the roller (where fitted) into the corner of the progression lever and adjust the spring loaded screws equally until the adjusting bar is in the centre of both forks.
5 Unscrew both throttle adjusting screws to permit the throttles to fully close, then screw them in until they *just* touch the throttle levers. Turn each one in a further half-turn to provide a datum setting.
6 Continue turning each screw by equal amounts until a balanced airflow is obtained together with an idle speed of 800rpm.
7 Hold the roller in the corner of the progression lever again (where fitted) and adjust the spring loaded screw to give a gap of 0.010 in (0.25 mm) between the forks and the adjusting bar (dimension A in Fig.3.11).
8 If necessary, re-adjust the idle speed to 800rpm.
9 Increase the engine speed to 1600rpm and recheck the airflow balance. Readjust, if necessary, by means of the throttle adjusting screws, then recheck the balance when idling.
10 Disconnect and plug the air pump outlet hose.
11 With the engine still at normal operating temperature, check that the idle speed is still 800rpm then use an exhaust gas analyser to check that the CO content is in accordance with the engine compartment emission control decal label.
12 If slight adjustment is required, the idle trim screw on each carburettor may be rotated by equal amounts until the correct CO reading is obtained.
13 If further adjustment is required, remove the piston damper from each carburettor.
14 Carefully insert the special tool 'S 353' into the dashpot until the outer part engages in the air valve, and the inner part engages in the hexagon of the needle adjuster plug.
15 Whilst preventing the outer part of the tool from moving (to prevent the diaphragm from rupturing) rotate the inner part either clockwise to richen the mixture or anticlockwise to weaken the mixture. Repeat this on the second carburettor, ensuring that the adjustment is made by the same amount. Remove the special tool.
16 Top-up the carburettor piston dampers using engine oil. Fully raise each piston using a finger, and add oil until it is ¼ in (6mm) below the top of the damper tube. Release the piston and refit the damper. **Screw down the plug, then raise and lower the piston to ensure correct location of the oil retaining cup in the damper tube.**
17 Recheck the CO content and idle speed as already described, then switch off the ignition. Unplug the air injection hose and reconnect it to the pump.

Fast idle speed - vehicles to Canadian specification
18 With the engine switched off but still at normal operating temperature, check that the mixture control cam lever on both carburettors returns to its stop.
19 Adjust the mixture control cable so that there is some free movement then pull out the control knob about ¼ in (6mm) until the fast idle cams are correctly engaged with the ball separators.
20 Loosen the fast idle screw locknuts, start the engine and rotate each screwhead against its cam until the engine speed reaches 1600rpm.
21 Adjust the fast idle screws if necessary to balance the airflow whilst maintaining the 1600rpm. This speed is approximately equal to 1300 rpm when the engine is cold.
22 Tighten the fast idle screw locknuts and push in the mixture control knob.

Deceleration bypass valves, Californian and Canadian vehicles - checking and adjusting
23 With the engine idling at normal operating temperature, disconnect the vacuum pipe from the distributor.
24 Plug the end of the pipe with the finger and check that the idle speed increases to around 1300rpm.
25 Should the speed increase to 2000/2500rpm, it indicates that one or both of the deceleration by-pass valves is/are floating off its seat. In this condition, if the throttle is momentarily opened, the additional increase in rpm will be slow to fall.
26 If adjustment is required, screw the by-pass valve adjusting screw fully anticlockwise onto its seat **on the carburettor not being adjusted.** (This procedure will not need to be repeated when adjusting the second carburettor.)
27 With the vacuum pipe still plugged, turn the by-pass valve adjusting screw clockwise until the speed increases abruptly to 2000/2500rpm, thereby causing the valve to float on its seat.
28 Turn the screw anticlockwise until the engine speed falls to around 1300rpm.
29 Momentarily open, then release, the throttle. The engine speed should increase then fall to around 1300rpm. If this does not occur, repeat the adjusting sequence.
30 Repeat the adjustment for the second carburettor then turn each by-pass adjusting screw ½ turn anticlockwise to seat the valves.
31 Reconnect the vacuum pipe; refit the air cleaner and fresh air duct.

11 Twin Zenith-Stromberg carburettors - repair operations

Starter assembly - removal and installation
1 Remove the fresh air duct as described in Chapter 12, Section 34, Paragraph 3.
2 Remove the air cleaner (refer to Section 3 if necessary).
3 Detach the mixture (choke) control cable from the appropriate carburettor.
4 Remove the two starter assembly securing screws.
5 If applicable, remove the EGR control valve from the rear carburettor.
6 Remove the starter assembly. No repairs can be carried out; if defective, a new unit must be obtained.
7 Installation is the reverse of the removal procedure, but first ensure that the mating faces are clean.

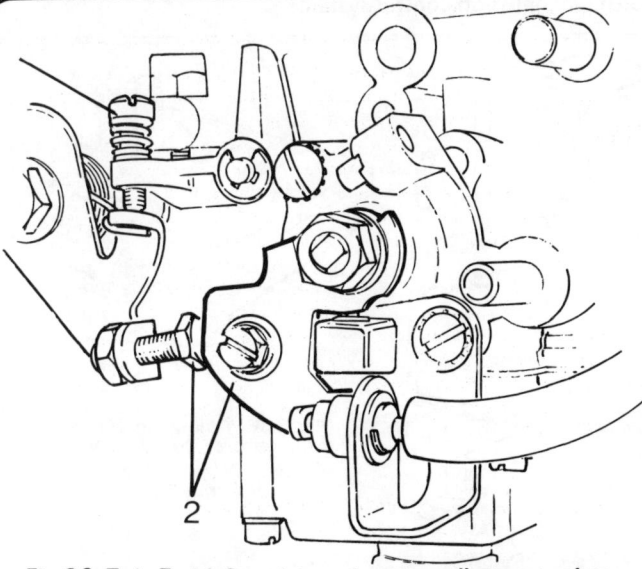

Fig. 3.9. Twin Zenith-Stromberg carburettors adjustment points

1　Throttle adjusting screw　　2　Fast idle screw

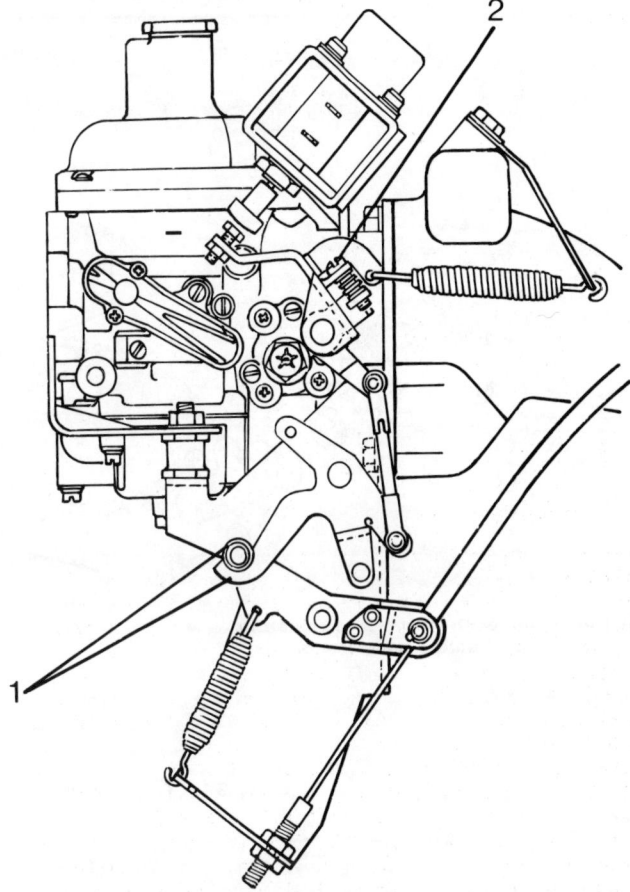

Fig. 3.10. Twin Zenith-Stromberg carburettors - tuning and adjusting

1　Roller in corner of progression lever
2　Spring loaded screws for adjusting bar

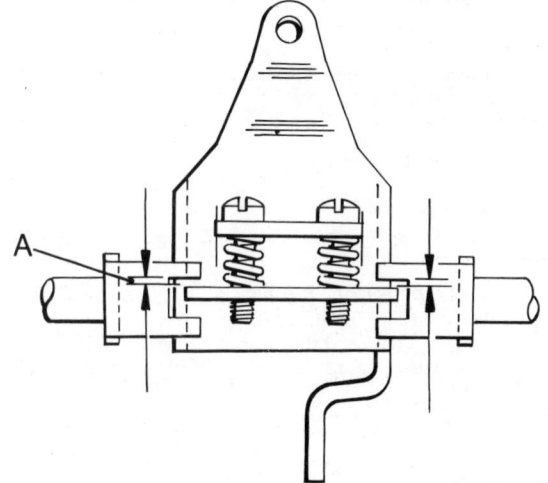

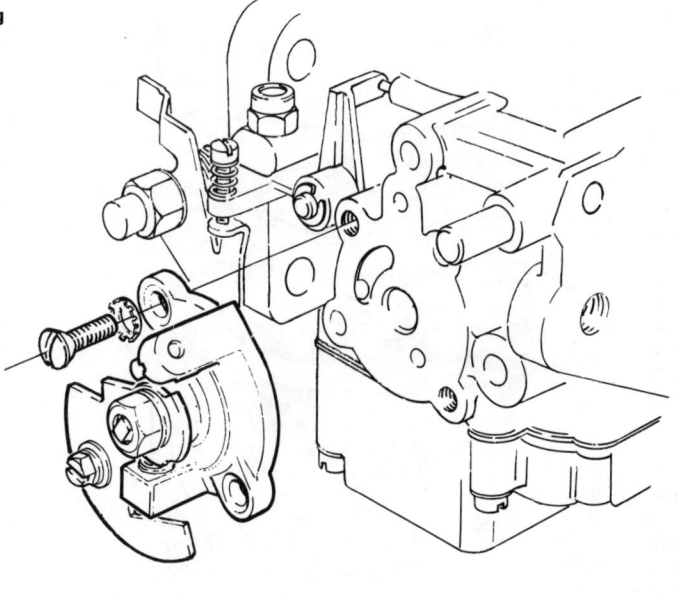

Fig. 3.11. Twin Zenith-Stromberg carburettors - setting the adjusting bar

A　Gap 0.010 in (0.25 mm)

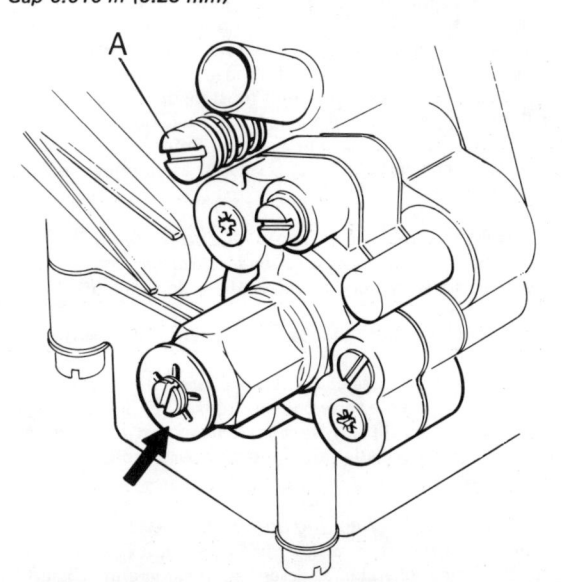

Fig. 3.12. Deceleration bypass valve (arrowed) on twin Zenith-Stromberg carburettors

A　—　Idle trim screw

Fig. 3.13. Twin Zenith-Stromberg carburettors - starter assembly

Fig. 3.14. Twin Zenith-Stromberg carburettors - temperature compensator

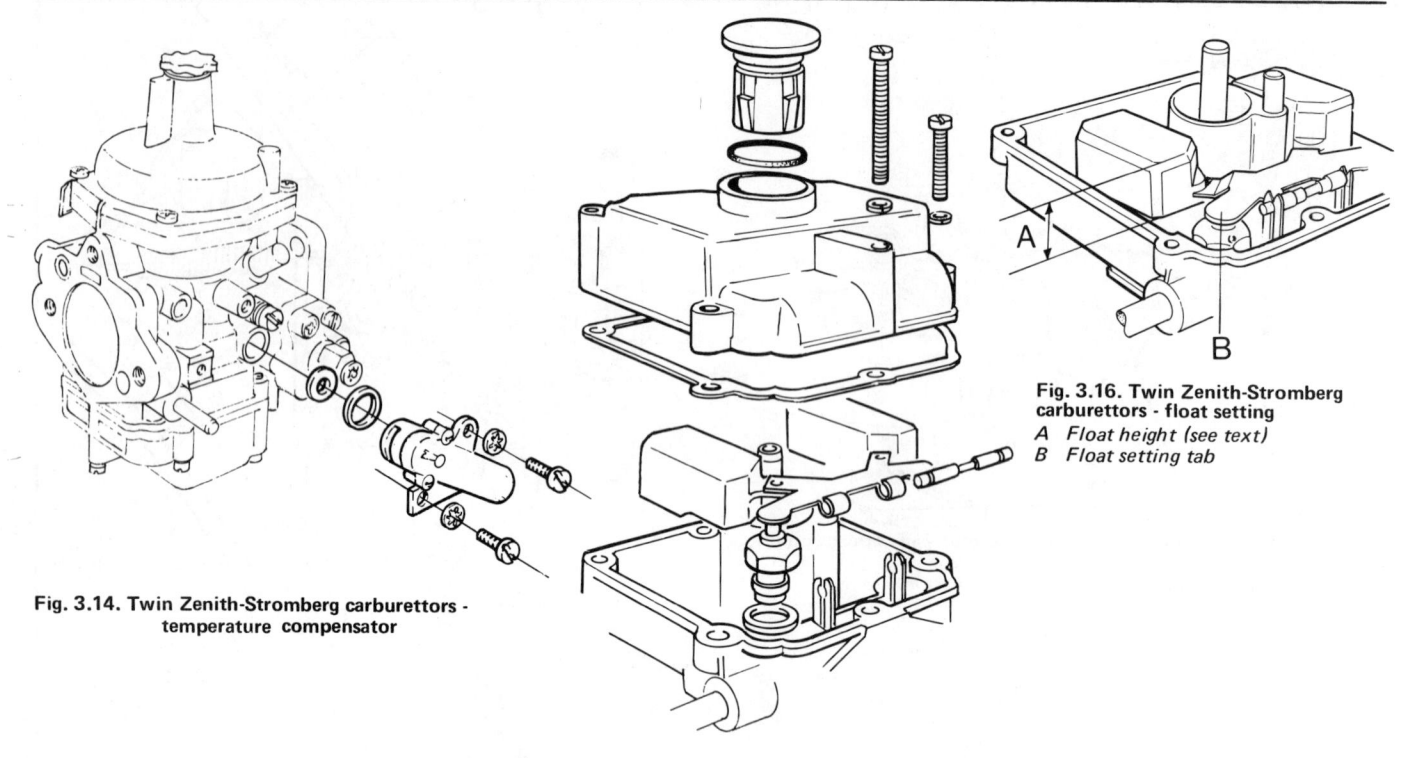

Fig. 3.15. Twin Zenith-Stromberg carburettors - float chamber parts

Fig. 3.16. Twin Zenith-Stromberg carburettors - float setting
A Float height (see text)
B Float setting tab

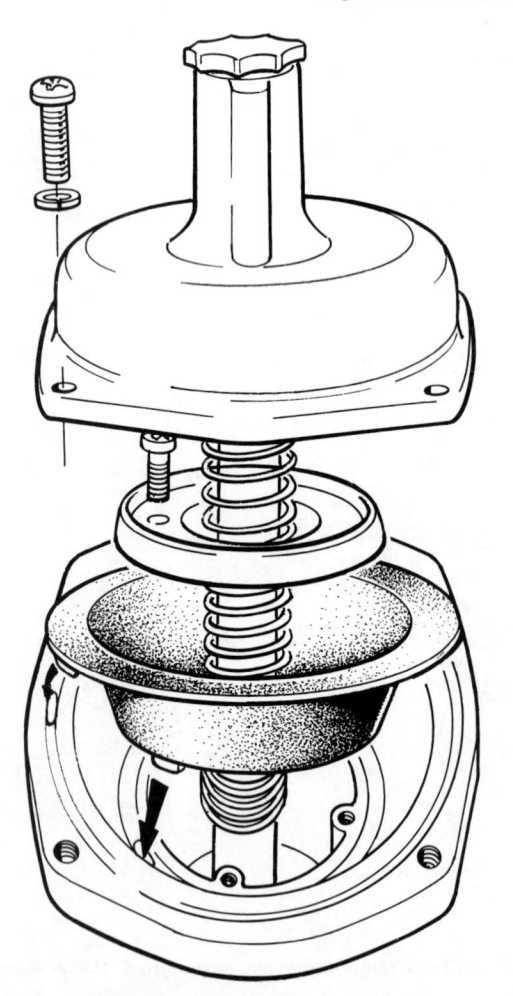

Fig. 3.17. Twin Zenith-Stromberg carburettors - diaphragm unit. The arrows show the location tags

Temperature compensator - removal and installation
8 Remove the air cleaner (refer to Section 3 if necessary).
9 Remove the two screws and shakeproof washers, and detach the compensator. No repairs can be carried out; if defective, a new unit must be obtained.
10 Installation is the reverse of the removal procedure. Ensure that the mating faces are clean and use new O-rings if the existing ones are hardened, distorted or are cracked.

Deceleration by-pass valve - removal and installation
11 Remove the carburettors as described in Section 7.
12 Remove the two cheesehead screws and the single countersunk head slotted screw. **Do not touch the countersunk head cross-slotted screws.**
13 Withdraw the valve assembly and remove the gasket. No repairs can be carried out; if defective, a new unit must be obtained.
14 Installation is the reverse of the removal procedure. Ensure that the mating faces are clean and use a new gasket. Before the air cleaner is fitted, check the deceleration by-pass valve(s) as described in Paragraphs 23 to 31 of Section 10.

Float chamber needle valve - removal and installation
15 Remove the carburettors as described in Section 7.
16 Remove the six screws and spring washers, and take off the float chamber and gasket.
17 Prise the spindle out of the locating clips, then remove the needle valve and washer.
18 Installation is the reverse of the removal procedure. Use a new washer on the float needle valve and ensure that the spindle is firmly secured in the locating clips. It is recommended that the float chamber level is checked as described in the following paragraphs before fitting the float chamber.

Float chamber level - checking and adjustment
19 Remove the carburettors as described in Section 7.
20 Remove the float chamber as described in Paragraphs 15 and 16.
21 With the carburettor in the inverted position, measure the distance between the carburettor body face to the highest point on the float. Bend the tab that contacts the needle valve, if necessary, to obtain a float height of 0.625 to 0.627in (16 to 17mm). Ensure that the tab

Chapter 3/Fuel and carburation, exhaust and emission control systems

remains at right angles to the valve for satisfactory operation.
22 Install the float chamber using a new gasket.

Diaphragm - removal and installation
23 Mark the installed position of the top cover then remove the four screws and spring washers.
24 Carefully lift off the cover, then remove the spring, retaining plate and diaphragm.
25 Installation is the reverse of the removal procedure, but ensure that the inner and outer tags of the diaphragm locate in the air valve and body recesses respectively.
26 Top up the damper oil level by removing the piston damper then following the procedure given in Section 10, Paragraph 16.

12 Single Zenith-Stromberg carburettor - tuning and adjusting

Before attempting to tune and adjust the carburettor, it is important to realize that certain items of equipment will be required. If these are not available, the job should not be attempted, but should be carried out by a suitably qualified Triumph dealer. For those motorists intending to do the job themselves, it is essential that service tool number S 353 is purchased beforehand to enable the jet needle to be repositioned. It is essential also that a non-dispersive infra-red exhaust gas analyser is used to check the CO content of the exhaust gas. Before any attempt is made to adjust the carburettor, the ignition timing, tappet clearances, spark plugs and distributor pick-up air gap should be checked and adjusted as necessary.
Note: *During the following procedure do not allow the engine to idle for longer than three minutes without purging for one minute's duration at 2000rpm.*
1 Run the engine until the normal operating temperature is attained, then turn the throttle adjusting screw, if necessary, to obtain an idle speed of 800 rpm. Stop the engine.
2 Disconnect and plug the outlet hose from the air pump.
3 Run the engine at idle speed then use an exhaust gas analyser to check that the CO content is in accordance with the engine compartment emission control decal label.
4 If slight adjustment is required, this can be done by turning the fine idle CO screw until the CO contact is satisfactory. Do not attempt to adjust the coarse idle CO screw (see Fig. 3.19).
5 If further adjustment is required, remove the piston damper from the carburettor.
6 Carefully insert the special tool 'S 353' into the dashpot until the outer part engages in the air valve, and the inner part engages in the hexagon of the needle adjuster plug.
7 Whilst preventing the outer part of the tool from moving (to prevent the diaphragm from rupturing) rotate the inner part either clockwise to richen the mixture or anticlockwise to weaken the mixture.
8 Remove the special tool. Remove the air cleaner and top-up the carburettor piston damper using engine oil. Fully raise the piston using a finger and add oil until it is ¼in (6mm) below the top of the damper tubes. Release the piston and refit the damper. **Screw down the plug, then raise and lower the piston to ensure correct location of the oil retaining cup in the damper tube.** Refit the air cleaner.
9 Recheck the CO content and idle speed as already described, then switch off the ignition. Unplug the air injection hose and reconnect it to the pump.
10 Run the engine again at normal operating temperature and disconnect the vacuum pipe from the distributor.
11 Plug the end of the pipe with a finger and check that the idle speed increases to around 1300rpm.
12 Should the speed increase to 2000/2500rpm, it indicates that the deceleration by-pass valve is floating off its seat.
13 If adjustment is required, keep the vacuum pipe plugged and turn the by-pass valve adjusting screw clockwise, then anticlockwise, until the engine speed falls to around 1300rpm.
14 Momentarily open, then release the throttle. The engine speed should increase then fall to around 1300rpm. If this does not occur, repeat the adjusting sequence.
15 When correctly set, turn the by-pass adjusting screw half a turn anticlockwise to seat the valve.
16 Reconnect the vacuum pipe to the distributor.

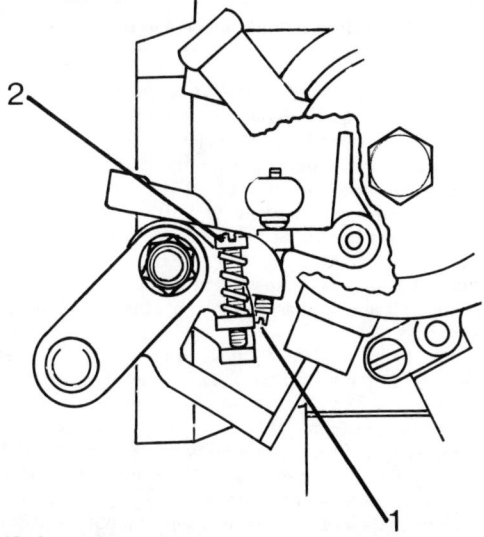

Fig. 3.18. Single Zenith-Stromberg carburettor - adjustment points

1 Throttle adjusting screw   2 Fast idle screw

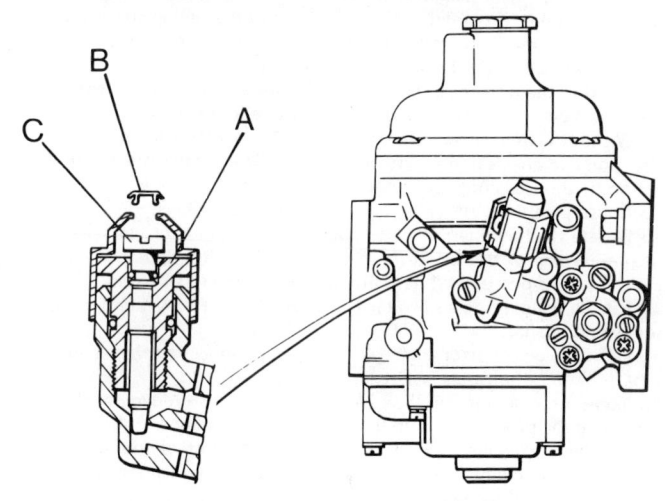

Fig. 3.19. Single Zenith-Stromberg carburettor - mixture adjustment

A Coarse idle CO screw (do not adjust)
B Plastic cap
C Fine idle CO screw

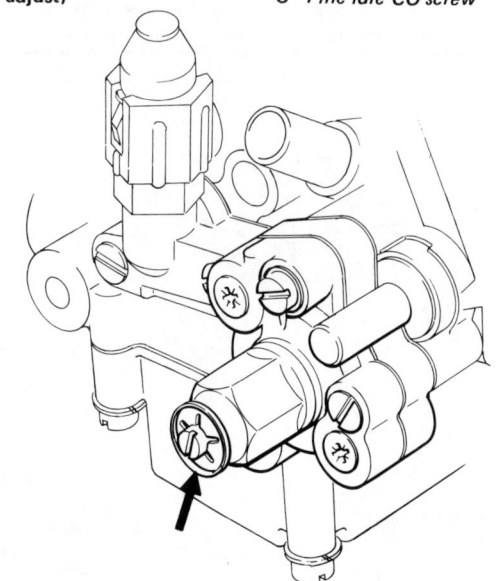

Fig. 3.20. Deceleration bypass valve (arrowed) on single Zenith-Stromberg carburettors

13 Single Zenith-Stromberg carburettor - repair operations

Automatic choke - removal, installation and setting-up

1. Remove the air cleaner from the carburettor.
2. Remove the carburettor from the engine.
3. Hold the throttle in the open position by inserting a suitable wooden plug between the throttle bore and butterfly.
4. Remove the three retaining screws and lift off the automatic choke and gasket. No repairs can be carried out; if defective, a new unit must be obtained.
5. Clean the carburettor and choke mating faces.
6. Remove the bolt and washer, and take off the water jacket and rubber sealing ring.
7. Remove the clamp ring (three screws and washers).
8. Carefully remove the heat mass, ensuring that the bi-metal coil is not strained.
9. Take off the heat insulator.
10. Using a new gasket, fit the choke body to the carburettor. Progressively and evenly tighten the screws to 40-45 lbf in (4.5-5.08 Nm), note that the lower screw is the shortest.
11. Adjust the fast idle screw to obtain a gap of 0.035 in (0.889mm) between the base circle of the cam and the fast idle pin (dimension A in Fig. 3.23).
12. Position the heat insulator so that the bi-metal lever protrudes through the slot. Provided that the choke is in the **ON** position, the insulator can only be fitted in one position; the back of it locates in the choke body when the three holes are aligned.
13. Position the heat mass with the ribs facing outwards so that the bi-metal rectangular loop fits over the bi-metal lever.
14. Without lifting the heat mass, rotate it 30 to 40° only, in each direction and check that it returns to the static position under spring action. If this does not occur, repeat Paragraph 13.
15. Loosely fit the clamp ring, screws and spring washers.
16. Rotate the heat mass anticlockwise 30 to 40° to align the scribed line on its edge with the datum mark on the insulator and choke body. Hold it in this position while tightening the clamp ring screws.
17. Fit the sealing ring and water jacket, but do not fully tighten the screw and washer.
18. Refit the carburettor to the engine. After connecting the water pipes, tighten the water jacket screw.
19. Top-up the damper oil level by removing the piston damper then following the procedure given in Section 12, Paragraph 8.
20. Fit the air cleaner, then run the engine up to normal operating temperature and adjust the idle speed to 800rpm by means of the throttle adjusting screw.
21. Stop the engine, allow it to cool then top-up the cooling system as necessary.

Deceleration by-pass valve - removal and installation

22. Remove the two cheesehead screws and the single countersunk head slotted screw. **Do not touch the countersunk head cross-slotted screws.**
23. Withdraw the valve assembly and remove the gasket. No repairs can be carried out; if defective, a new unit must be obtained.
24. Installation is the reverse of the removal procedure. Ensure that the mating faces are clean and use a new gasket. Check the deceleration by-pass valve as described in Paragraphs 10 to 16 of Section 12.

Float chamber needle valve - removal and installation

25. Remove the carburettor then follow the procedure given in Paragraphs 16 to 18 of Section 11.

Float chamber level - checking and adjustment

26. Remove the carburettor then follow the procedure given in Paragraphs 16, 21 and 22 of Section 11.

Diaphragm - removal and installation

27. Follow the procedure given in Paragraphs 23 to 26 of Section 11, but note that topping-up of the damper oil is dealt with in Section 12, Paragraph 8.

14 Throttle pedal assembly - removal and installation

1. Detach the battery earth lead.
2. Working from inside the car, remove the spring clip and slide out the cable.
3. From under the bonnet remove the two bolts, spring washers and flat washers, and remove the pedal assembly from the car.
4. If necessary, remove the split pin and washer to dismantle the pedal assembly.
5. Reassembly and installation are the reverse of the removal procedure.

15 Mixture (choke) control cable - removal and installation

Note: *The procedure described is for right-hand drive cars. Slight differences may occur for left-hand drive models.*

1. Detach the inner cable from the carburettor trunnion.
2. From inside the car, pull the choke knob and withdraw the complete inner cable.
3. Pull off the connector from the choke switch.
4. Loosen the outer cable retaining nut then unscrew the bezel.
5. Remove the complete outer cable; loosen the grubscrew and detach the switch.
6. Installation is the reverse of the removal procedure. If necessary, refer to Chapter 10, Section 32 for further information.

Fig. 3.21. Single Zenith-Stromberg carburettor - automatic choke assembly

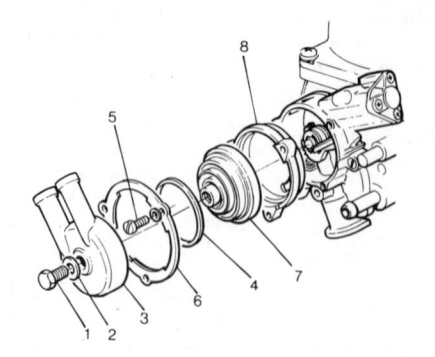

Fig. 3.22. Exploded view of automatic choke

1. Bolt
2. Washer
3. Water jacket
4. Sealing ring
5. Screw
6. Clamp ring
7. Heat mass
8. Heat insulator

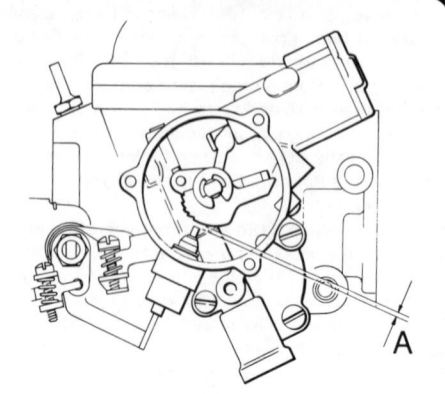

Fig. 3.23. Single Zenith-Stromberg carburettor - fast idle adjustment

A Gap - see text for dimension

Chapter 3/Fuel and carburation, exhaust and emission control system

16 Fuel filler cap, filler assembly and hose - removal and installation

1 Turn the filler cap anticlockwise to remove it.
2 Remove the filler tube assembly (three screws).
3 From inside the boot remove the access panel (four screws) (photo).
4 Loosen the top hose clip and withdraw the filler tube; now remove the finisher.
5 Where applicable, remove the hoses from the vapour separator then pull the separator sideways from its retaining clip.
6 Release the hose clip at the tank end and pull off the hose.
7 Installation is the reverse of the removal procedure. Where applicable, connect the tank breather hose to the left-hand vapour separator connection, and the absorption canister pipe-run hose to the right-hand vapour separator connection.

17 Fuel tank - removal and installation

Note: *Wherever possible, reduce the contents of the fuel tank by normal usage before attempting to remove it, thereby reducing its overall weight.*

1 Detach the battery earth lead.
2 Raise the car to a suitable working height for access to the fuel tank and rear suspension. Support it on the body frame side members using axle stands or blocks. Chock the front wheels for safety.
3 Remove the fuel filler cap assembly as described in the previous Section.
4 Either unclip the breather hose from the left-hand side rear chassis member or disconnect the vapour separator pipe from the tank, as applicable.
5 Disconnect the dampers at their lower attachment points (refer to Chapter 11 if necessary).
6 Detach the leads from the tank unit.
7 Pull off the main fuel line connection.
8 Disconnect the right-hand radius rod from the body bracket; remove the left-hand one completely. Refer to Chapter 11 for further information if necessary.
9 Whilst supporting the axle weight, remove both springs. Take care that the brake hose is not stretched during this operation.
10 Remove the left-hand rear wheel and lower the axle, still taking care that the brake hose is not stretched.
11 Remove the tail pipe and silencer assembly, referring to Section 20 if necessary.
12 Carefully prise out the left-hand bump stop.
13 Remove the four nuts whilst supporting the tank. Take off the straps and remove the tank from the left-hand side of the car.
14 If necessary, remove the filler hose, breather hose and tank unit (refer to Chapter 10 for this item).
15 Installation is basically the reverse of the removal procedure, but the following points should be noted:-

(a) *Fit new adhesive cushion strips if necessary, the thicker ones going on the top of the tank.*
(b) *The elongated holes in the straps should be towards the rear of the tank.*
(c) *Check for fuel leaks on completion.*

18 Inlet manifold - removal and installation

Note: *At the time of compilation of this Manual, no details were available for removal of the single carburettor inlet manifold.*

Inlet manifold - twin SU carburettors

1 Disconnect the battery earth lead.
2 Drain the cooling system; refer to Chapter 2 if necessary.
3 Remove the fresh air duct as described in Chapter 12, Section 34, Paragraph 3.
4 Remove the air cleaner as described in Section 3.
5 Detach the top water hoses from the engine, and disconnect the water temperature sender lead.
6 Detach the fuel feed to the carburettors.
7 Detach the brake servo hose from the manifold.
8 Detach the mixture (choke) control cable from the front carburettor.
9 Pull off the spill pipe from the rear carburettor float chamber.
10 Detach the throttle cable from the linkage.
11 Pull off the camshaft cover breather hose.
12 Remove the distributor cap, then disconnect the distributor vacuum pipe from the rear carburettor.
13 Remove the six manifold attachment bolts and the engine lifting eye bracket (photo).
14 Disconnect the heater pipe at the inlet manifold union, then lift off the manifold (photos).
15 Remove the O-ring between the manifold and the cylinder head, then remove the gasket (photo).
16 If necessary, take out the water pump connecting tube.
17 If necessary, the carburettors can be removed. Further information will be found in Section 7.
18 Installation is the reverse of the removal procedure. Before fitting the gasket, O-ring and water pump connecting tube, ensure that the mating faces are clean. A little non-setting gasket sealant is recommended for the water pump connecting tube sealing rings. Tighten the bolts progressively to the specified torque (photo).

Inlet manifold - twin Zenith-Stromberg carburettors

19 Follow the procedure given in Paragraphs 1 to 8 of this Section. In addition detach the emission control hoses (see Fig. 3.24).
20 Detach the mixture (choke) control and throttle cables from the carburettors.
21 Take off the distributor cap.
22 Remove the six inlet manifold attachment bolts.
23 Disconnect the heater pipe at the union, then lift off the manifold complete with carburettors.
24 Remove the O-ring between the manifold and the cylinder head, then remove the gasket.
25 If necessary, take out the water pump connecting tube.
26 Installation is the reverse of the removal procedure. Before fitting the gasket, O-ring and water pump connecting tube, ensure that the mating faces are clean. A little non-setting gasket sealant is recommended for the water pump connecting tube sealing rings. Tighten the bolts progressively to the specified torque. (Note that early types with a moulded connecting tube have no O-rings. The tube should be refitted by removing the thermostat housing and thermostat, and pushing the tube home).

19 Exhaust manifold - removal and installation

1 Detach the battery earth lead, then raise the left-hand side of the car for access to the front exhaust pipe/manifold flange.
2 Remove the three flange bolts.
3 Remove the four lower manifold retaining bolts.
4 Lower the car to the ground. Where applicable, remove the hot air hose from the manifold.
5 Remove the three top manifold retaining bolts and lift away the manifold. Where applicable, remove the hot air collection pressing.
6 Installation is the reverse of the removal procedure. Tighten the bolts progressively to the specified torque. It is recommended that a new gasket is used for the flange joint; note that no gasket is used at the manifold/cylinder head joint (photo).

20 Exhaust system - removal and installation

Front pipe

1 Raise the car for access to the front pipe.
2 Loosen the silencer-to-downpipe clamp bolt then unhook the silencer from the body hanger.
3 Unhook the tailpipe from its hanger then drive the silencer from the downpipe.
4 Remove the nut(s) and bolt(s) securing the rear end of the downpipe, and the two nuts and bolts retaining the front end of the pipe (photos).
5 Remove the three flange bolts and lower away the pipe complete with gasket.
6 Installation is the reverse of the removal procedure. It is recommended that a new gasket is used for the flange joint (photo).

Front silencer

7 Raise the car for access to the silencer.
8 Loosen the front and rear silencer clamps.
9 Disconnect the tailpipe from the body, then withdraw it from the silencer. The silencer can now be withdrawn.

16.3 Access panel for filler tube

18.13 Engine lifting eye on manifold

18.14a Manifold heater pipe connection

18.14b Manifold heater pipe arrangement

18.15 Inlet manifold O-ring

18.18 Inlet manifold gasket in position

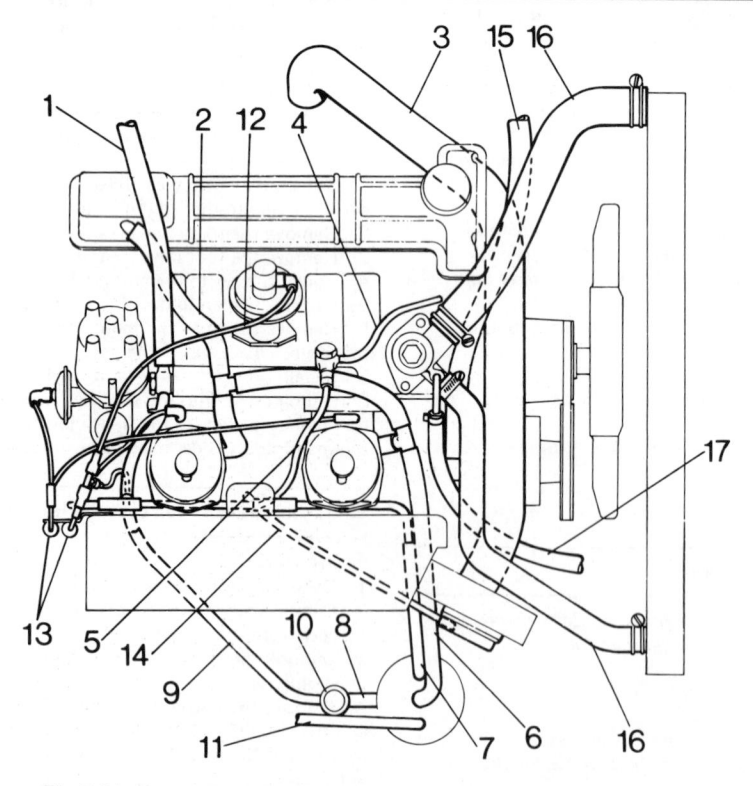

Fig. 3.24. Hoses to be detached when removing the inlet manifold where twin Zenith-Stromberg carburettors are installed

1 Brake servo hose
2 Engine breather hose
3 Hot air hose
4 Vacuum hose to diverter and relief valve
5 One-way valve hose
6 Canister purge hose
7 Float chamber vent hose
8 Anti run-on valve hose
9 Manifold vacuum hose
10 Anti run-on valve
11 Fuel tank vent hose
12 EGR vacuum hose
13 Vapour traps
14 Temperature sensor hose
15 Check valve air hose
16 Coolant top hoses
17 Coolant expansion hose

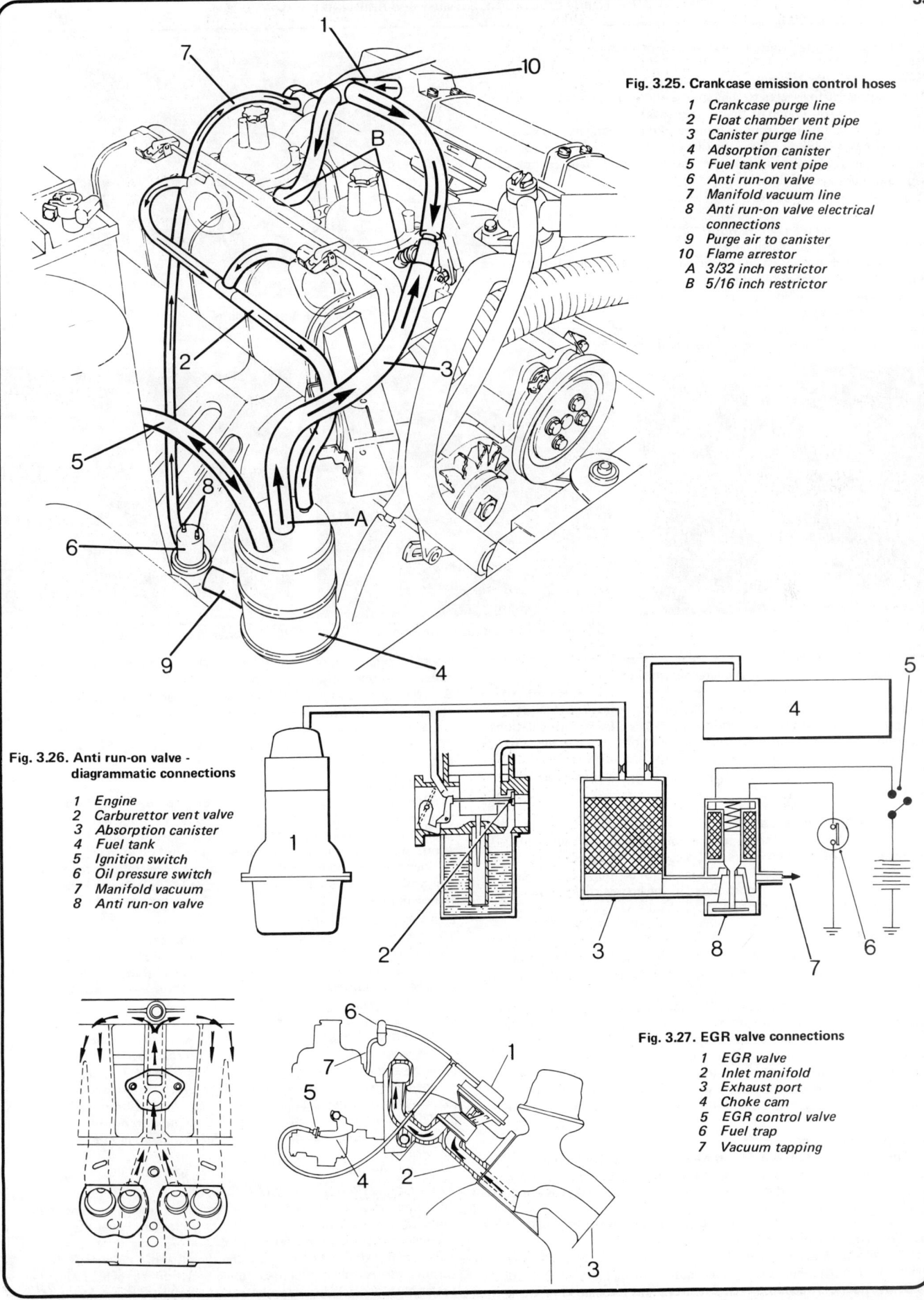

Fig. 3.25. Crankcase emission control hoses

1. Crankcase purge line
2. Float chamber vent pipe
3. Canister purge line
4. Adsorption canister
5. Fuel tank vent pipe
6. Anti run-on valve
7. Manifold vacuum line
8. Anti run-on valve electrical connections
9. Purge air to canister
10. Flame arrestor
A. 3/32 inch restrictor
B. 5/16 inch restrictor

Fig. 3.26. Anti run-on valve - diagrammatic connections

1. Engine
2. Carburettor vent valve
3. Absorption canister
4. Fuel tank
5. Ignition switch
6. Oil pressure switch
7. Manifold vacuum
8. Anti run-on valve

Fig. 3.27. EGR valve connections

1. EGR valve
2. Inlet manifold
3. Exhaust port
4. Choke cam
5. EGR control valve
6. Fuel trap
7. Vacuum tapping

Chapter 3/Fuel and carburation, exhaust and emission control system

10 Installation is the reverse of the removal procedure.

Tailpipe and rear silencer
11 Raise the car for access to the tailpipe and rear silencer.
12 Loosen the rear clamp of the front silencer.
13 Unhook the tailpipe from its forward and rear hangers (photo).
14 Withdraw the tailpipe from the silencer, then remove the two nuts and bolts securing the hanger bracket assembly to the tailpipe bracket. Separate the two assemblies.
15 Installation is the reverse of the removal procedure.

Catalytic converter (where applicable)
16 Detach the battery earth lead.
17 Raise the car for access to the converter. Ensure that it is not too hot to handle before attempting to remove it.
18 Remove the three bolts at each end, then ease out the converter and extract the olives.
19 Installation is the reverse of the removal procedure, but ensure that the deflection plate is facing towards the front of the car.

20.4a Attachment points ...

20.4b ... for front exhaust pipe

20.6 Flange joint gasket for front exhaust pipe

20.13 Forward hanger for tailpipe

21 Emission control systems - general description

1 The emission control system enables cars manufactured for the North American market to conform with all 1976 Federal Regulations governing the emission of hydrocarbons, carbon monoxide, nitric oxide and fuel vapours from the crankcase, exhaust and fuel systems.
2 With the exception of the catalytic converter, the items described are fitted to all models.

Crankcase breather system
3 Crankcase breathing and removal of 'blow-by' vapours are achieved by making use of the inherent partial vacuum in the carburettors. In this way, emissions are burnt in the engine combustion process. An oil separator/flame trap is incorporated in the engine rocker cover - see Fig. 3.25 for the hose routing.

Anti-run-on valve
4 This valve prevents the engine from 'running on' after the ignition has been switched off when, due to engine heat, a compression-ignition situation could occur.
5 When the ignition is switched off, a solenoid valve is activated and seals off the inlet to the bottom of the charcoal canister. A connection to the inlet manifold thus applies a partial vacuum to the canister and consequently to the float chamber(s) of the carburettor(s). This vacuum is sufficient to prevent fuel being drawn into the engine. When the engine has stopped and the oil pressure has dropped to zero, the solenoid valve is de-activated and the engine is thus capable of being restarted.

Air injection system
6 This system is used to reduce the emission of hydrocarbons, nitric oxide and carbon monoxide in the exhaust gases, and comprises an air pump, a combined diverter and relief valve, a check valve and an air manifold.
7 The rotary vane type pump is belt driven from the engine and delivers air to each of the four exhaust ports.
8 The diverter and relief valve (earlier models only) diverts air from the pump to atmosphere during deceleration, being controlled in this mode by manifold vacuum. Excessive pressure is discharged to atmos-

phere by operation of the relief valve. Note that later models have a relief valve incorporated into the air pump.
9 The check valve is a diaphragm-spring operated non-return valve. Its purpose is to protect the pump from exhaust gas pressures both under normal operation and in the event of the drive belt failing.
10 The air manifold is used to direct the air into the engine exhaust ports.

Air intake temperature control
11 Refer to Section 2 for further information on the air cleaner and temperature sensor.
12 The control system for the air cleaner (twin carburettor installation) incorporates a one-way valve so that full vacuum influence on the flap valve is maintained during sudden acceleration when the manifold vacuum is temporarily destroyed. The valve is installed in the vacuum line.

Evaporative control system
13 This system uses an activated adsorption canister through which the fuel tank is vented, and incorporates the following features:
(a) *The carburettor float chamber(s) is/are vented to the engine when the throttle is open, and to the adsorption canister when the throttle is closed.*
(b) *The carburettor constant depression is used to induce a purge condition through the canister via the anti-run-on valve. The crankcase breathing is also coupled into this system.*
(c) *A separator tank is used to prevent fuel surges from reaching the canister which could otherwise saturate the system.*
(d) *A sealed filler cap is used to prevent loss by evaporation.*
(e) *The fuel filler tube extends into the fuel tank to prevent complete filling; this permits the fuel to expand in hot weather.*
(f) *Twin adsorption canisters are fitted to some models.*

Exhaust gas recirculation (EGR) system
14 To minimise nitric oxide exhaust emissions, the peak combustion temperatures are lowered by recirculating a metered quantity of exhaust gas through the inlet manifold.
15 A control signal is taken from the throttle edge tapping of the carburettor. At idle or full load no recirculation is provided, but under part load conditions a controlled amount of recirculation is

Chapter 3/Fuel and carburation, exhaust and emission control system

provided according to the vacuum signal profile of the metering valve. The EGR valve is mounted on the inlet manifold.

16 An EGR control valve cuts the vacuum signal when the choke is in operation by opening an air bleed into the vacuum line.

Catalytic converter

17 To further reduce the emission of carbon monoxide and hydrocarbons, a catalytic converter is installed in the exhaust system of certain models.

18 The following precautions should be taken with cars equipped with a catalytic converter:

(a) Avoid heavy impacts to the casing which can cause damage to the internal ceramic material.
(b) Never use anything but unleaded fuel, or the emission control system efficiency will be seriously impaired.
(c) The catalytic converter becomes extremely hot during operation of the car. Avoid contacting combustible materials such as long dry grass when motoring, and allow the converter to cool before touching it during any repair or maintenance operations.

19 If misfiring of the engine occurs, the cause must be traced and rectified immediately to prevent damage to the catalytic converter.

22 Emission control systems - repair and maintenance operations

Vapour separator - removal and installation
1 This operation is included in Section 16.

Absorption canister - removal and installation
2 Disconnect the bottom pipe leading to the anti-run-on valve.
3 Disconnect the three top pipes, noting their installed positions.
4 Loosen the clamping screw and lift out the canister.
5 Installation is the reverse of the removal procedure.

Air pump - removal and installation
6 Detach the battery earth lead.
7 Disconnect the air pump hose.
8 Remove the adjusting nut and bolt, and the pivot nut and bolt.
9 Draw the pivot bolt towards the radiator and remove the pump.
10 Installation is the reverse of the removal procedure. Refer to the following Paragraph for belt tensioning.

Air pump drive belt - removal, installation and tensioning
11 Detach the battery earth lead.
12 Remove the fan guard and fan blades, referring to Chapter 2 if necessary.
13 Loosen the adjusting nut and bolt, and the pivot nut and bolt.
14 Remove the drive belt from the pulleys.
15 Installation of the belt is the reverse of the removal procedure. Do not connect the battery lead or tighten the adjusting and pivot bolt nuts, until the belt has been tensioned.
16 If the tension is to be adjusted where the belt has not just been installed as described in Paragraphs 11 to 15, initially carry out the procedure of Paragraphs 11 and 13.
17 Carefully lever against the pump to obtain a total deflection of 3/8in (9.5mm) along the longest belt run under firm thumb pressure.
18 Tighten the adjusting and pivot bolt nuts whilst maintaining this tension.
19 Connect the battery earth lead.
20 Where a new belt has been installed, recheck the tension after about 150 miles (250 km) of travelling.

Air distribution manifold - removal and installation
21 Disconnect the air hose at the diverter and relief valve.
22 Unscrew the four union nuts, and withdraw the complete manifold (and check valve) from the exhaust ports.
23 If necessary, hold the manifold in a vice, and unscrew the check valve.
24 Installation is the reverse of the removal procedure.

Check valve - removal, testing and installation
25 Disconnect the air hose at the check valve.

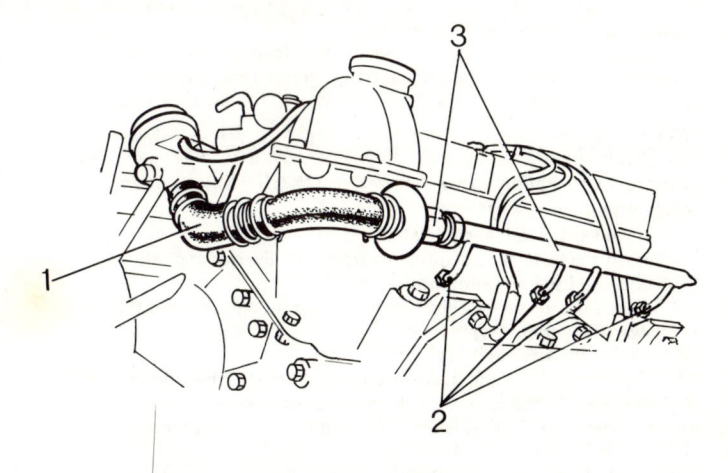

Fig. 3.28. Air distribution manifold

1 Hose
2 Union nuts
3 Check valve union

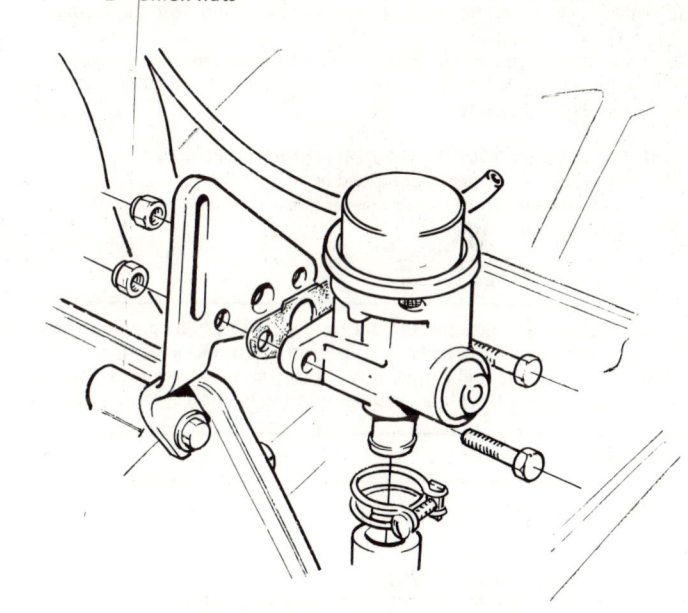

Fig. 3.29. Diverter and relief valve removal

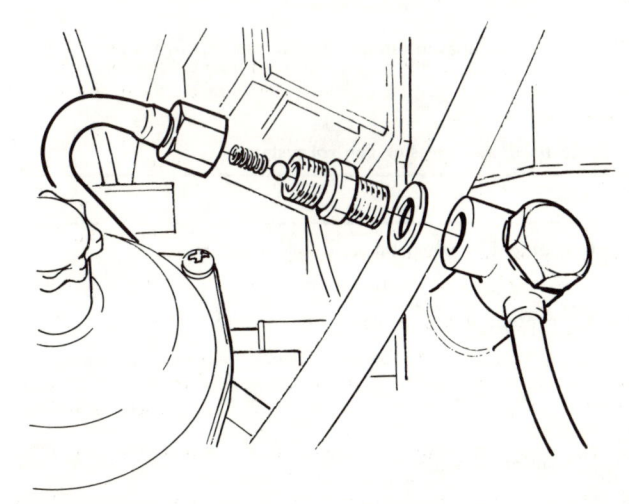

Fig. 3.30. One-way valve connections

26 Using two open ended spanners, unscrew the check valve whilst preventing strain on the air manifold.
27 If necessary the valve can be checked by blowing air (by mouth only) through the valve. Air should pass through from the hose connection end but not from the manifold end. Renew a defective valve.
28 Installation is the reverse of the removal procedure.

Diverter and relief valve (earlier models) - removal and installation
29 Detach the battery earth lead.
30 Detach the hose and vacuum line from the diverter and relief valve.
31 Disconnect the air pump hose.
32 Remove the air pump adjusting nut and bolt, and the bolt securing the valve/pump adjusting bracket.
33 Remove the valve and bracket.
34 Separate the valve and gasket from the bracket (2 nuts and bolts).
35 Installation is the reverse of the removal procedure; it is recommended that a new gasket is used between the valve and bracket.

One-way valve - removal and installation
36 Detach the pipe connection from the valve.
37 Unscrew the valve, complete with the union nut, from the banjo connection.
38 Separate the union nut from the valve body; do not lose the spring and ball.
39 Clean the valve parts, and make sure that the passageway is clear.
40 Reassemble the valve then install it using a new fibre washer.
41 Reconnect the pipe to the valve.

Temperature sensor unit - removal and installation
42 Remove the air cleaner as described in Section 3.
43 Prise off the clip from the air cleaner back-plate, and remove the sensor unit and felt washer.
44 Installation is the reverse of the removal procedure. Ensure that the pipe from the one-way valve is fitted to the connection nearest to the back-plate and that the pipe from the air inlet temperature control unit is connected furthest from the back-plate.

Anti-run-on valve - removal and installation
45 Detach the solenoid electrical leads.
46 Detach the vacuum pipe and the hose leading from the canister.
47 Twist the valve out of its retaining bracket.
48 Installation is the reverse of the removal procedure.

Exhaust gas recirculation (EGR) valve - removal and installation
49 On earlier models, detach the vacuum pipe.
50 Remove the two bolts, then lift the valve and gasket from the manifold.
51 Clean the valve-to-manifold mating faces and the valve port.
52 Installation is the reverse of the removal procedure. Use a new gasket if possible; note that one side is marked 'manifold side'.
53 On later models, detach the vacuum pipe and the asbestos lagged pipe from the valve.
54 Unscrew the valve.
55 To refit, reverse the removal procedure. Check that the valve is properly sealed to the manifold.

Exhaust gas recirculation (EGR) control valve - removal and installation
56 Disconnect the pipe from the valve, then remove the screw and shakeproof washer securing the valve bracket to the rear carburettor.
57 Remove the valve and bracket together.
58 Installation is the reverse of the removal procedure.

Catalytic converter - removal and installation
59 Refer to Section 20 for this procedure.

23 Fault diagnosis - fuel system

Unsatisfactory engine performance and excessive fuel consumption are not necessarily the fault of the fuel system or carburettor. In fact they more commonly occur as a result of ignition faults. Before acting on the fuel system it is necessary to check the ignition system first. Even though a fault may lie in the fuel system it will be difficult to trace unless the ignition is correct.
The table below therefore, assumes that the ignition system is in order.

Symptom	Reason/s
Smell of petrol when engine is stopped	Leaking fuel lines or unions. Leaking fuel tank.
Smell of petrol when engine is idling	Leaking fuel line unions between pump and carburettor. Overflow of fuel from float chamber due to wrong level setting, ineffective needle valve or punctured float.
Excessive fuel consumption for reasons not covered by leaks or float chamber faults	Worn needle. Sticking needle.
Difficult starting, uneven running, lack of power, cutting out	Incorrectly adjusted carburettor. Float chamber fuel level too low or needle sticking. Fuel pump not delivering sufficient fuel. Intake manifold gaskets leaking, or manifold fractured.

24 Fault diagnosis - emission control systems

Symptom	Reason/s
Low CO content of exhaust gases (weak or lean mixture)	Fuel level incorrect in carburettor. Incorrectly adjusted carburettor.
High CO content of exhaust gases (rich mixture)	Incorrectly adjusted carburettor. Choke sticking. Absorption canister blocked. Fuel level incorrect in carburettor. Air injection system fault.
Noisy air injection pump	Belt tension incorrect. Relief valve faulty. Diverter faulty. Check valve faulty.

Chapter 4 Ignition system

For modifications, and information applicable to later models, see Supplement at end of manual

Contents

Condenser (AC Delco) - removal, testing and installation 4	Distributor (Lucas) - removal and installation 7
Contact breaker points (AC Delco) - adjustment 2	Distributor - lubrication 10
Contact breaker points (AC Delco) - removal, cleaning and installation 3	Fault diagnosis - contact breaker type ignition system 13
	Fault diagnosis - Lucas Opus electronic ignition system 14
Distributor (AC Delco) - dismantling, overhaul and reassembly 8	General description 1
	Ignition timing - checking and adjustment 11
Distributor (AC Delco) - removal and installation 5	Pick-up air gap (Lucas) - adjustment 6
Distributor (Lucas) - dismantling, overhaul and reassembly ... 9	Spark plugs and HT leads 12

Specifications

UK and general market models

Distributor
Type	AC Delco D302
Rotation	Anticlockwise at rotor
Contact breaker gap	0.014 to 0.016 in (0.34 to 0.41 mm)
Dwell angle	$39 \pm 1°$
Moving contact spring tension	19 to 24 oz (540 to 680 gm)
Condenser capacity	0.18 to 0.23 mfd

Spark plugs
Type	Champion N12Y
Electrode gap	0.024 to 0.026 in (0.61 to 0.66 mm)

Ignition coil
Type	Lucas 15C6
Primary resistance	1.3 to 1.45 ohms
Ballast wire resistance	1.3 to 1.5 ohms

Ignition timing
Static	$10°$ btdc
Dynamic (vacuum pipe connected)	$10°$ btdc at 650/850 rpm
Locations of timing mark	Notch on pulley and scale on front cover

Engine firing order ... ... 1, 3, 4, 2 (no 1 cylinder nearest radiator)

Torque wrench settings
	lbf ft	Nm
Spark plugs	20	27

USA models

Distributor
Type	Lucas 47DE4 (Opus electronic) or Delco Remy series D302
Rotation	Anticlockwise at rotor
Pick-up air gap	0.014 to 0.016 in (0.34 to 0.41 mm)
Drive resistor	Lucas 10 ohms ± 5%

Spark plugs
Type:	
Early models	Champion N11Y
Late models	Champion N12Y
Electrode gap	0.024 to 0.026 in (0.61 to 0.66 mm)

Ignition coil (Lucas distributor)
Type	Lucas 15C6
Primary resistance	1.3 to 1.45 ohm
Ballast wire resistance	1.3 to 1.5 ohm

Ignition coil (Delco Remy distributor)
Type	AC Delco

Ignition timing
	Static	Dynamic at idle
Federal specification (carburettor engines)	10° btdc	10° btdc
Californian specification (carburettor and fuel injection engines)	10° btdc	2° atdc
Location of timing marks	Notch on pulley and scale on front cover	

Engine firing order
1, 3, 4, 2 (no 1 cylinder nearest radiator)

Torque wrench settings
	lbf ft	Nm
Spark plugs	20	27

1 General description

In order that the engine may run correctly, it is necessary for an electrical spark to ignite the fuel/air charge in the combustion chamber at exactly the right moment in relation to engine speed and load. The ignition system is based on supplying low tension voltage from the battery to the ignition coil where it is converted to high tension voltage by virtue of contact breaker operation. The high tension voltage is powerful enough to jump the spark plug gap in the cylinder many times a second under high compression pressure, providing that the ignition system is in good working order and that all adjustments are correct.

The ignition system comprises two individual circuits known as the low tension circuit and the high tension circuit, and the operation is as described below.

Systems with AC Delco contact breaker type distributor

The low tension (or primary) circuit comprises the lead from the positive terminal of the 12 volt battery, the ignition/starter switch, a ballast resistor wire, the primary winding of the 6volt ignition coil, the contact breaker points of the distributor (which are bridged by the condenser) and an earth connection. Since the negative terminal of the battery is also earthed, current will flow in the low tension circuit when the distributor contacts are closed and a magnetic field will be set up in the primary winding of the coil.

The high tension (or secondary) circuit comprises the secondary winding of the ignition coil (one end of thich is connected internally to the output terminal of the primary winding), the heavily insulated ignition lead from the centre of the coil to the centre of the distributor cap, the rotor arm, the spark plug leads and the spark plugs.

The complete ignition system operation is as follows: Low tension voltage from the battery is changed within the ignition coil to high tension voltage by the opening and closing of the contact breaker points in the low tension circuit. High tension voltage is then fed via the carbon brush in the centre of the distributor cap to the rotor arm of the distributor. The rotor arm revolves inside the distributor cap and each time it comes into line with one of the four metal segments in the cap, these being connected to the spark plug leads, the opening and closing of the contact breaker points causes the high tension voltage to build up, jump the gap from the rotor arm to the appropriate metal segment and so, via the spark plug lead, to the spark plug where it finally jumps the gap between the two spark plug electrodes, one being connected to the earth system.

During the engine starting sequence, the ballast resistance - wire, in series with the coil primary winding, is by-passed so that the full battery voltage (which will be low anyway due to the high current drawn by the starter motor) is passed to the 6 volt coil. In this way a bigger secondary voltage will be induced and therefore a bigger spark will result.

The ignition timing is advanced and retarded automatically to ensure the spark occurs at just the right instant for the particular load at the prevailing engine speed.

The ignition advance is controlled by a mechanical and vacuum operated system. The mechanical governor mechanism comprises two weights which move out under centrifugal force from the central distributor shaft as the engine speed rises. As they move outwards they rotate the cams relative to the distributor shaft, and so advance the spark. The weights are held in position by two springs, and it is the tension of the springs which is largely responsible for correct spark advancement.

The vacuum control comprises a diaphragm, one side of which is connected, via a small bore tube, to the carburettor, and the other side to the contact breaker plate. Depression in the induction manifold and carburettor, which varies with engine speed and throttle opening, causes the diaphragm to move, so moving the contact breaker plate and advancing or retarding the spark. A fine degree of control is achieved by a spring in the vacuum assembly.

Systems with Lucas 47DE4 Opus electronic type distributors

This system differs from that previously described, principally by the type of distributor used. The conventional cam, contact breaker points and capacitor are replaced by an oscillator, timing rotor, pick-up, amplifier and power transistor. These components are housed in a standard Lucas distributor body.

During operation, the oscillator supplies pulses to the pick-up in a continuous sequence. As the timing rotor, which is driven by the distributor shaft and has a ferrite rod for each 'lobe' passes the pick-up, one of its ferrite rods catches one of the pulses and applies it to the amplifier. This amplified pulse causes the power transistor to switch off; this results in a collapse of the ignition coil primary current and a high secondary voltage is induced as for a conventional system.

Where the conventional centrifugal advance and retard unit is retained, the vacuum advance unit is no longer used. The vacuum retard unit operates during deceleration or closed throttle conditions, its purpose being to delay the spark operation, resulting in cleaner exhaust emissions. On certain models, the retard unit also is omitted.

A separately mounted drive resistor is used as part of the control circuit for the amplifier. This is mounted alongside the ignition coil, and is not to be confused with the ballast resistor which is of the resistance wire type.

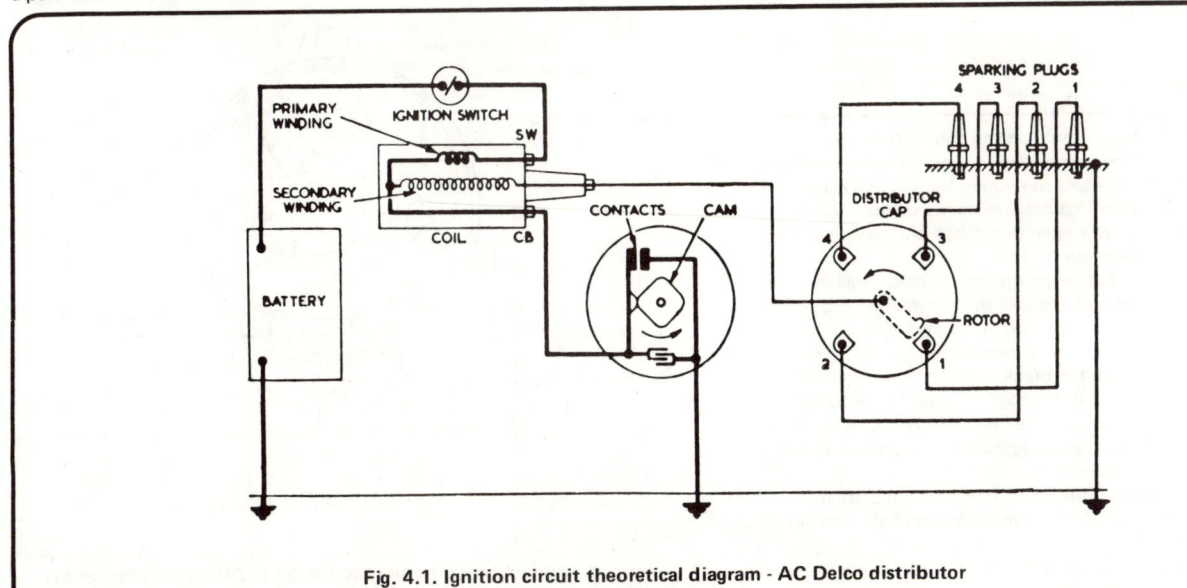

2 Contact breaker points (AC Delco) - adjustment

1 To adjust the contact breaker points to the correct gap, first remove the fresh air duct as described in Chapter 12, Section 34, paragraph 3.
2 Prise back the two clips securing the distributor cap to the body. Lift away the cap and then clean it with a dry cloth.
3 If the electrodes in the cap are severely burnt, the cap will need to be renewed. Otherwise, light corrosion or scoring can be removed with a penknife or small screwdriver.
4 Grasp the distributor rotor firmly and turn it in the direction of rotation then check that it returns to its original position under the action of its own springs.
5 Now take off the distributor rotor. It is held in place by two screws.
6 Put the car into third gear and push it backwards a little to get the contact breaker activating heel on one of the peaks of the cam. On automatic models it will be necessary to turn the crankshaft from the pulley.
7 Now prise the contacts open a little further with a screwdriver and examine the condition of their faces. If they are rough, dirty or pitted they will need to be removed for refacing, or renewed, as described in Section 3.
8 Assuming that the contacts are satisfactory, or have already been renewed, check the gap using a feeler gauge. A 0.014 to 0.016 in (0.34 to 0.41 mm) feeler gauge should just slide between the contact faces. If the feeler is tight or loose, adjustment will be necessary.
9 When adjusting the gap, first slacken the two screws which secure the contact plate then insert a screwdriver in the notched hole at the end of the plate. Turn the screwdriver as necessary to obtain the correct gap, then retighten the two screws.
10 Finally recheck the gap then fit the rotor, distributor cap, electrical connections and fresh air duct.
Note: the rotor has square and round alignment holes to prevent incorrect assembly.

3 Contact breaker points (AC Delco) - removal, cleaning and installation

1 If the contact breaker points are burnt, pitted or badly worn, they must be removed and either renewed or their faces must be rubbed smooth. For access to the distributor, first remove the fresh air duct as described in Chapter 12, Section 34, paragraph 3.
2 To remove the points, first remove the distributor cap and rotor as described in Section 2. Now press the moving contact spring towards the cam to release the wires from the terminal post then remove the two contact plate securing screws. Take out the contact securing screw and lift out the contact assembly.
3 To reface the points, rub the faces on a fine carborundum stone, or on fine emery paper. It is important that the faces are kept flat and parallel, or very slightly domed, in order to minimise wear when bedding in.
4 Installation of the contacts is a straightforward reversal of the removal procedure, but don't forget to reconnect the electrical connections.
5 Before refitting the rotor and distributor cap, set the gap correctly as described in Section 2.

Fig. 4.1. Ignition circuit theoretical diagram - AC Delco distributor

Chapter 4/Ignition system

Fig. 4.2. Ignition circuit - Lucas Opus electronic distributor

1. Battery
2. Ignition/starter switch
3. Ballast resistor wire
4. Start supply from interlock starter motor relay
5. Ignition coil
6. Oscillator
7. Timing rotor
8. Pick up
9. Amplifier
10. Power transistor
11. Drive resistor
12. HT circuit

Colour coding
- N Brown
- W White
- KW Pink/White
- WY White/Yellow
- WS White/Slate
- WU White/Blue
- B Black ident
- R Red ident
- U Blue ident

4 Condenser (AC Delco) - removal, testing and installation

1 The purpose of the condenser (sometimes known as a capacitor) is to ensure that when the contact breaker points open there is no sparking across them which would waste voltage and cause wear. It also boosts the high tension voltage.

2 The condenser is fitted in parallel with the contact breaker points. If it develops a short circuit, it will cause ignition failure as the points will be prevented from interrupting the low tension circuit.

3 If the engine becomes very difficult to start or begins to 'miss' after several miles running and the breaker points show signs of excessive burning, then the condition of the condenser must be suspect.

4 Without special test equipment, the only sure way to diagnose condenser trouble is to replace a suspected unit with a new one and note if there is any improvement.

5 To remove the condenser from the distributor, take off the distributor cap and rotor arm. Now remove the condenser fixing screw and the lead connection.

6 Installation of the condenser is the reverse of the removal procedure, but take care that the flexible lead cannot contact the moving parts or become trapped when the cap is replaced.

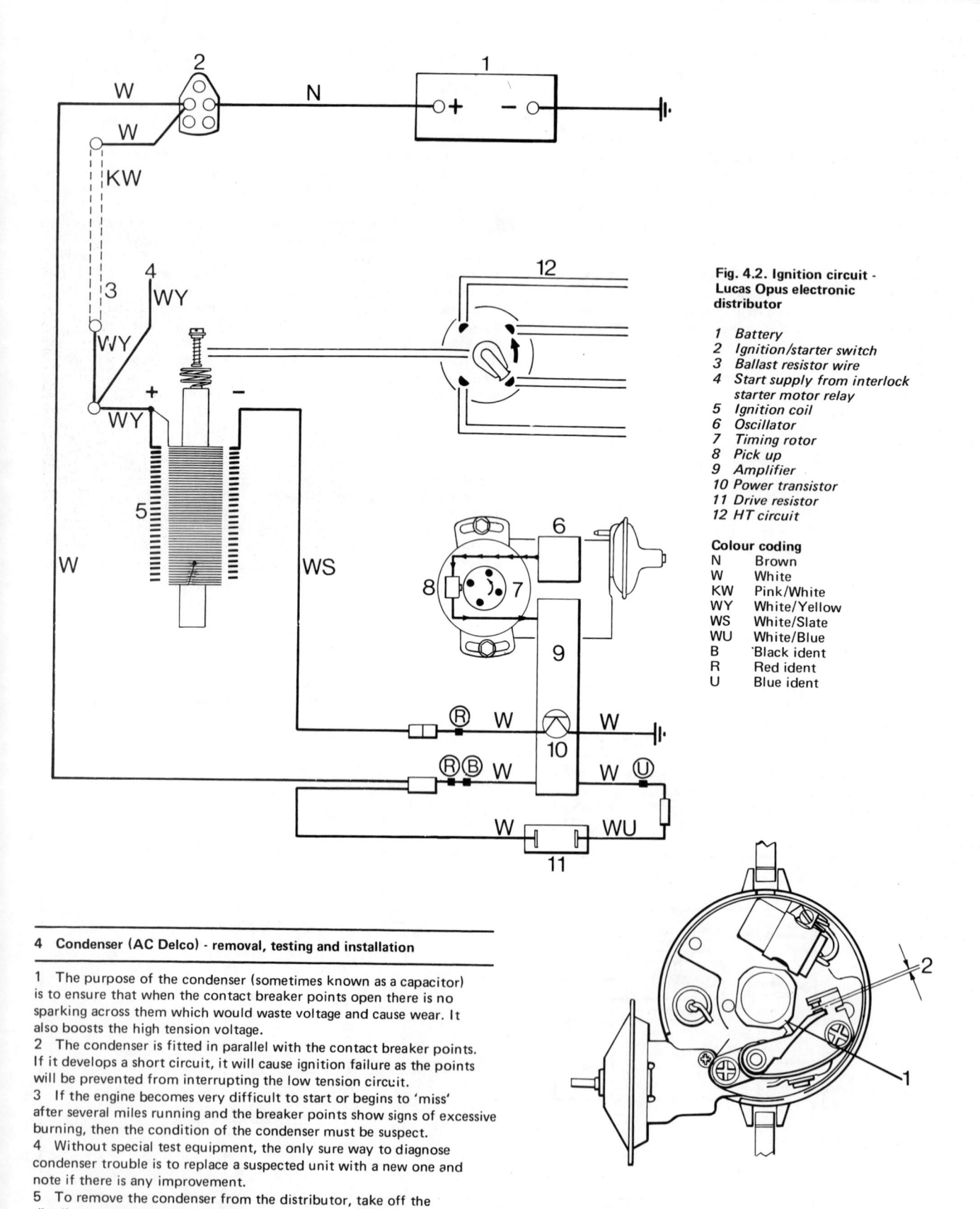

Fig. 4.3. Contact breaker points adjustment - AC Delco distributor

1 Cam peak   2 Points gap

Chapter 4/Ignition system

5 Distributor (AC Delco) - removal and installation

1 Remove the fresh air duct as described in Chapter 12, Section 34, paragraph 3.
2 Remove the distributor cap, low tension connections and vacuum connection.
3 Rotate the crankshaft clockwise until the notch on the pulley aligns with the O mark on the scale on the timing chain housing (top-dead-centre) (photo).
4 Mark the fitted position of the distributor in relation to the cylinder block and also note the position of the rotor arm in relation to the distributor body (see Chapter 13, Fig.13.1).
5 Remove the two mounting bolts and washers and carefully ease the distributor out.
6 If the same distributor is not to be refitted, make a sketch of the marking and distributor position so that they can be transferred to the new one.
7 When refitting, first align the pulley marking as described in paragraph 2.
8 Refit the distributor making sure that the marks are aligned and the rotor is in the correct position.
9 Tighten the mounting bolts, then fit the vacuum connection and low tension connections.
10 Recheck the ignition timing as described in Section 11.
11 Finally refit the distributor cap and fresh air duct.

6 Pick-up air gap (Lucas) - adjustment

Caution: Do not attempt to check the pick-up air gap with the ignition switched on.
1 Detach the battery earth lead.
2 Remove the fresh air duct as described in Chapter 12, Section 34, paragraph 3.
3 Spring back the clips and remove the distributor cap, rotor and anti-flash cover.
4 Slide a 0.014 to 0.016 in (0.34 to 0.41 mm) feeler gauge between the timing rotor and the pick-up coil contact. It should be a firm sliding fit, if too tight or too loose, adjustment will be required.
Note: *The angular position of the rotor is not important.*
5 To adjust, slacken the pick-up retaining screws and reposition as necessary. Recheck the gap after the screws have been tightened.
6 Whilst the distributor cap is removed, wipe it clean with a clean dry cloth. If the electrodes in the cap are badly burnt the cap will need to be renewed. Otherwise, light corrosion or scoring can be removed with a penknife or small screwdriver.
7 Push the carbon brush (located in the centre of the distributor cap) inwards a few times and make sure that it is free to move.
8 Install the anti-flash cover, rotor and distributor cap.
9 Install the fresh air duct and connect the battery earth lead.

7 Distributor (Lucas) - removal and installation

1 The procedure is essentially as described for the AC Delco distributor, but the following points should be noted:

   a) Initially, detach the battery earth lead.
   b) There are three wire connectors to be detached.
   c) Do not reconnect the wires incorrectly during installation, or connect them directly to the battery with the distributor removed, or irreparable damage may be caused.

8 Distributor (AC Delco) - dismantling, overhaul and reassembly

1 Remove the distributor from the car as previously described, and take off the cap.
2 Remove the two screws and spring washers, and take off the rotor.
3 Note the position of the two wires on the terminal post, then push the moving contact away to release the wire eyelets.
4 Remove the crosshead screws and spring washers to release the contact assembly, moving the weights as necessary.
5 Remove the screw and take out the condenser.

5.3 The crankshaft pulley and scale plate timing marks

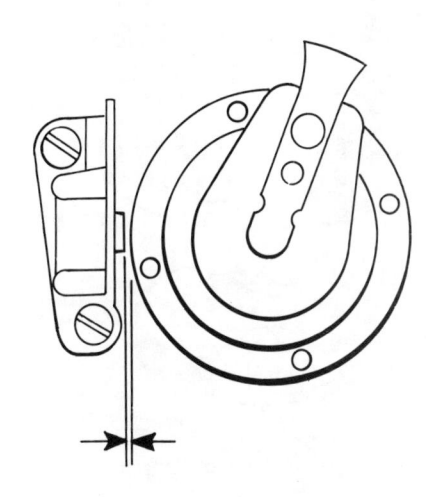

Fig. 4.4. Pick-up air gap measurement - Lucas distributor

6 Squeeze together the sides of the cam lubrication post; withdraw the post and remove the foam pad (where applicable).
7 Drive out the roll pin, and remove the gear and thrust washer. Ensure that there are no burrs, then withdraw the shaft.
8 Remove the retaining clip and lift off the plate.
9 Remove the two screws to release the earth lead and vacuum unit.
10 Carefully prise out the felt pad.
11 Remove the grommet and take out the wire assembly.
12 Taking care that they are not distorted, remove the control springs; also remove the weights.
13 Remove the rubber O-ring from the body, then clean all the mechanical parts carefully in petrol.
14 Check the fit of the balance weights on the distributor shaft. If the pivots are loose or the holes excessively worn, the relevant parts must be renewed. The springs are best renewed anyway.
15 Examine the driving gear teeth for wear and renew if necessary.
16 Check the fit of the driveshaft in the housing. If excessive wear is present the parts must be renewed.
17 Check that the vacuum unit is working correctly by sucking on the tube and checking that the linkage moves.
18 Check the metal contact on the distributor rotor for security of fixing and burning. Small burning marks can be dressed out with a smooth file or very fine emery paper, but if anything else is wrong the rotor should be renewed. Look also for cracks in the plastic moulding.
19 Check the condition of the points, and renew or clean them as necessary.
20 Check the distributor cap in a similar manner to the rotor, renewing if the condition warrants it.
21 Reassembly is essentially the reverse of the dismantling procedure, but the following points should be noted:

a) Lubricate the shaft with engine oil.
b) Saturate the felt washer with engine oil.
c) Lubricate the plate bearing with engine oil.
d) Lubricate the foam pad (where fitted) by working in some petroleum jelly.
e) Apply one drop of engine oil to the weight and spring pivots.
f) Adjust the points as described in Section 2.

9 Distributor (Lucas) - dismantling, overhaul and reassembly

1 Remove the distributor from the car as previously described, and take off the cap.
2 Remove the rotor and anti-flash cover, and withdraw the felt lubrication pad.
3 Remove the two screws, lock washers and flat washer to release the pick-up.
4 Hold the amplifier module in position and remove the three retaining screws and washers (two long screws, one short screw).
5 Carefully manoeuvre the module to unhook the retard unit link from the moving plate pin. Also remove the distributor cap clips.
6 Pull out the wire grommet, then remove the module and pick-up.
7 Tap out the pin and withdraw the retard unit.
8 Remove the circlip, flat washer and O-ring, then remove the timing rotor.
9 Remove the two screws and take out the base plate.
10 Drive out the retaining pin, and remove the gear and thrust washer. Ensure that there are no burrs and then withdraw the shaft.
11 Remove the control springs.
13 Remove the O-ring from the distributor body. No further dismantling should be attempted.
14 Examine the parts for wear and damage by reference to paragraphs

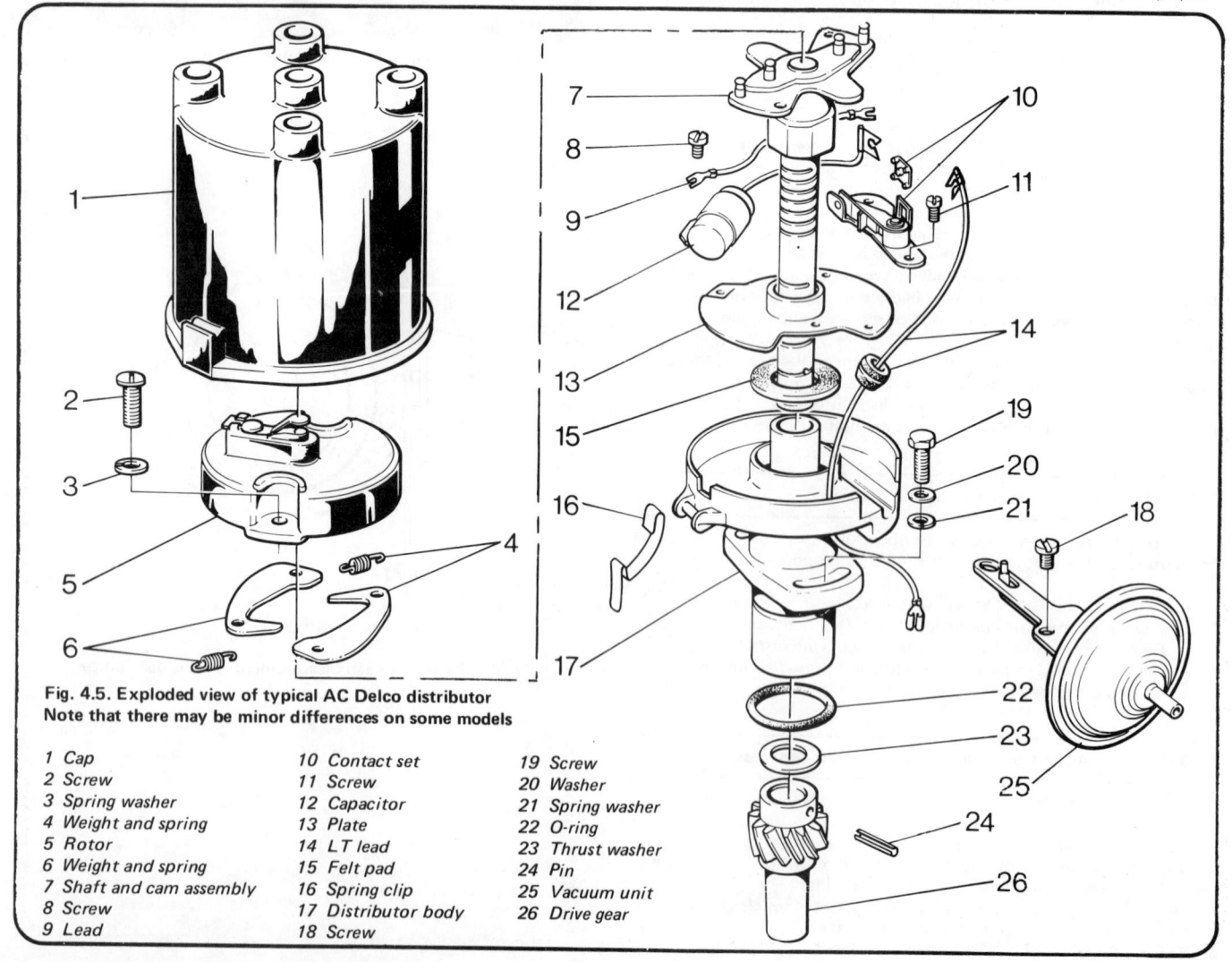

Fig. 4.5. Exploded view of typical AC Delco distributor
Note that there may be minor differences on some models

1 Cap
2 Screw
3 Spring washer
4 Weight and spring
5 Rotor
6 Weight and spring
7 Shaft and cam assembly
8 Screw
9 Lead
10 Contact set
11 Screw
12 Capacitor
13 Plate
14 LT lead
15 Felt pad
16 Spring clip
17 Distributor body
18 Screw
19 Screw
20 Washer
21 Spring washer
22 O-ring
23 Thrust washer
24 Pin
25 Vacuum unit
26 Drive gear

Chapter 4/Ignition system

14 to 18 and 20 of Section 8.
15 Reassembly is essentially the reverse of the dismantling procedure, but the following points should be noted:

a) Lubricate the weight assembly with a dry lubricant such as Moly-pad.
b) Lubricate the shaft with a dry lubricant such as Moly-pad.
c) Lubricate the moving plate pin with a dry lubricant such as Moly-pad.
d) Adjust the pick-up coil air gap as described in Section 6.
e) Lubricate the distributor as described in Section 10.

10 Distributor - lubrication

1 During routine maintenance, and where otherwise specified in this Chapter, the distributor should be lubricated as follows:

AC Delco distributor
2 Remove the distributor cap and rotor as described in Section 2.
3 Squeeze together the sides of the cam lubrication post; withdraw the post and remove the foam pad (where applicable).
4 Lubricate the foam pad by working in some petroleum jelly, then refit the pad and lubrication post. Where no pad is fitted, lightly grease the cam with petroleum jelly.
5 Inject a few drops of engine oil through the lubrication hole to lubricate the top bearing.
6 Apply one drop of engine oil to each 'cam action' position and centrifugal weight pivot post.
7 Install the rotor, cap and fresh air duct.

Lucas distributor
8 Remove the distributor cap, rotor and anti-flash cover as described in Section 6.
9 Apply a few drops of engine oil to the felt pad in the cavity below the rotor.
10 Inject a few drops of engine oil through the apertures to lubricate the centrifugal advance mechanism.
11 Apply one drop of engine oil to each of the two lubrication apertures in the moving plate bearing.
12 Install the anti-flash cover, rotor, cap and fresh air duct.

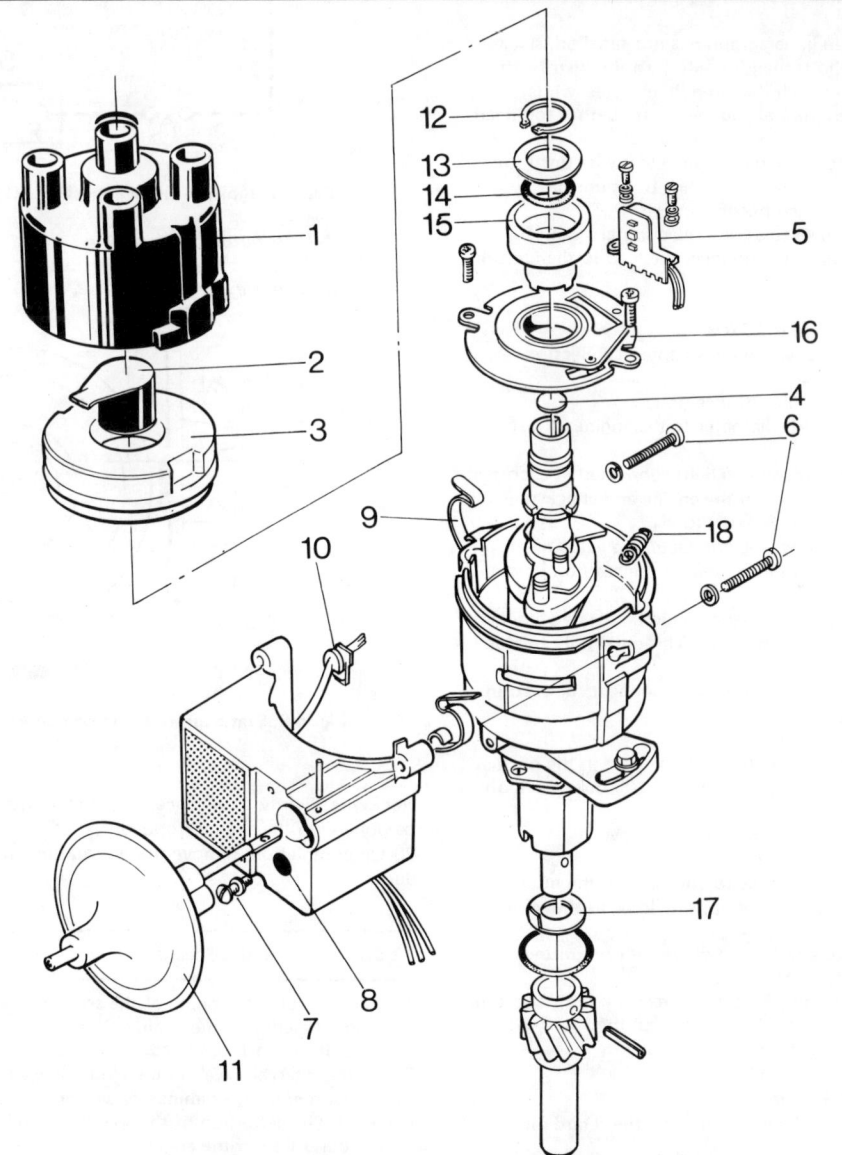

Fig. 4.6. Exploded view of Lucas Opus electronic distributor

1 Cap
2 Rotor
3 Anti-flash cover
4 Lubrication pad
5 Pick-up
6 Long screws
7 Short screw
8 Amplifier unit
9 Retaining clip
10 Grommet
11 Retard unit (where fitted)
12 Circlip
13 Flat washer
14 O-ring
15 Timing rotor
16 Base plate
17 Thrust washer
18 Spring

11 Ignition timing - checking and adjustment

1 In this Section, procedures are given for Static and Dynamic checking. For most practical purposes, the Static setting is satisfactory for the AC Delco distributors. This is not the case with the Lucas distributors; here, the Static setting may only be regarded as a setting up point when the dynamic setting point has been lost.

Static setting - AC Delco distributor

2 Initially check the contact breaker gap, and adjust or clean the contacts as necessary (See Sections 2 and 3).
3 Isolate the battery, and disconnect the distributor lead at the negative terminal of the coil.
4 Using a 12volt bulb of up to 5 watts rating (eg, sidelamp or instrument panel lamp) connect it between the distributor/coil lead and the battery positive terminal.
5 Rotate the crankshaft in the normal direction of rotation, and check that as the notch in the crankshaft pulley is moving towards the O marking on the scaleplate, ie, top-dead-centre (tdc), the lamp is on until 10° before top-dead-centre is reached, and then goes off (photo).
6 If the requirement described in paragraph 5 is not satisfied, slacken the two distributor mounting bolts then initially turn the distributor anticlockwise past the point at which the lamp illuminates, whilst maintaining the notch on the crankshaft pulley at 10° before top-dead-centre.
7 Now turn the distributor slowly and carefully clockwise until the lamp just goes off; then tighten the two mounting bolts on the distributor without allowing the distributor to move.
8 The setting can now be re-checked as described in paragraph 5.
9 Finally, remove the test lamp and reconnect the lead to the ignition coil.

Dynamic setting - AC Delco distributor

10 Remove the fresh air duct as described in Chapter 12, Section 34, paragraph 3.
11 Pull off the vacuum pipe from the advance unit.
12 If possible connect an external tachometer to the engine; if not available, use the car tachometer.
13 Using a proprietary stroboscopic timing light connected in accordance with the manufacturer's instructions, run the engine and check that the centrifugal advance is as given in the Specifications.
Note: *The static setting must be added to the tabulated value to give the actual amount of advance as indicated by the stroboscopic light.*
14 If adjustment is required, the distributor should be rotated clockwise slightly to advance, or anticlockwise to retard. This must only be done with the engine stopped.
15 On completion, remove the test equipment, reconnect the vacuum line and install the fresh air duct.

Static setting - Lucas distributor

16 With the battery disconnected, rotate the crankshaft in the normal direction to align the notch on the pulley with the 10° btdc mark on the scale.
17 Remove the distributor cap, rotor and anti-flash cover as described in Section 6.
18 If the static timing is correctly set the relationship of the nearest ferrite rod in the timing rotor to the pick-up, should be as shown in Fig. 4.8.
19 If correction is required, loosen the two distributor mounting bolts and rotate the distributor body as necessary.
20 Tighten the distributor mounting bolts, then recheck the static timing.
21 Fit the anti-flash cover, rotor and distributor cap, then carry out the dynamic setting check as described below.

Dynamic setting - Lucas distributor

22 Connect the battery earth lead, and ensure that the retard pipe is connected (where relevant).
23 If possible, connect an external tachometer to the engine; if not available, use the car tachometer.
24 Using a proprietary stroboscopic timing light connected in accordance with the manufacturer's instructions, run the engine at idle speed and check that the ignition timing is as given in the Specifications.
25 If adjustment is required, the distributor should be rotated

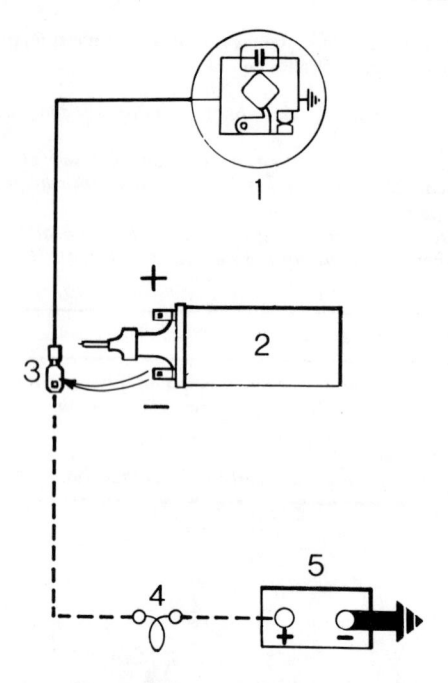

Fig. 4.7. Ignition timing checks - AC Delco distributor

1 Distributor
2 Coil
3 Negative coil lead
4 Test lamp, 5 watts maximum
5 Battery

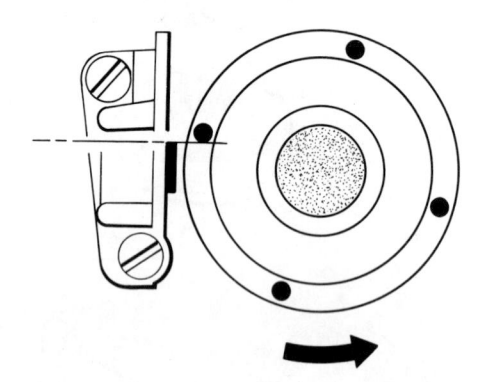

Fig. 4.8. Static ignition timing checks - Lucas distributor

clockwise slightly to advance, or anticlockwise to retard. This must only be done with the engine stopped.
26 On completion, remove the test equipment and install the fresh air duct.

12 Spark plugs and HT leads

1 The correct functioning of the spark plugs is vital for the proper running efficiency of the engine. Plugs of the recommended type (see Specifications), or approved equivalents only, should be used.
2 At the intervals given in the Routine Maintenance Section the plugs should be removed, examined and cleaned, and if worn excessively, renewed. The condition of the spark plug will also tell much about the overall condition of the engine.
3 If the insulator nose of the spark plug is clean and white with no deposits, this is indicative of a weak mixture, or too hot a plug (a hot plug transfers heat away from the electrode slowly - a cold plug ;fers transfers heat away quickly).
4 If the top and insulator nose is covered with hard black-looking deposits, then this is indicative that the mixture is too rich. Should the plug be black and oily, then it is likely that the engine is fairly worn, as

Measuring plug gap. A feeler gauge of the correct size (see ignition system specifications) should have a slight 'drag' when slid between the electrodes. Adjust gap if necessary

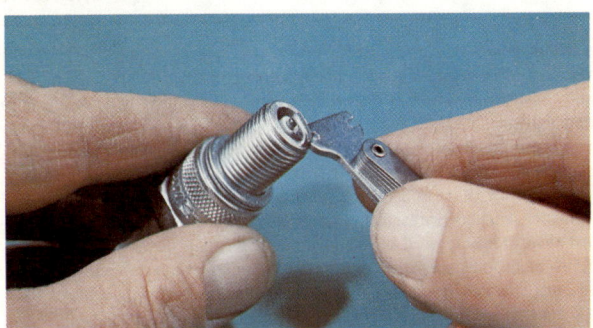

Adjusting plug gap. The plug gap is adjusted by bending the earth electrode inwards, or outwards, as necessary until the correct clearance is obtained. Note the use of the correct tool

Normal. Grey-brown deposits, lightly coated core nose. Gap increasing by around 0.001 in (0.025 mm) per 1000 miles (1600 km). Plugs ideally suited to engine, and engine in good condition

Carbon fouling. Dry, black, sooty deposits. Will cause weak spark and eventually misfire. Fault: over-rich fuel mixture. Check: carburettor mixture settings, float level and jet sizes; choke operation and cleanliness of air filter. Plugs can be re-used after cleaning

Oil fouling. Wet, oily deposits. Will cause weak spark and eventually misfire. Fault: worn bores/piston rings or valve guides; sometimes occurs (temporarily) during running-in period. Plugs can be re-used after thorough cleaning

Overheating. Electrodes have glazed appearance, core nose very white – few deposits. Fault: plug overheating. Check: plug value, ignition timing, fuel octane rating (too low) and fuel mixture (too weak). Discard plugs and cure fault immediately

Electrode damage. Electrodes burned away; core nose has burned, glazed appearance. Fault: pre-ignition. Check: as for 'Overheating' but may be more severe. Discard plugs and remedy fault before piston or valve damage occurs

Split core nose (may appear initially as a crack). Damage is self-evident, but cracks will only show after cleaning. Fault: pre-ignition or wrong gap-setting technique. Check: ignition timing, cooling system, fuel octane rating (too low) and fuel mixture (too weak). Discard plugs, rectify fault immediately

well as the mixture being too rich.
5 If the insulator nose is covered with light tan to greyish brown deposits, then the mixture is correct and it is likely that the engine is in good condition.
6 If there are any traces of long brown tapering stains on the outside of the white portion of the plug, then the plug will have to be renewed as this shows that there is a faulty joint between the plug body and the insulator, and compression is being allowed to leak away.
7 Plugs should be cleaned by a sand blasting machine which will free them from carbon better than cleaning by hand. The machine will also test the condition of the plugs under compression. Any plug that fails to spark at the recommended pressure should be renewed.
8 The spark plug gap is of considerable importance. If it is too large or too small the size of the spark and its efficiency will be seriously impaired. The spark plug gap is given in the Specifications.
9 To set it, measure the gap with a feeler gauge, and then bend open, or close the outer plug electrode until the correct gap is achieved. The centre electrode should never be bent as this may crack the insulation and cause plug failure if nothing worse.
10 When installing the plugs, remember to connect the leads from the distributor in the correct firing order, which is 1, 3, 4, 2 (No 1 cylinder is the nearest one to the radiator).
11 The plug leads require no routine attention other than being kept clean and wiped over regularly. During routine maintenance however, pull each lead off the plugs in turn and remove them from the distributor cap. Water can seep down into these joints giving rise to a white corrosive deposit which must be carefully removed from the end of each cable.

13 Fault diagnosis - contact breaker type ignition system

Engine fails to start

1 If the engine fails to start and the car was running normally when it was last used, first check there is fuel in the fuel tank. If the engine turns over normally on the starter motor and the battery is evidently well charged, then the fault may be in either the high or low tension circuits. If any dismantling has been done, in addition to check given below, also check the ignition timing. First check the HT circuit.
Note: *If the battery is known to be fully charged; the ignition light comes on, and the starter motor fails to turn the engine check the tightness of the leads on the battery terminal and also the secureness of the earth lead to its connection to the body. It is quite common for the leads to have worked loose, even if they look and feel secure. If one of the battery terminal posts gets very hot when trying to work the starter motor this is a sure indication of a faulty connection to that terminal.*
2 One of the commonest reasons for bad starting is wet or damp spark plug leads and distributor. Remove the distributor cap; if condensation is visible internally, dry the cap with a rag and also wipe over the leads. Refit the cap.
3 If the engine still fails to start, check that voltage is reaching the plugs by disconnecting each plug lead in turn at the spark plug end, and hold the end of the cable about 1/8 in (5 mm) away from the cylinder block. Spin the engine on the starter motor.
4 Sparking between the end of the cable and the block should be fairly strong with a regular blue spark. (Hold the lead with a dry cloth or rubber glove to avoid electric shocks). If current is reaching the plugs, remove them, and clean and regap them. The engine should now start.
5 If there is no spark at the plug leads, take off the HT lead from the centre of the distributor cap and hold it to the block as before. Spin the engine on the starter once more. A rapid succession of blue sparks between the end of the lead and the block indicates that the coil is in order and that the distributor cap is cracked, the rotor arm faulty, or the carbon brush in the top of the distributor cap is not making good contact with the spring on the rotor arm. Possibly the points are in bad condition. Clean and reset them as described in this Chapter.
6 If there are no sparks from the end of the lead from the coil, check the connections at the coil end of the lead. If it is in order start checking the low tension circuit.
7 Use a 12v voltmeter, or a 12v bulb and two lengths or wire. With the ignition switched on and the points open test between the low tension wire to the coil connection + (positive) and earth. No reading indicates a break in the supply from the ignition switch. Check the connections at the switch to see if any are loose. Refit them and the engine should run. A reading shows a faulty coil or condenser, or broken lead between the coil and the distributor.
8 Take the condenser wire off the points assembly and with the points open, test between the moving points and earth. If there now is a reading, then the fault is in the condenser. Fit a new one and the fault is cleared.
9 With no reading from the moving point to earth take a reading between earth and the —(negative) terminal of the coil. A reading here shows a broken wire which will need to be replaced between the coil and distributor. No reading confirms that the coil has failed and must be replaced, after which the engine will run once more. Remember to refit the condenser wire to the points assembly. For these tests it is sufficient to separate the points with a piece of dry paper while testing with the points open.

Engine misfires

10 If the engine misfires regularly, run it at a fast idling speed. Pull off each of the plug caps in turn and listen to the note of the engine. Hold the plug cap in a dry cloth or with a rubber glove as additional protection against a shock from the HT supply.
11 No difference in engine running will be noticed when the lead from the defective circuit is removed. Removing the lead from one of the good cylinders will accentuate the misfire.
12 Remove the plug lead from the end of the defective plug and hold it about 1/8 in (5 mm) away from the block. Restart the engine. If the sparking is fairly strong and regular, the fault must lie in the spark plug.
13 The plug may be loose, the insulation may be cracked, or the points may have burnt away giving too wide a gap for the spark to jump. Worse still, one of the points may have broken off. Either renew the plug, or clean it, reset the gap, and then test it.
14 If there is no spark at the end of the plug lead, or if it is weak and intermittent, check the ignition lead from the distributor to the plug. If the insulation is cracked or perished, renew the lead. Check the connections at the distributor cap.
15 If there is still no spark, examine the distributor cap carefully for tracking. This can be recognised by a very thin black line running between two or more electrodes or between an electrode and some other part of the distributor. These lines are paths which conduct electricity across the cap thus letting it run to earth. The only answer is a new distributor cap.
16 Apart from the ignition timing being incorrect, other causes of misfiring have already been dealt with under the section dealing with the failure of the engine to start. To recap - these are that:

   a) *the coil may be faulty giving an intermittent misfire;*
   b) *there may be a damaged wire or loose connection in the low tension circuit;*
   c) *the condenser may be defective;*
   d) *there may be a mechanical fault in the distributor (broken driving spindle or contact breaker spring).*

17 If the ignition timing is too far retarded, it should be noted that the engine will tend to overheat, and there will be a quite noticeable drop in power. If the engine is overheating and the power is down and the ignition timing is correct, then the carburettor should be checked, as it is likely that this is where the fault lies.

14 Fault diagnosis - Lucas Opus electronic ignition system

1 Refer to the previous Section, but ignore Paragraphs 7, 8 and 9.
2 If the foregoing does not isolate the fault, check all the lead connections in the LT circuit. If these are satisfactory, there are two other checks which can be carried out if an ohmmeter is available. First detach the white/yellow lead from the positive (+) terminal of the ignition coil, and both leads from the drive resistor.
3 Check for a resistance of 1.3 to 1.5 ohms, between the coil white/yellow lead and the drive resistor white lead; if this is not obtained, either the ballast resistor wire or its connections in the circuit are faulty.
4 If this is satisfactory, check the resistance of the drive resistor. This should be 10 ohms ± 5%; if not obtained, a new drive resistor must be obtained. Remake the connections on completion.
5 If these tests do not isolate a fault, the nearest Triumph or Lucas dealer should be contacted for further diagnostic checking.

Chapter 5 Clutch

Contents

Clutch pedal box - removal, overhaul and installation ... 9	General description ... 1
Clutch - removal, inspection and installation ... 7	Master cylinder - overhaul ... 4
Clutch release bearing assembly - removal, overhaul and installation ... 8	Master cylinder - removal and installation ... 3
	Slave cylinder - overhaul ... 6
Clutch system - bleeding ... 2	Slave cylinder - removal and installation ... 5
Fault diagnosis - clutch ... 10	

Specifications

Type ...	Hydraulically operated single diaphragm spring
Make ...	Borg and Beck
Clutch plate diameter ...	8.5 in (216 mm)
Facing material ...	2124 F

Clutch plate damper springs

Number of springs ...	6
Spring colour ...	Red and violet
Release bearing ...	Ball journal
Clutch fluid specification ...	SAE J1703d or DOT 3(FMV SS 116)

Torque wrench settings:

	lb f ft	Nm
Clutch housing-to-rear engine plate dowel bolts ...	32	43
Clutch housing-to-rear engine plate bolts ...	20	27
Clutch-to-flywheel bolts ...	22	30
Clutch housing-to-cylinder block bolts ...	20	27
Clutch master cylinder bolts ...	21	28
Sump coupling plate-to- clutch housing bolt ...	37	50

1 General description

The clutch is fitted in order that the engine may run without being mechanically connected to the transmission. It enables the engine torque to be progressively applied to the gearbox so that the car can move off gradually from rest, and then for the gears to be changed easily as the speed increases or decreases.

The main parts of the clutch assembly are, the driven plate assembly, the cover assembly and the release bearing assembly. When the clutch is in use, the driven plate assembly, being splined to the input shaft, is sandwiched between the flywheel and the pressure plate by the diaphragm spring. Engine torque is, therefore, transferred from the flywheel to the clutch driven plate assembly and then to the input shaft.

By depressing the clutch pedal, the piston in the master cylinder pressurises hydraulic fluid through the clutch hydraulic pipe to the slave cylinder, the piston of which moves forward on the entry of the fluid and actuates the pivoted clutch release lever. The release bearing assembly is pushed against the diaphragm centre, releasing its pressure on the driven plate assembly, thus breaking the drive between the engine and gearbox.

When the clutch pedal is released, the pressure plate diaphragm spring forces the pressure plate into contact with the friction linings on the clutch driven plate, at the same time forcing the clutch driven plate assembly against the flywheel and so taking up the drive.

As the friction linings on the clutch driven plate wear, the pressure plate automatically moves closer to the driven plate to compensate. This makes the centre of the diaphragm spring move nearer to the release bearing, so decreasing the release bearing clearance.

2 Clutch system - bleeding

Whenever the clutch hydraulic system has been overhauled, a part renewed, or the level in the reservoir is too low, air will have entered the system necessitating its bleeding. During this operation the level of hydraulic fluid in the reservoir should not be allowed to fall below half full, otherwise air will be drawn in again.

1 Obtain a clean, dry, glass jar, a length of plastic or rubber tubing which will fit the bleed nipple of the clutch slave cylinder and which is about 12 in (300 mm) long, a supply of the correct type of fluid and the services of an assistant.

2 Check that the master cylinder reservoir is full and, if not, fill it to within ¼ in (6.5 mm) of the top. Also add about one inch of fluid to the jar.
3 Remove the rubber dust cap from the slave cylinder bleed nipple, wipe the nipple clean then attach the bleed tube.
4 With the other end of the tube immersed in the fluid in the jar (which can be supported on the subframe if required), and the assistant ready inside the car, unscrew the bleed nipple one full turn.
5 The assistant should now pump the clutch pedal up and down until the air bubbles cease to emerge from the end of the tubing. Check the reservoir frequently to ensure that the hydraulic fluid does not drop too far, so letting air into the system.
6 When no more air bubbles appear, tighten the bleed nipple on a downstroke.
7 Install the rubber dust cap over the bleed nipple.
8 One-man bleeding devices, as sold by most motor stores, may be found helpful for the bleeding operation.
Note: *Never use the fluid bled from the hydraulic system immediately for topping up the master cylinder, but allow to stand for at least twenty-four hours in a sealed air tight container to allow the minute air bubbles held in suspension to escape.*

3 Master cylinder - removal and installation

1 Remove the split pin, washer and clevis pin from the clutch pedal-to-pushrod attachment.
2 Hold a suitable sized container beneath the master cylinder fluid pipe, then unscrew the gland nut. Pull out the pipe and allow the fluid to drain into the container. Do not allow the fluid to spill on any body panel as it will destroy the paint finish.
3 Remove the two nuts, spring washers and bolts, and withdraw the master cylinder.
4 Installation is the reverse of the removal procedure. On completion, top up the cylinder with the correct type of fluid then bleed the system as described in Section 2.

4 Master cylinder - overhaul

1 With the master cylinder removed, slide the rubber boot along the pushrod so that the circlip can be removed.
2 Apply air from a low pressure airline to the pressure port to extract the piston and associated parts.
3 Refer to Fig. 5.1 or 5.2 for the component parts of the alternative types of master cylinder in use. Dismantling of the type shown in Fig. 5.1 is straightforward; for the type shown in Fig. 5.2 the prong of the spring thimble has to be raised to detach the piston, then the valve stem can be released from the keyhole slot in the thimble.
4 Discard all the rubber parts, then clean the metal parts in methylated spirit, isopropyl alcohol or clean brake fluid. Check for wear, scoring and corrosion, obtaining new parts as necessary.
5 Assembly of the master cylinder is basically the reverse of the dismantling procedure. All rubber parts should be renewed, and the moving parts should be lubricated with clean brake fluid. The following points should be noted:

Master cylinder shown in Fig. 5.1
  a) *Insert the spring into the master cylinder, larger diameter first.*
  b) *Insert the spring retainer button, front up (cup lips leading), dished spring washer and piston (complete with rear cup).*
  c) *Smear the ball end of the pushrod with rubber grease or disc brake lubricant.*
  d) *Use a new retaining circlip if the original one is damaged.*

Master cylinder shown in Fig. 5.2
  e) *After fitting a new seal, assemble the spacer, spring and thimble to the valve stem.*
  f) *Fit the piston seal with the lip towards the spring.*
  g) *After installing the thimble on the piston, press down the prong to retain it.*
  h) *Smear the ball end of the pushrod with rubber grease or disc brake lubricant.*
  i) *Use a new retaining circlip if the original one is damaged.*

5 Slave cylinder - removal and installation

1 Apply the handbrake then raise the right-hand side of the car for access to the slave cylinder. Support the car on the bodyframe sidemember; chock the wheels remaining on the ground for safety.
2 Clean off any mud and dirt from around the slave cylinder fluid pipe and nipple. Disconnect the fluid pipe, then plug the pipe and port to prevent ingress of dirt.
3 Remove the attachment nuts and bolts, and withdraw the slave cylinder. **Do not attempt to move the operating pushrod in a forward direction as this may cause the clutch release lever to be dislodged; this will necessitate removing the gearbox for the lever to be refitted.**
4 Installation is the reverse of the removal procedure. Ensure that the nipple is above the fluid pipe, and bleed the system on completion (see Section 2).

6 Slave cylinder - overhaul

1 With the slave cylinder removed from the car, remove the rubber boot.
2 Take out the wire retaining clip, piston, cup and spring.
3 Discard the rubber boot and piston, then clean the remaining parts in methylated spirit, isopropyl alcohol or clean brake fluid. Check for wear, scoring and corrosion, obtaining new parts as necessary.
4 Assembly is basically the reverse of the dismantling procedure, all internal parts being lubricated with clean brake fluid. **Note:** *The smaller diameter of the spring is fitted to the piston.* After assembly, apply rubber grease or disc brake lubricant to the exposed end of the piston and the ball end of the pushrod before the rubber boot is installed.

7 Clutch - removal, inspection and installation

1 Remove the gearbox from the engine as described in Chapter 6.
2 Mark the clutch cover and flywheel so that the clutch may be refitted in its original position, unless it is to be renewed. The clutch cover, pressure plate and diaphragm spring assembly must be replaced as a unit if it is found to be faulty. Only the driven plate is able to be replaced as a separate entity.
3 Progressively slacken the six pressure plate to flywheel bolts, a turn at a time, so releasing them evenly. As they are being released, check that the pressure plate flange is not binding on the dowels, otherwise it could fly off causing an accident.
4 Lift away the six bolts and spring washers, followed by the cover assembly and drive plate. **Note which way round the driven plate is fitted.** The longer boss is facing towards the gearbox.
5 Using a stiff-brush or clean rags, clean the face of the flywheel, the pressure plate assembly and the driven plate. Note that the dust is harmful to the lungs as it contains asbestos, so do not inhale it.
6 It is important that neither oil nor grease comes into contact with the clutch facings, and that absolute cleanliness is observed at all times.
7 Inspect the friction surfaces of the driven plate and, if worn, a complete new assembly must be fitted. The linings are completely worn out when the faces of the rivets are flush with the lining face. There should be at least 1/32 in (0.8 mm) of lining material left clear of the rivet faces, or the driven plate is not worth refitting. Check that the friction linings show no signs of heavy glazing or oil impregnation. If evident, a new assembly must be fitted. If a small quantity of lubricant has found its way onto the facing, due to heat generated by the resultant slipping, it will be burnt off. This will be indicated by darkening of the facings. This is not too serious provided that the grain of the facing material can be clearly identified. Fit a new assembly if there is any doubt at all. It is important that if oil impregnation is present, the cause of the oil leak is found and rectified to prevent recurrence.
8 Carefully inspect the driven plate and flywheel contact faces for signs of overheating, distortion, cracking and scoring; if any serious evidence of scoring exists, then it will probably be necessary to have the flywheel skimmed; if you simply renew the driven plate, you could very soon be faced with the same faulty condition.
9 Mount the driven plate onto the input shaft and check for looseness or wear on the hub splines. Also check the driven plate damper springs for damage or looseness.

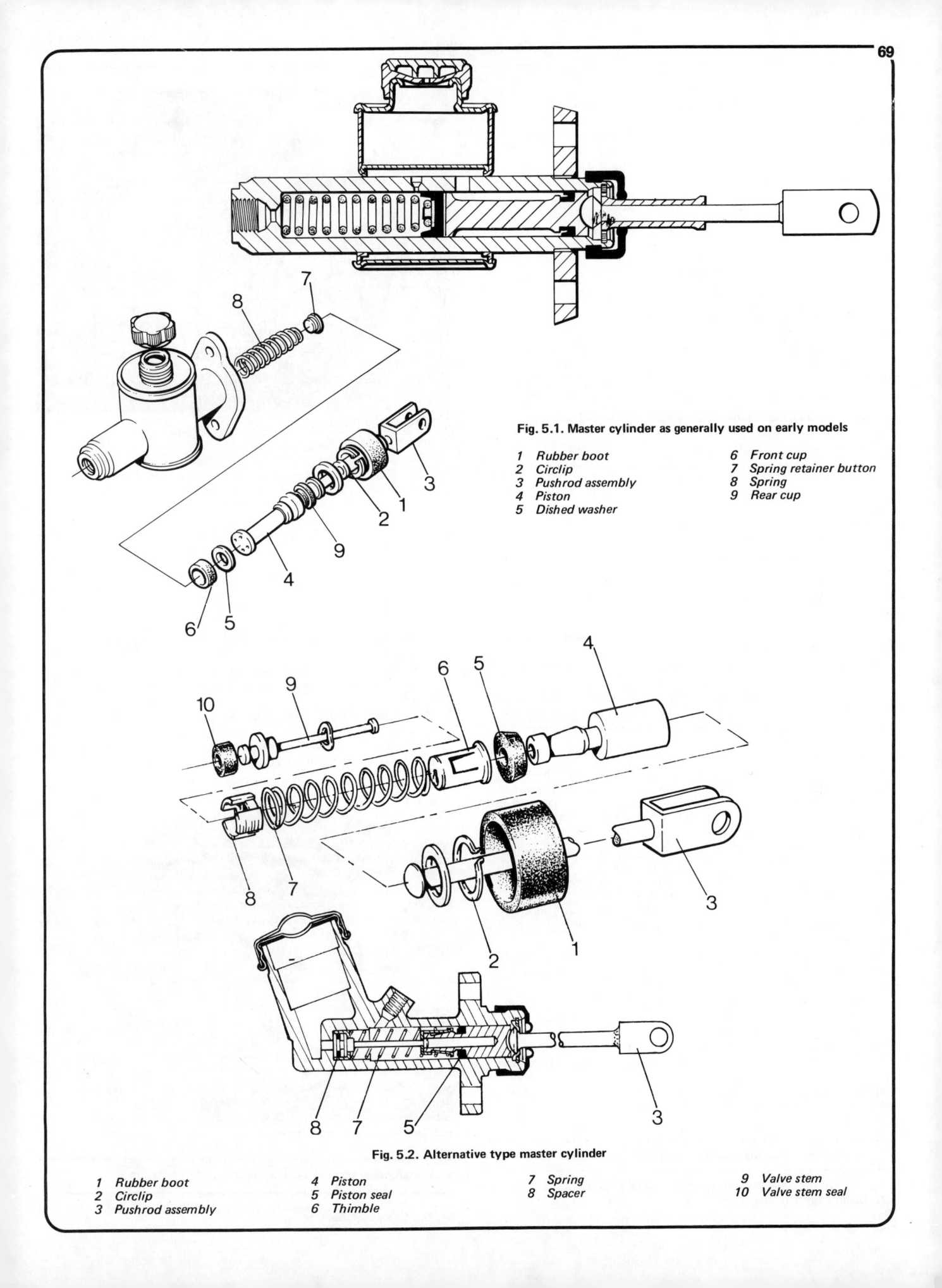

Fig. 5.1. Master cylinder as generally used on early models

1 Rubber boot
2 Circlip
3 Pushrod assembly
4 Piston
5 Dished washer
6 Front cup
7 Spring retainer button
8 Spring
9 Rear cup

Fig. 5.2. Alternative type master cylinder

1 Rubber boot
2 Circlip
3 Pushrod assembly
4 Piston
5 Piston seal
6 Thimble
7 Spring
8 Spacer
9 Valve stem
10 Valve stem seal

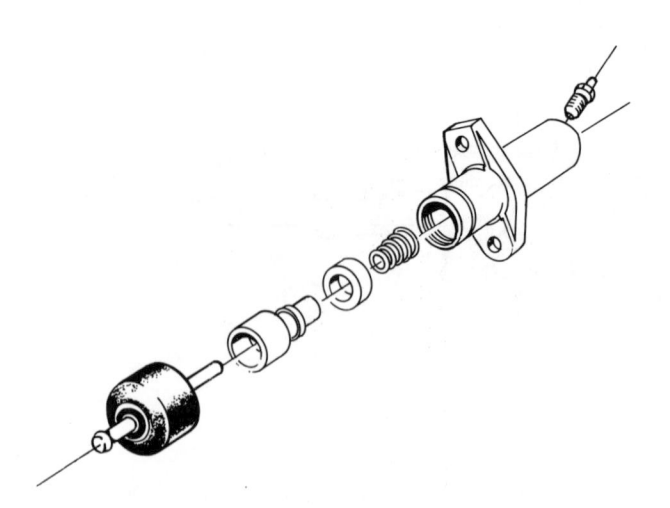

Fig. 5.3. Clutch slave cylinder - exploded view

7.13 Installing the clutch

10 Inspect the clutch release mechanism for wear, and obtain a new assembly if evident (Section 8).
11 Inspect the release bearing for wear as indicated by roughness or sloppiness between the inner and outer tracks.
12 At this point it is a good idea to check the condition of the clutch pilot bearing (spigot bush). Further information on this will be found in Chapter 1 in the Section entitled Miscellaneous components - examination and renovation.
13 When installing the clutch, offer up the clutch assembly and friction plate to the flywheel, aligning the scribed marks (where applicable). Do not forget that the larger boss of the friction plate is towards the gearbox, as is the disc of the spring cushioning mechanism. Install the bolts and washers, and lightly tighten them (photo).
14 The clutch friction plate must now be centralised so that when the engine and gearbox are mated, the gearbox input shaft splines will pass through the splines in the centre of the driven plate hub.
15 Centralisation can be carried out quite easily by inserting a round bar or long screwdriver through the hole in the centre of the clutch, so that the end of the bar rests in the small hole in the end of the crankshaft containing the input shaft bearing bush. Ideally an old input shaft should be used.
16 Using the input shaft bush (pilot bearing) as a fulcrum, moving the bar sideways or up and down will move the clutch disc in whichever direction is necessary to achieve centralisation.
17 Tighten the clutch bolts firmly in a diagonal sequence to ensure that the cover plate is pulled down evenly and without distortion of the flange. It is best to prevent the flywheel from turning by using a locally manufactured bracket which can be fitted into the teeth of the starter ring at one end and against one of the starter motor bolts at the other end.
18 Smear a little general purpose grease on the nose of the input shaft, then install the gearbox as described in Chapter 6.

8 Clutch release bearing assembly - removal, overhaul and installation

1 Remove the gearbox from the engine as described in Chapter 6.
2 Using a suitably cranked spanner, unscrew the release lever pivot bolt.
3 Withdraw the release lever and pivot bolt, and detach the release bearing (photo).
4 If the bearing is unserviceable it can be pressed or driven off the sleeve; it cannot be repaired or cleaned, and a new one must be obtained.
5 A new bearing can be pressed on, but pressure must only be applied to the inner race. Ensure that the release face is outermost.
6 Apply a little general purpose grease to the pivot bolt and the release lever slippers, then install the assembly by following the reverse of the removal procedure.

8.3 The release bearing in position

9 Clutch pedal box - removal, overhaul and installation

1 See Chapter 9, Section 13.

Chapter 5/Clutch

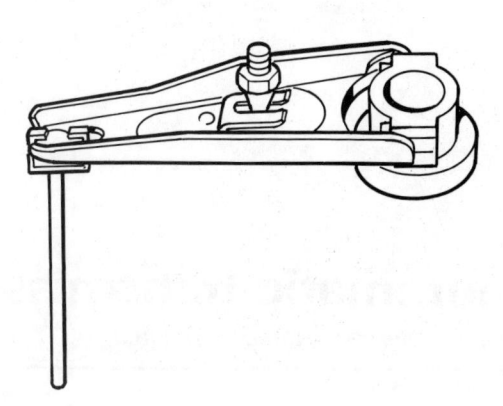

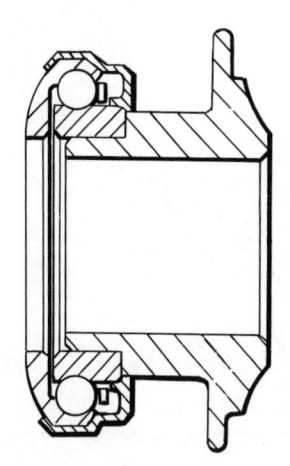

Fig. 5.4. Release lever and bearing

Fig. 5.5. Sectional view of release bearing and sleeve

10 Fault diagnosis - clutch

Symptom	Reason/s
Judder when taking up drive	Loose engine mountings. Worn or oil contaminated driven plate friction linings. Worn splines on driven plate hub or first input shaft. Worn crankshaft spigot bush (pilot bearing).
Clutch slip	Damaged or distorted pressure plate assembly. Driven plate linings worn or oil contaminated.
Noise on depressing clutch pedal	Dry-worn or damage clutch release bearing. Excessive play in input shaft splines.
Noise as clutch pedal is released	Distorted driven plate. Broken or weak driven plate hub cushion coil springs. Distorted or worn input shaft. Release bearing loose.
Difficulty in disengaging clutch for gearchange	Fault in master cylinder or slave cylinder. Air in hydraulic system. Driven plate hub splines rusted on shaft.

Chapter 6 Manual gearbox and automatic transmission

For modifications, and information applicable to later models, see Supplement at end of manual

Contents

Part 1 - Gearbox
Fault diagnosis - manual gearbox	13
Five-speed gearbox - dismantling	8
Five-speed gearbox - examination	9
Five-speed gearbox - reassembly	11
Five-speed gearchange remote control assembly - dismantling and reassembly	12
Five-speed mainshaft - dismantling and reassembly	10
Four-speed gearbox - dismantling	3
Four-speed gearbox - examination	4
Four-speed gearbox - reassembly	7
Four-speed input shaft - dismantling and reassembly	5
Four-speed mainshaft - dismantling and reassembly	6
Gearbox - removal and installation	2
General description	1

Part 2 - Automatic transmission
Automatic transmission - fault diagnosis procedure	33
Automatic transmission - removal and installation	15
Down shift cable - pressure test	30
Fluid level - checking	29
Front brake band - adjustment	17
Front servo - removal, overhaul and installation	19
General description	14
Governor - removal, overhaul and installation	23
Kickdown cable - adjustment	16
Rear brake band - adjustment	18
Rear extension - removal and installation	21
Rear extension oil seal - renewal	22
Restrictor valve and bypass pipe - removal, cleaning and installation	24
Rear Servo - removal, overhaul and installation	20
Road test to be carried out in conjunction with the fault diagnosis procedure (Section 33)	32
Selector rod - adjustment	25
Stall test	31
Starter inhibitor/reverse lamp switch - removal and refitting	28
Transmission sump - draining and refilling	26
Transmission sump - removal and refitting	27

Specifications

MANUAL GEARBOX
Type ... 4-speed or 5-speed. Synchromesh on all forward gears.

Gearbox ratios
4-speed
Top (fourth)	1.00 : 1
Third	1.25 : 1
Second	1.78 : 1
First	2.65 : 1
Reverse	3.98 (early cars), 3.05 (later cars)

5-speed
Top (fifth)	0.83 : 1
Fourth	1.00 : 1
Third	1.40 : 1
Second	2.09 : 1
First	3.32 : 1
Reverse	3.43 : 1

Gearbox lubricant
Type
4-speed	SAE90EP gear oil
5-speed	SAE75EP gear oil (SAE80EP may be used for topping-up)

Quantity (approx)
4-speed	2.0 Imp pint/2.4 US pint (1.1 litres)
5-speed	3.7 Imp pint/4.4 US pint (2.1 litres)

AUTOMATIC TRANSMISSION

Type ... Borg Warner model 65

Transmission conversion range
Third ... 1.00 to 1.91
Second ... 1.45 to 2.77
First ... 2.39 to 4.57
Reverse ... 2.09 to 3.99

Shift speeds

Throttle position	Selection	Shift	Speed mph (km/h)
Closed	1	2 - 1	30 to 38 (48 to 61)
Light throttle	2	1 - 2	8 to 12 (13 to 19)
	D	2 - 3	12 to 16 (23 to 26)
Part throttle	D	3 - 2	30 to 40 (48 to 64)
Kick down	D	1 - 2	38 to 44 (61 to 71)
	D	2 - 3	65 to 71 (105 to 114)
	D	3 - 2	58 to 66 (93 to 106)
	1	2 - 1	42 to 50 (68 to 81)
	D	3 - 1	30 to 38 (48 to 61)
	2	1 - 2	38 to 44 (61 to 71)
	2	2 - 1	30 to 38 (48 to 61)

Automatic transmission fluid ... Castrol TQF or similar specification *(NOT Dexron type)*

Automatic transmission fluid capacity (including oil cooler) ... 9.51 Imp pints/11.4 US pints (5.4 litres)

Torque wrench settings

	lbf ft	Nm
Flange to mainshaft	120	163
Drain filler plugs	25	34
Sump coupling plate to clutch housing	37	50

PART 1: MANUAL GEARBOX

1 General description

4-speed gearbox

The gear change lever is mounted on the extension housing and operates the selector mechanism in the gearbox by a long shaft. When the gear change lever is moved sideways the shaft is rotated so that the pins in the gearbox end of the shaft locate in the appropriate selector fork. Forward or rearward movement of the gear change lever moves the selector fork, which in turn moves the synchromesh unit outer sleeve until the gear is firmly engaged. When reverse gear is selected, a pin on the selector shaft engages with a lever and this in turn moves the reverse idler gear into mesh with the laygear reverse gear and mainshaft. The direction of rotation of the mainshaft is thereby reversed.

The gearbox input shaft is splined and it is onto these splines that the clutch driven plate is located. The gearbox end of the input shaft is in constant mesh with the laygear cluster, and the gears formed on the laygear are in constant mesh with the gears on the mainshaft with the exception of the reverse gear. The gears on the mainshaft are able to rotate freely which means that when the neutral position is selected the mainshaft does not rotate.

When the gear change lever moves the synchromesh unit outer sleeve via the selector fork, the synchromesh cup first moves and friction, caused by the conical surfaces meeting, takes up initial rotational movement until the mainshaft and gear are both rotating at the same speed. This condition achieved, the sleeve is able to slide over the dog teeth of the selected gear and thereby giving a firm drive. The synchromesh unit inner hub is splined to the mainshaft and because the outer sleeve is splined to the inner hub engine torque is passed to the mainshaft and propeller shaft.

5-speed gearbox

This gearbox was developed for the Rover 3500 model and is a completely new design. The most unusual feature is the use of taper roller bearings for the mainshaft/input shaft and layshaft, which provide for a less bulky gearbox using standard off-the-shelf bearings, and also simplifies assembly.

The gearbox is built up on a centre plate, a rigid iron casting between the tailpiece and the main housing. Since taper rollers come apart easily the shafts and gears can be simply placed in position and jiggled around until the gears mesh: All that is needed is for the shaft and play to be adjusted which is a relatively straightforward operation.

A single shaft selector mechanism is used for gear selection. Gears 1 to 4 are arranged in the usual H-pattern, but reverse and 5th are left-forward and right-forward respectively. The single shaft has two operating dowels, the rear one engaging reverse and the front one engaging the forward gears, via bosses on the selector forks. An interlock arrangement prevents engagement of more than one gear at the same time.

The selector mechanisms are mounted on the centre plate. A simple externally mounted hairpin-type spring gives the gear lever a natural bias to the 3/4 plane. This bias has to be overcome in progressive steps to obtain the 1/2 plane, 5th gear and reverse gear in order, the latter having a separate biasing arrangement.

With the exception of fifth gear, which overhangs the centre plate to the rear, the gear layout and mode of operation is conventional.

2 Gearbox - removal and installation

Note: *Wherever possible, it is recommended that the engine and gearbox are lifted out together. If this is not possible, the gearbox may be removed from below as described in this Section.*

1 Detach the battery earth lead.
2 With neutral selected, unscrew and remove the gear lever knob.
3 Take out the retaining screws and remove the gear lever gaiter and top panel assembly.

Chapter 6/Manual gearbox and automatic transmission

4 Remove the retaining screws and lift away the draught excluder and flange assembly (photo).
5 *4-speed gearbox:* Remove the bayonet cap and lift out the gear lever. Do not lose the strong anti-rattle spring and nylon plunger.
6 *5-speed gearbox:* Remove the single bolt and lift off the ball cap. Using a large screwdriver, prise the ends of the hairpin spring aside, and press them down to retain them below the bolt heads. Lift out the gear lever, taking care that the strong anti-rattle spring and nylon plunger are not lost (photos).
7 Take off the radiator fan guard, if fitted.
8 Raise the car and support it on blocks or similar items beneath the bodyframe sidemembers, for access to the gearbox area.
9 Mark the installed position of the propeller shaft flange, then detach it from the gearbox. Support the propeller shaft so that it does not obstruct removal of the gearbox.
10 Remove the exhaust downpipe; refer to Chapter 3 for further information if necessary.
11 Disconnect the speedometer cable drive from the gearbox.
12 Disconnect the electrical leads to the seat belt interlock switch (where applicable) and the reverse lamp switch.
13 Where applicable, remove the engine tie bar, restraint cable and engine stabilizer.
14 Place a jack with a wood block packer beneath the engine sump for support. Raise the jack to *just* take the engine weight.
15 Remove the starter motor; refer to Chapter 10 for further information if necessary.
16 Detach the clutch slave cylinder and tie it to one side.
17 Remove the bolts from the sump stiffening plate.
18 Remove the bolts retaining the rear mounting to the body underframe.
19 Carefully lower the jack for access to the clutch housing bolts. Make sure that no cables and hoses are stretched during this operation.
20 Loosen, then remove, all the bellhousing bolts. *Note the position of the dowel bolt near the clutch slave cylinder.*
21 Whilst suitably supporting the gearbox, withdraw it from the engine taking care that load is not taken on the input shaft. The gearbox is a rather heavy unit, and help from an assistant is recommended.
22 Where necessary, remove the vibration damper and bracket (where applicable) and the mounting bracket.
23 Installation of the gearbox is basically the reverse of the removal procedure, but the following points should be noted where applicable:
(a) Adjust the stabilizer as described in Chapter 1, Section 49.
(b) *Slacken the front nut at the rear end of the restraint cable, then tighten the rear nut to settle the cable in position. Slacken the rear nut whilst holding the cable taut, to provide a clearance of 1/16 to 1/32in (0.5 to 1.5mm) between the nut and the cable bracket. Finally, tighten the front nut.*

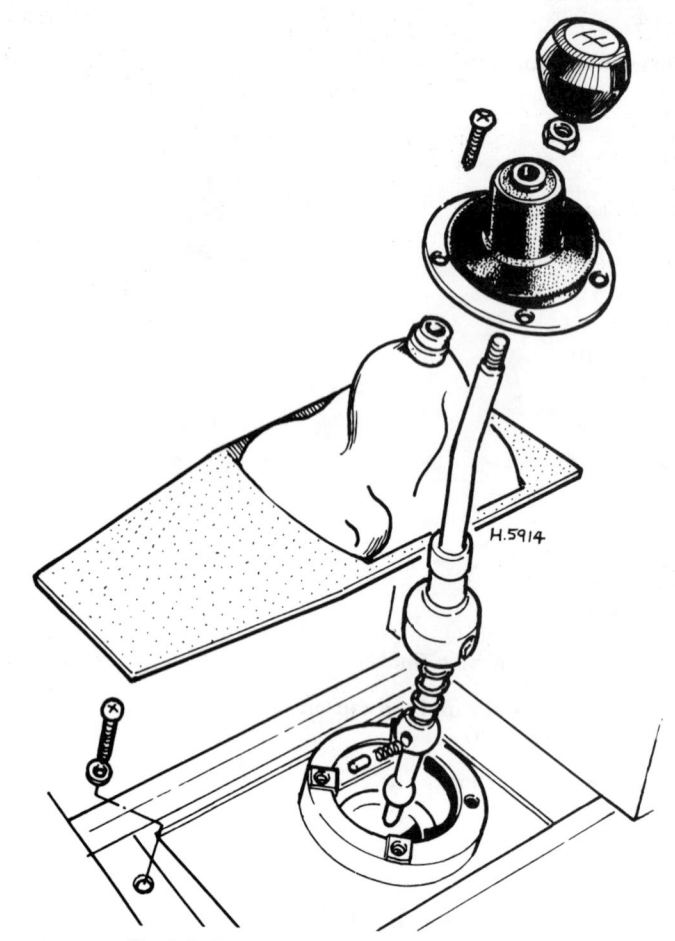

Fig. 6.1. Gearchange lever - four-speed type shown

3 Four-speed gearbox - dismantling

1 Before commencing to dismantle the gearbox, clean the exterior with a water-soluble solvent. This will make the gearbox more easy to handle, and possibly prevent dirt from contaminating the internal parts. Drain the gearbox oil, then refit the plug.
2 Remove the clutch release bearing and lever as described in Chapter 5.
3 Remove the bolts and washers, and withdraw the bellhousing. *Note that a copper washer is used for the lowermost bolt, whereas the others have spring washers.*
4 Remove the three layshaft preload springs, and the bellhousing gasket.
5 Remove the gearbox top cover and the interlock spool plate.
6 Support the gearbox in a vice with the jaws firmly clamping the drain plug.
7 Remove the reverse lamp switch (and seat belt interlock switch if applicable). Refer to Chapter 10 for further information if necessary.
8 Drive out the roll pin from the selector rod.
9 Release the speedometer cable clamp plate and remove the pinion housing.
10 Pass a long bolt through the mainshaft flange so that it can wedge against the rear extension housing, then unscrew the flange nut and washer.
11 Pull off the mainshaft flange.

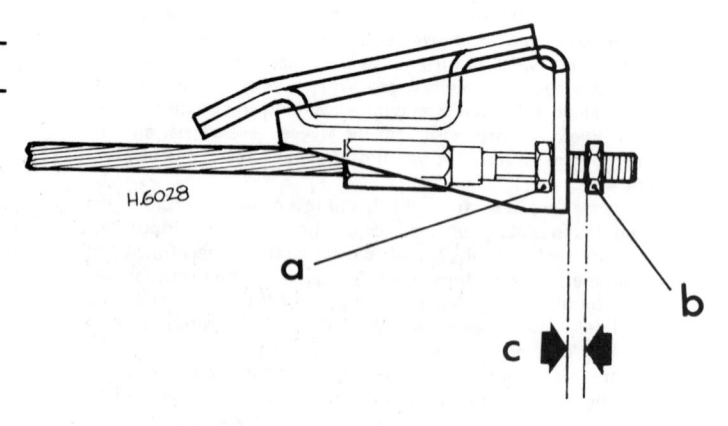

Fig. 6.2. Restraint cable adjustment

a Front nut
b Rear nut
c Clearance between nut and cable bracket

2.4 Remove the draught excluder

2.6a Remove the single bolt

2.6b This photo shows the hairpin spring

2.6c The anti-rattle spring and nylon plunger

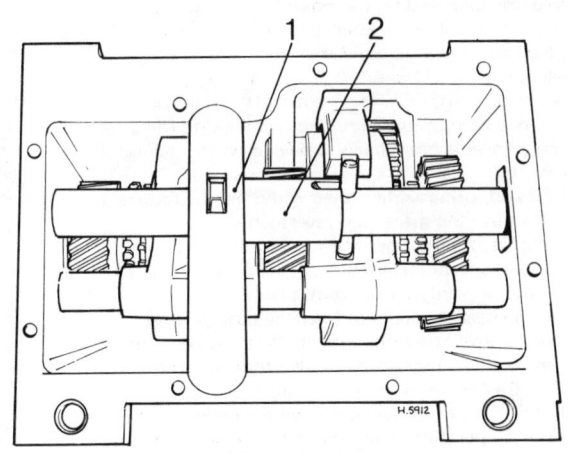

Fig. 6.3. Interlock spool plate (1) and selector spool (2)

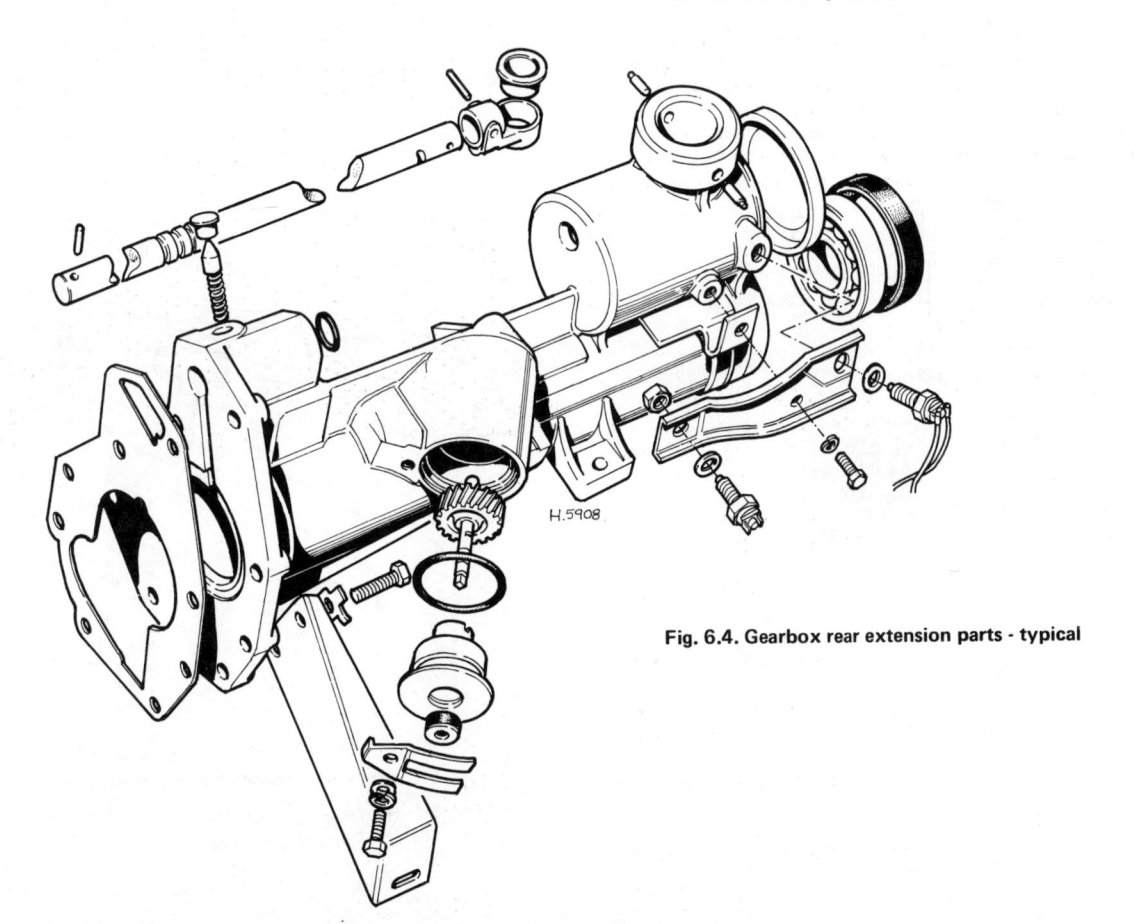

Fig. 6.4. Gearbox rear extension parts - typical

12 Move the gear selector to engage reverse gear; ensure that the selector shaft pins clear the interlock spool and selector forks.
13 Remove the bolts and spring washers, and detach the exhaust bracket and rear extension. Ensure that the selector pins do not foul, and that the layshaft does not move.
14 Lift out the interlock spool, then remove the rear extension gasket.
15 Remove the distance washer from the end of the mainshaft.
16 Slide the gear selector shaft off the rear extension rearwards to contact the blanking plug, then gently push out the plug using the shaft.
17 Slide the shaft rearwards, then drive out the rollpin securing the yoke. Withdraw the yoke.
18 Push the selector shaft forwards, keeping the rollpin hole horizontal to prevent the plunger from trapping the shaft.
19 Remove the shaft. Take out the nylon plug, plunger, spring and O-ring.
20 Remove the selector shaft and forks from the gearbox.
21 Using a round bar of suitable diameter, (eg, an old layshaft), drive out the layshaft and allow the laygear cluster to fall to the bottom of the gearbox.
22 Using a suitable small drift, drive the input shaft and bearing forwards out of the gearbox. Ensure that load is applied to the outer race only. If the caged needle bearing has remained on the mainshaft spigot, remove this also.
23 Remove the bolt and spring washer, then withdraw the reverse idler gear spindle. Remove the spacer and reverse idler gear.
24 Release the circlip securing the bearing to the mainshaft.
25 Carefully tap off the speedometer gear.
26 Tap the mainshaft rearwards a little so that two large screwdrivers can be used behind the bearing circlip to lever the bearing out.
27 With the bearing removed from the mainshaft shoulder, lift the mainshaft out of the top of the gearbox. Collect the circlip, selective washer, bearing and thrust washer as this is being done.
28 Remove the laygear and thrust washers from the casing.
29 Remove the reverse operating lever.
30 Remove the 25 needle rollers from each end of the laygear. The retaining rings need not be removed unless renewal is found to be necessary.

4 Four-speed gearbox - examination

1 The gearbox has been stripped, presumably, because of wear or malfunction, possibly excessive noise, ineffective synchromesh or failure to stay in a selected gear. The cause of most gearbox ailments is failure of the ball bearings on the input or mainshaft and wear on the synchro rings, both the cup surfaces and dogs. The nose of the mainshaft which runs in the needle roller bearing in the input shaft is also subject to wear. This can prove very expensive as the mainshaft would need replacement, and this represents about 20% of the total cost of a new gearbox.
2 Examine the teeth of all gears for signs of uneven or excessive wear and, of course, chipping. If a gear on the mainshaft requires replacement check that the corresponding laygear is not equally damaged. If it is the whole laygear may need replacing also.
3 All gears should be a good running fit on the shaft with no signs of rocking. The hubs should not be a sloppy fit on the splines.
4 Selector forks should be examined for signs of wear or ridging on the faces which are in contact with the operating sleeve.
5 Check for wear on the selector rod and interlock spool.
6 The ball bearings may not be obviously worn but if one has gone to the trouble of dismantling the gearbox it would be shortsighted not to renew them. The same applies to the four synchronizer rings although for these the mainshaft has to be completely dismantled for the new ones to be fitted.
7 Examine the bush in the reverse idler gear for wear. If any is found, press out the old bush, and then press in a new one so that it is flush with the boss opposite the collar of the operating lever. Ream the bush to a diameter of 0.6585/0.6592in (16.7279/16.8011mm).
8 If worn, the reverse lever pivot pin can be pressed out of the casing and a new one inserted.
9 It is recommended that new oil seals are fitted in the input shaft guide counterbore of the clutch bellhousing and in the rear extension. These should be fitted with the lips towards the gearbox.
Note: *If the rear extension bearing is to be renewed, this should be done before the oil seal is refitted.*
10 Before finally deciding to dismantle the mainshaft and replace parts, it is advisable to make enquiries regarding the availability of parts and their cost. It may still be worth considering an exchange gearbox even at this stage. You should reassemble the old gearbox before exchanging it.

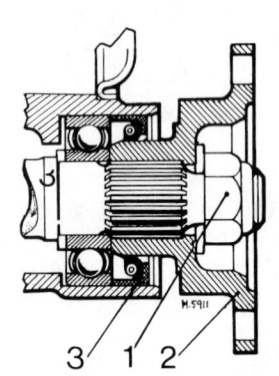

Fig. 6.5. Mainshaft flange and oil seal

1 Nut
2 Flange
3 Oil seal

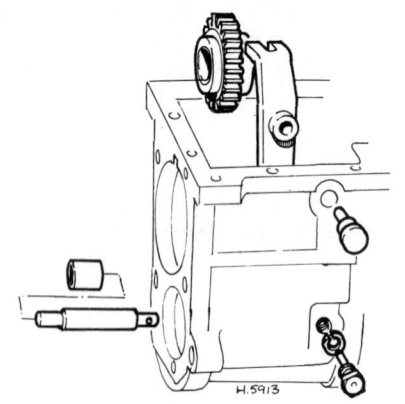

Fig. 6.7. Reverse gear and associated parts

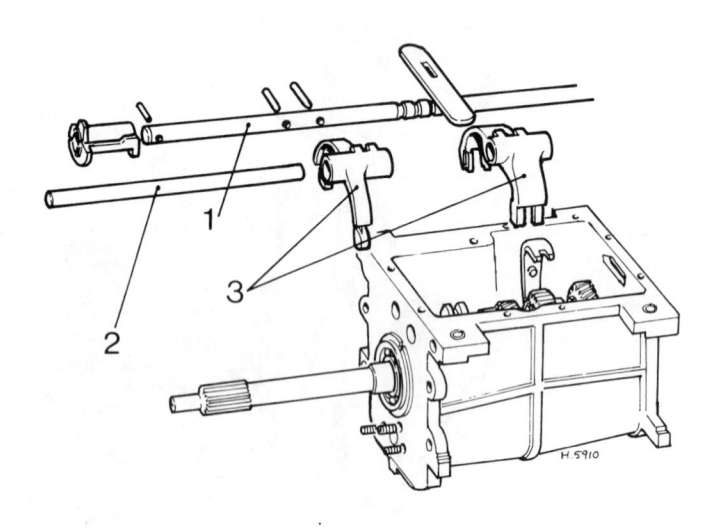

Fig. 6.6. Selectors and associated parts

1 Gear selector shaft
2 Selector fork shaft
3 Selector forks

Fig. 6.8. Exploded view of gearbox parts

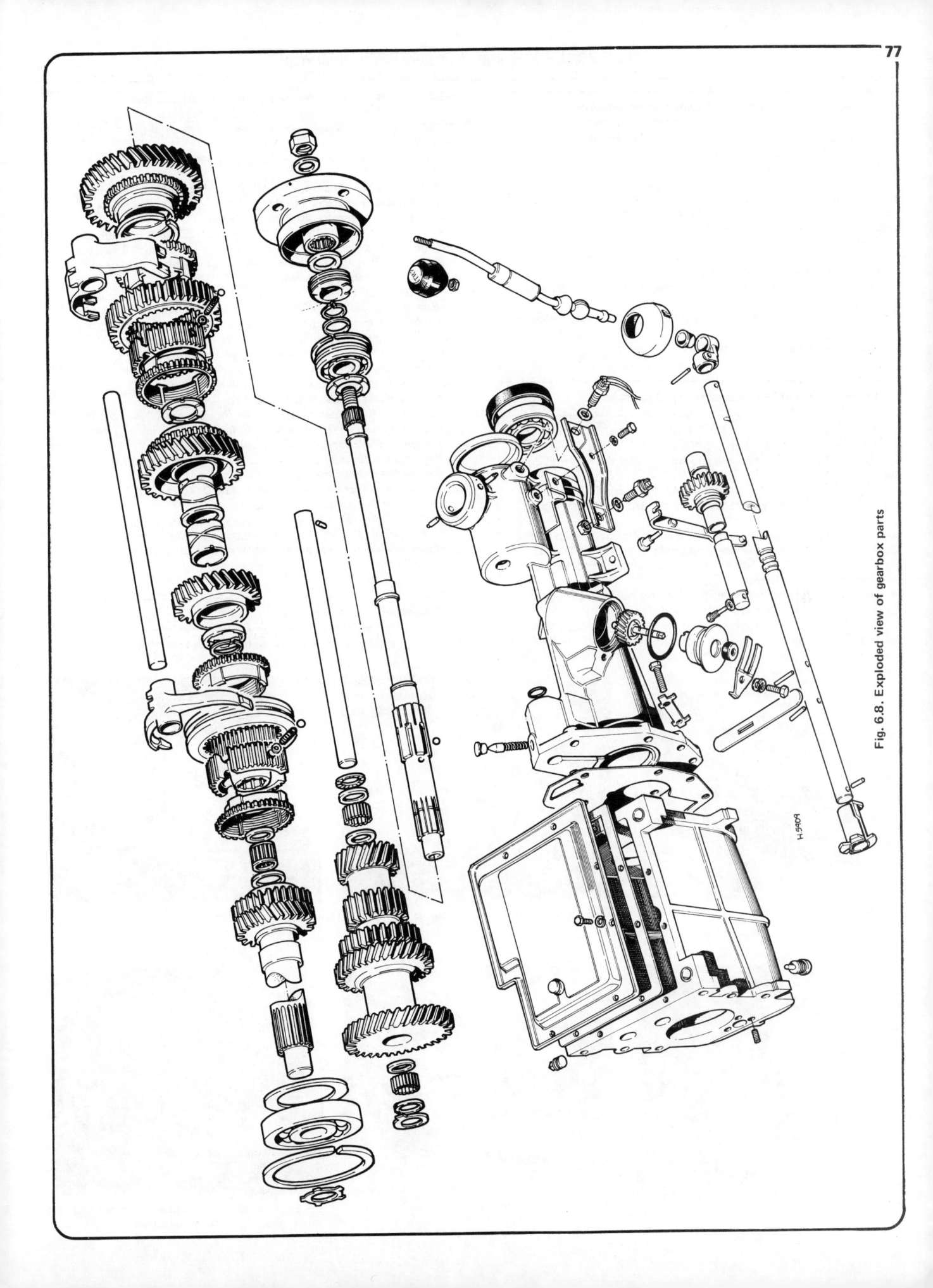

5 Four-speed input shaft - dismantling and reassembly

1 Place the input shaft in a vice, splined end upwards, and with a pair of circlip pliers, remove the circlip which retains the ball bearing in place.
2 With the bearing resting on top of the open jaws of the vice and splined end upwards, tap the shaft through the bearing with a soft faced hammer. Note: *The offset circlip groove in the outer track of the bearing is towards the front of the input shaft.*
3 Lift away the oil flinger.
4 Remove the caged needle roller bearing from the centre of the rear of the input shaft if it is still in place.
5 Remove the circlip from the old bearing outer track and transfer it to the new bearing.
6 Refit the oil flinger and with the aid of a block of wood and vice tap the bearing into place. Make sure it is the right way round.
7 Finally refit the bearing retaining circlip.

6 Four-speed mainshaft - dismantling and reassembly

1 Remove the 1st gear (and thrust washer if not already removed) from the rear of the mainshaft (photo).
2 Remove the 1st gear synchro cup followed by the two split collars (photos).
3 Remove the 3rd/4th synchro hub and sleeve assembly from the front end of the mainshaft, followed by the 3rd gear synchro cup (photos).
4 Carefully prise open the ends of the retaining circlip then lever off the 3rd gear, bush, thrust washer and circlip (photo).
5 Remove the 2nd gear and bush, followed by the selective washer and synchro cup (photos).
6 Using a small magnet, extract the selective washer locating ball from the mainshaft drilling.
7 Withdraw the 1st/2nd synchro hub and sleeve assembly (photo).
8 Index mark the sleeve and synchro hub assemblies of the 1st/2nd and 3rd/4th gears to ensure assembly in their original locations. Separate the sleeves from the hubs, ensuring that the three balls and springs are not lost. Note: *In some cases shims may be fitted below the springs.*
9 Clean all the parts in petrol or paraffin, and dry them with a lint-free cloth.
10 Reassemble the synchro assemblies ensuring that the teeth of the outer members are adjacent to the longer boss of the synchro hub. Where shims were previously fitted, refit them.
11 Check that a load of 19 to 27 lb f (8.7 to 12.2 kg f) is required to shift the 1st/2nd sleeve in either direction, and a load of 19 to 21 lb f (8.7 to 9.5 kg f) is required to shift the 3rd/4th sleeve. Add or remove shims to obtain this requirement.
12 The mainshaft can now be assembled following the reverse of the removal procedure. Note: *The second gear bush flange and the rim of the 3rd gear thrust washer are towards the front of the gearbox.* When installing the retaining circlip ensure that the inclined end faces forwards and the clip aligns with the edge of the mainshaft spline. Ensure that the larger boss of the 3rd gear synchro assembly is towards the front of the gearbox. During the assembly procedure, the following clearances must be checked:
(a) *1st gear endfloat measured between the split collars and thrust washer, should be 0.004 to 0.013in (0.102 to 0.33mm). Renew the split collars and/or thrust washer as necessary.*
(b) *2nd gear endfloat on the bush should be 0.002in (0.051mm). Renew the flanged bush if necessary.*
(c) *3rd gear endfloat on the bush should be 0.002in (0.051mm). Renew the flanged bush if necessary.*
(d) *2nd and 3rd gear bushes should have a clearance of 0 to 0.006in (0 to 0.15mm). If adjustment is required obtain a new selective washer; these are available in steps of 0.003in (0.076mm).*

7 Four-speed gearbox - reassembly

1 Using a general purpose grease to hold them in position, fit the

6.1 Removing the 1st gear

6.2a Removing the synchro cup ...

6.2b ... and split collars

6.3a Removing the 3rd/4th synchro hub and sleeve assembly ...

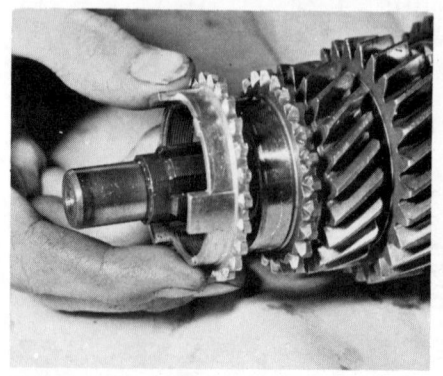

6.3b ... and the synchro cup

6.4 3rd gear being removed after circlip and thrust washer have been removed

6.5a Removing the 2nd gear ...

6.5b ... 2nd gear bush ...

6.5c ... selective washer ...

6.5d ... and synchro cup

6.7 Removing the 1st/2nd synchro hub and sleeve assembly

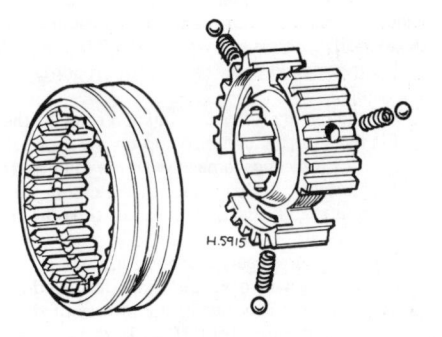

Fig. 6.9. Synchro assembly exploded view

Fig. 6.10. Installation depth for needle roller retaining rings in laygear

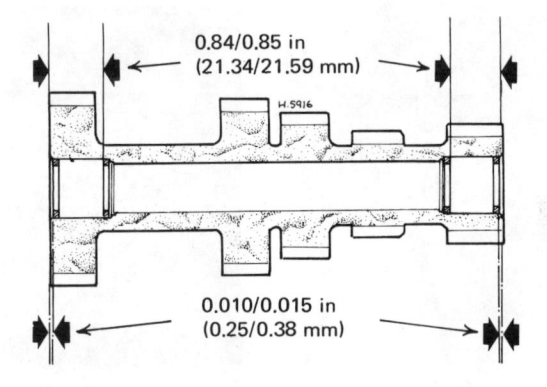

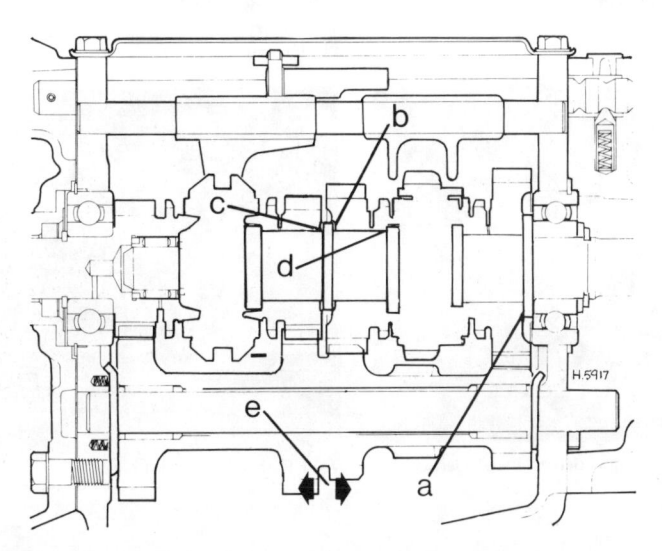

0.84/0.85 in (21.34/21.59 mm)

0.010/0.015 in (0.25/0.38 mm)

Fig. 6.11. Setting-up dimensions for mainshaft and laygear

a 1st gear endfloat
b 2nd gear endfloat on bush
c 3rd gear endfloat on bush
d 2nd and 3rd gear bush clearances
e Laygear endfloat

needle rollers into each end of the layshaft, and put the outer retaining ring in position (photo).
2 Locate the laygear thrust washers in the casing with their respective tabs in the casing slots.
3 Hold the laygear in position and check for an endfloat of 0.007 to 0.015in (0.178 to 0.381mm). If necessary, obtain new thrust washers to achieve this dimension.
4 Take out the laygear and insert the dummy layshaft used when dismantling. Place the laygear assembly into the casing again so that the large gear is towards the front.
5 Tilt the assembled mainshaft into the casing, then place reverse gear in the bottom of the casing and fit the reverse operating lever (photo).
6 Fit the circlip to the casing rear bearing, then slide the bearing onto the mainshaft (photo).
7 Whilst supporting the front of the mainshaft with a suitable metal bar, drive the rear bearing into position in the casing and on the mainshaft. A suitable size metal tube can be used, but apply loads to the bearing inner race only (photos).
8 Ensure that the bearing is fully home then fit a washer and selective circlip so that mainshaft endfloat does not exceed 0.002in (0.051mm). Circlips are available in 0.003in (0.076mm) steps.
9 Fit the speedometer gear so that it contacts the shoulder on the mainshaft.
10 Fit the needle roller bearing and rings to the mainshaft spigot.
11 Fit the 4th gear synchro cup on the input shaft, then push the input shaft into engagement with the mainshaft so that the bearing circlip contacts the front face of the casing (photo).
12 Push in the layshaft, at the same time driving out the dummy shaft (photos).
13 Fit the reverse gear, shaft and spacer.
14 Fit the selector forks and shaft, followed by the selector mechanism. This is basically the reverse of the removal procedure as described in Paragraphs 16 to 19 of Section 3 (photos). Also refer to Fig. 6.6.
15 Ensure that the mating faces are clean, then position a new gasket on the rear face of the casing.
16 Fit the distance washer to the mainshaft then offer up the rear extension, guiding the selector rail into position. Do not forget to fit the selector interlock spool (photos).
17 Fit the bolts and washers, together with the exhaust bracket and lockplate.
18 Smear some gearbox oil around the oil seal lip and drive flange surface, then install the flange, washer and nut. Tighten the nut to the specified torque whilst preventing the flange from turning.
19 Fit the rollpin to the front end of the selector rail so that it is positioned centrally (photo).
20 Refit the speedometer drive pinion using a new O-ring.
21 Fit the reverse light switch (and seat belt interlock switch, if applicable). Refer to Chapter 10 for further information if necessary.
22 Fit the three layshaft thrust springs.
23 Fit the clutch housing gasket, clutch housing, release fork and bearing, following the reverse of the removal procedure. Take care that the input shaft splines do not damage the oil seal.
24 Fit the gearbox drain plug, and top up to the level plug hole using the specified type of gear oil. Fit the filler/level plug.
25 Fit the spool interlock plate, a new gasket and the top cover (photos).

8 Five-speed gearbox - dismantling

1 Before commencing to dismantle the gearbox, clean the exterior with a water-soluble solvent. This will make the gearbox more easy to handle and possibly prevent dirt from contaminating the internal parts. Drain the gearbox oil, then refit the plug.
2 Remove the clutch release bearing and lever as described in Chapter 5.
3 If required, remove the six bolts, flat and spring washers and take off the bellhousing.
4 Make up a suitable bracket which can be bolted to the mainshaft flange, which will wedge against the remote control extension, then unscrew the flange nut and washer (photo).
5 Pull off the mainshaft flange.
6 Remove the nut and pin which are used to connect the remote

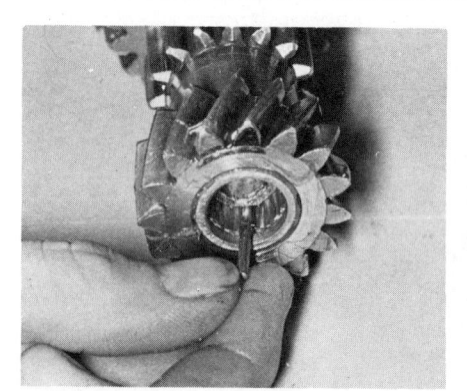

7.1 Installing the needle rollers

7.5 Fitting the mainshaft

7.6 Fitting the casing rear bearing onto the mainshaft

7.7a Supporting the mainshaft with a metal bar ...

7.7b ... as the bearing is driven into position

7.11 Fitting the input shaft

7.12a Installing the layshaft

7.12b Note that the pin aligns with the groove

7.14a Fitting the 3rd/4th selector fork

7.14b Fitting the 1st/2nd selector fork

7.14c Inserting the shaft through the forks

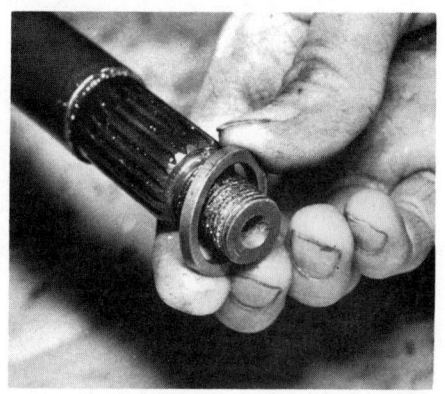

7.16a Fitting the distance washer

7.16b Fitting the interlock spool to the selector rail

7.19 Fitting the rollpin to the selector rail

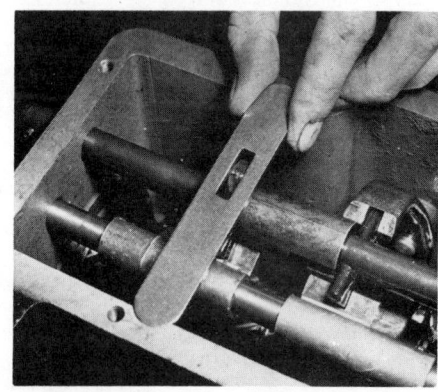

7.25a Fitting the interlock spool plate

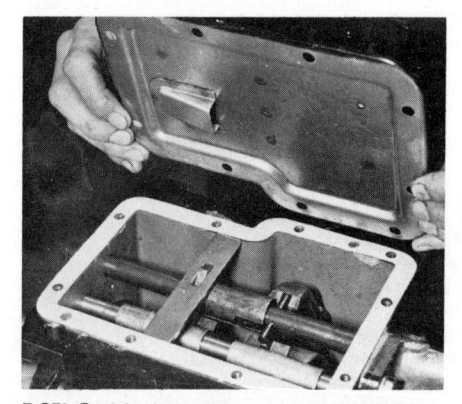

7.25b Positioning the top cover on a new gasket

8.4 Removing the flange nut

8.6 Selector shaft nut (arrowed)

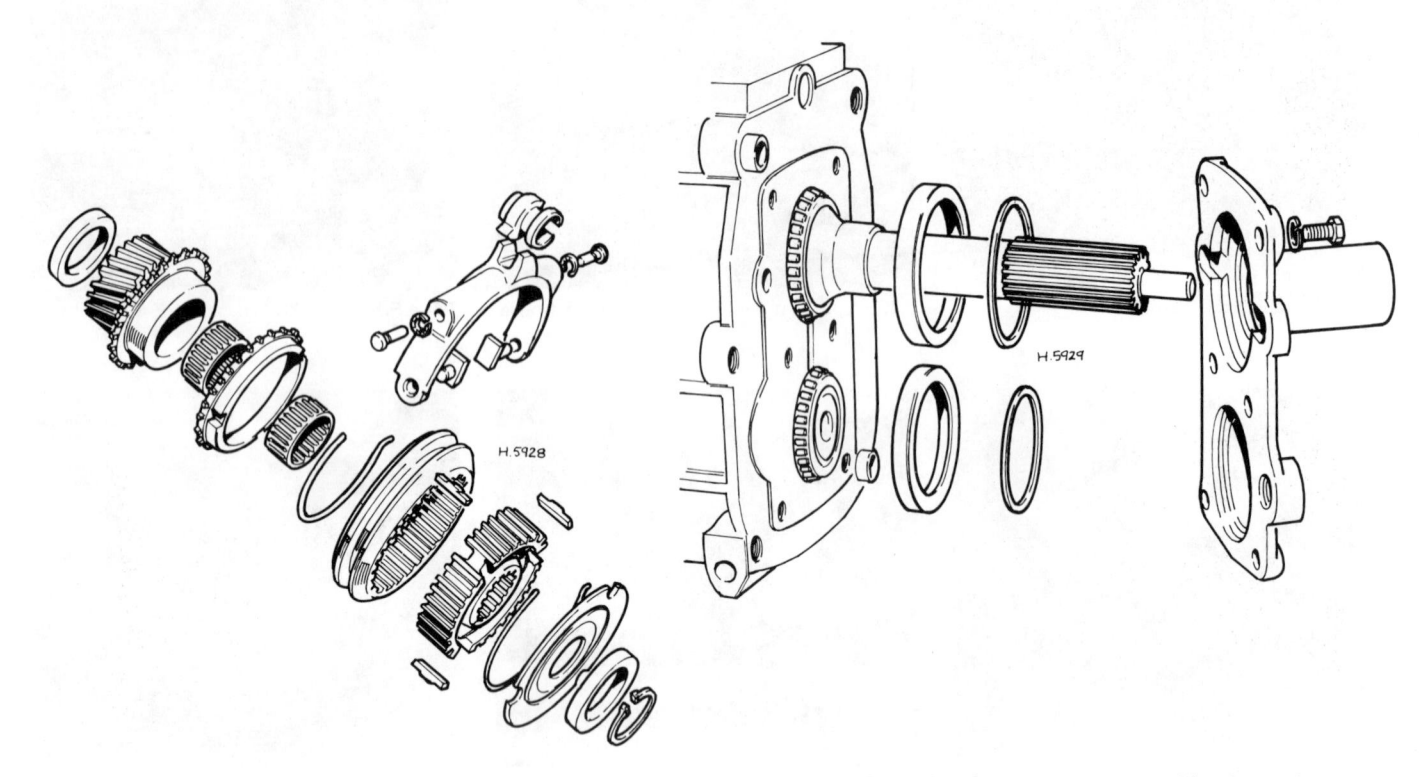

Fig. 6.12. 5th gear synchro and associated parts

Fig. 6.13. Gearbox front cover, spacers and bearing tracks

control shaft to the selector shaft (photo).
7 Remove the four bolts, spring and flat washers securing the remote control housing to the gearbox rear cover. *Note the positions of the plastic bushes and tubular spacers when the housing is removed* (photo).
8 Remove the clamp plate, and withdraw the speedometer driven gear and housing (photo).
9 Remove the two bolts and spring washers, and withdraw the locating boss for the 5th gear (rear) selector spool.
10 Remove the ten bolts, spring and flat washers, and withdraw the rear cover and gasket. *Note the positions of the tubular dowels.*
11 Remove the oil pump drive shaft from the 5th gear or oil pump so that it is not lost (photo).
12 Remove the two bolts and spring washers, and take off the 5th gear selector fork and bracket (photo).
13 Remove the circlip from the selector shaft and withdraw the 5th gear selector spool. *Note that the longer cam is pointing downwards* (photo).
14 Remove the circlip, and withdraw the 5th gear synchro assembly, 5th gear and spacer from the mainshaft.
15 Remove the circlip retaining the layshaft 5th gear, then use a 2- or 3-prong extractor to pull the gear off (photo).
16 Remove the six bolts and spring washers, and remove the gearbox front cover.
17 Remove the selective washers from the input shaft and layshaft.
18 Remove the two bolts and spring washers, and withdraw the locating boss for the front selector spool (photo).
19 Remove the selector plug, spring and ball from the drilling in the centre plate.
20 Support the gearbox on the centre plate and pull off the main casing.
21 Take off the input shaft and the first gear synchro cone.
22 Withdraw the layshaft gear cluster from the centre plate.
23 Using protective jaw clamps, support the centre plate in a vice.
24 Remove the retaining circlip and take out the reverse lever pivot pin. The reverse lever and slipper pad can now be taken out (photo).
25 Slide the reverse shaft rearwards so that the reverse gear spacer, mainshaft, selector shaft, selector shaft fork and spool can be drawn forwards and away from the centre plate.
26 Withdraw the selector fork and spool. *Note that the shorter cam is pointing towards the bottom of the gearbox.*
27 Remove the nut and spring washers, if necessary, and remove the reverse gear pivot shaft.
28 If subsequent inspection shows the input shaft bearing to be defective, the outer race can be driven out of the casing using a soft drift, and the inner race and bearing can be pulled or levered off.
29 If subsequent inspection shows the layshaft bearings to be faulty, the outer races can be driven out of the casing and centre plate, and the inner race and bearings pulled or levered off (photo).
30 Remove the bolts and spring washers, and take off the oil pump and intake pipe (photo).
31 Remove the oil pump gears (photo).
32 Remove the rear cover oil seal, bearing, speedometer gear, circlip and sleeve, and oil sleeve.
33 If subsequent inspection shows the mainshaft bearing to be defective, the outer race can be driven out of the centre plate. Removal of the inner race and bearing is dealt with in Section 10.

9 Five-speed gearbox - examination

1 The essential points to be noted when checking the gearbox parts, are dealt with in Section 4. However, the following should also be checked:
(a) Examine the oil pump gears for wear and damaged teeth, renewing parts as necessary.
(b) Examine the oil pump drive shaft for wear and damage, renewing as necessary.

10 Five-speed mainshaft - dismantling and reassembly

1 From the front end of the mainshaft remove the pilot bearing and spacer.
2 Remove the mainshaft bearing circlip, then use a lead hammer or similar item to drive the mainshaft out of the 1st gear. In this way the gear, bush and bearing can be removed together (photo).
3 From the rear end of the mainshaft, pull off the 3rd gear and the 3rd/4th synchronizer hub and sleeve (photo).

8.7 Note the plastic bushes and tubular spacers

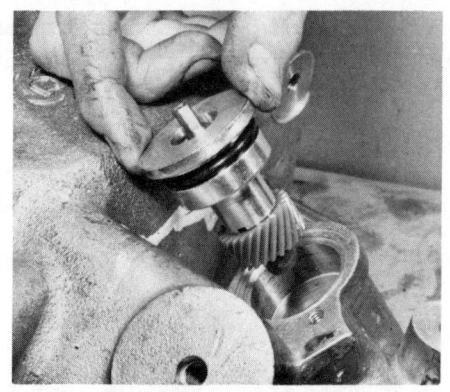

8.8 Removing the speedometer driven gear

8.11 Oil pump, showing the driveshaft

8.12 Removing the selector fork and bracket

8.13 Removing the selector shaft circlip

8.15 Removing the layshaft 5th gear

8.18 Removing the front selector spool locating boss

8.24 Reverse lever pivot pin (arrowed)

8.29 The layshaft showing the bearings

8.30 Removing the oil pump cover and intake pipe

8.31 The oil pump gears in the cover

10.2 Driving the mainshaft out of the 1st gear

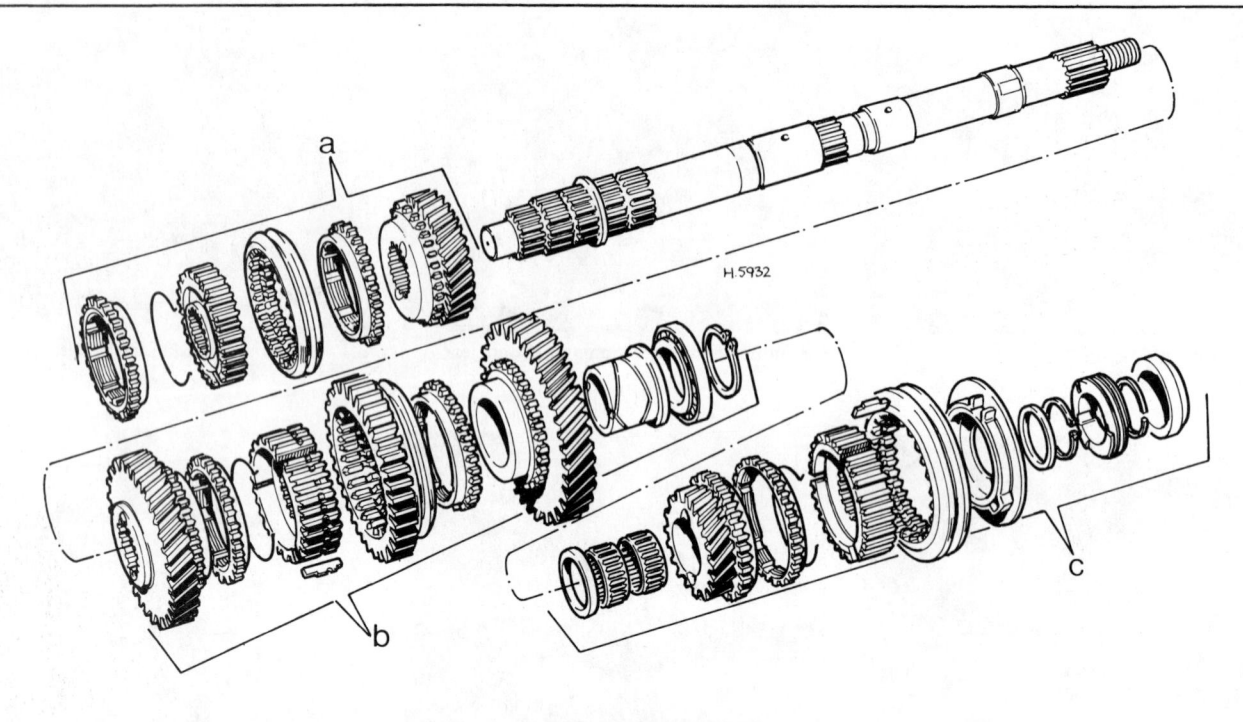

Fig. 6.14. Exploded view of mainshaft gears

a 3rd/4th synchro hub and sleeve, and 3rd gear
b 1st/2nd gear assembly, including reverse gear
c 5th gear assembly

4 Now take off the 1st/2nd gear hub, sleeve, synchro cones and 2nd gear (this assembly incorporates the reverse gear).
5 Clean all the parts in petrol or paraffin, and dry them with a lint-free cloth.
6 On the assembled synchro assemblies, check that a load of 18 to 22 lb f (8.2 to 10 kg f) is required to push the synchro hub through the outer sleeve in either direction.
7 Manufacture a spacer to the dimensions shown in Fig. 6.15, so that the 1st gear bush endfloat can be checked as described in Paragraphs 8 to 11 below.
8 Fit 2nd gear, 1st/2nd synchro hub and 1st gear to the mainshaft; then fit the spacer.
9 Using an old circlip and a feeler gauge, check the clearance between the spacer and circlip. This should be 0.0002 to 0.002 in (0.005 to 0.055 mm).
10 If necessary, select a new bush with a collar thickness which will give this dimension.
11 Remove the circlip, spacer, bush, synchro hub and 2nd gear from the mainshaft.
12 Check the 5th gear endfloat as described in Paragraphs 13 to 16 below.
13 Fit the 5th gear assembly to the mainshaft (this comprises the front spacer, 5th gear, synchro hub, rear plate and spacer).
14 Using an old circlip and a feeler gauge, check the endfloat. This should be 0.0002 to 0.002 in (0.005 to 0.055 mm).
15 If necessary, select a new rear spacer to provide this clearance.
16 Remove the circlip, spacer and 5th gear assembly from the mainshaft.
17 Ensure that the 1st/2nd synchro is assembled with the short splines on the inner member towards 2nd gear.
18 Fit 3rd gear, baulk ring, and the synchro sleeve and hub, so that the longer boss of the hub is towards the front of the gearbox. Fit the spacer and bearing (photos).
19 Fit the 2nd gear, baulk ring, synchro hub and sleeve (selector fork annulus towards the rear of the gearbox), baulk ring, 1st gear and selective bush, bearing and a new circlip. **During fitment, the circlip internal diameter must not be opened out beyond 1.272 in (32.3mm)** (photo).

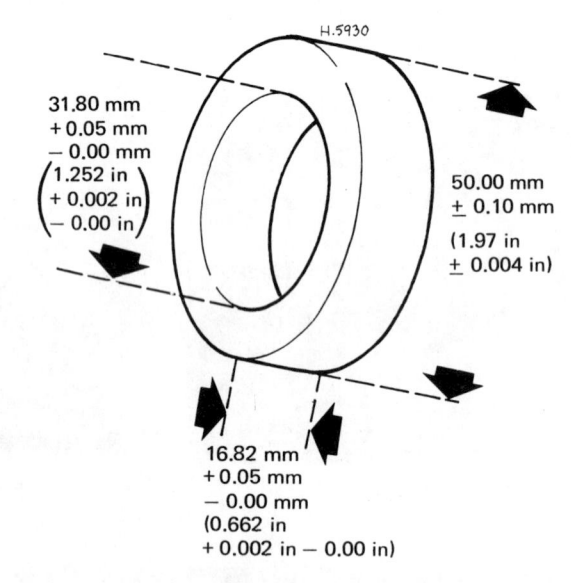

Fig. 6.15. Spacer dimensions for checking 1st gear bush endfloat

31.80 mm
+ 0.05 mm
− 0.00 mm
(1.252 in
+ 0.002 in
− 0.00 in)

50.00 mm
± 0.10 mm
(1.97 in
± 0.004 in)

16.82 mm
+ 0.05 mm
− 0.00 mm
(0.662 in
+ 0.002 in − 0.00 in)

11 Five-speed gearbox - reassembly

1 Commence reassembly by installing new layshaft bearings if the original ones were removed.
2 At this point it is also convenient to fit new bearing tracks in the front of the casing and in the centre plate if the original ones were removed.
3 Using protective jaw clamps, mount the centre plate in a vice.
4 Engage the forks of the assembled selector shaft (1st/2nd selector fork, front spool and 3rd/4th selector fork) in their respective synchro sleeves on the mainshaft. At the same time, engage the selector shaft and mainshaft assemblies in the centre plate (photos).

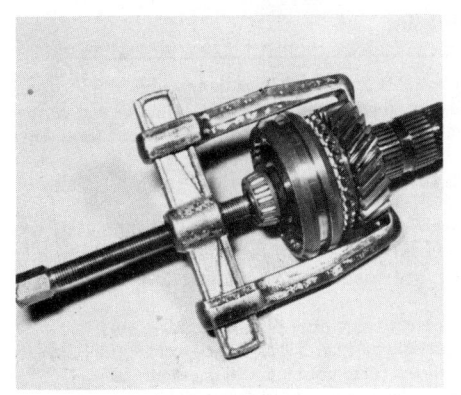

10.3 Pulling off the 3rd gear

10.18a Fitting the baulk ring ...

10.18b ... and synchro hub and sleeve

10.18c ... fitting the input shaft/mainshaft pilot bearing

10.19a Fit the 2nd gear and baulk ring ...

10.19b ... synchro hub and sleeve (including reverse gear), and baulk ring ...

10.19c ... 1st gear and selective bush ...

10.19d ... bearing ...

10.19e ... and circlip

11.4a, b Two views of the mainshaft and layshaft assembled to the centre plate

5 Fit the spacer, 5th gear, baulk ring, synchro hub and sleeve, endplate, selective spacer (see paragraph 15 of Section 10) and a new circlip. **During fitment, the circlip internal diameter must not be opened out beyond 1.088 in (27.63 mm).**
6 Fit the layshaft to the centre plate, and fit the 5th gear, the spacer, and a new circlip. **During fitment the circlip internal diameter must not be opened out beyond 0.886 in (22.5 mm).**
7 Fit the reverse gear so that the lip for the slipper pad is towards the front of the gearbox; fit the front and rear spacers, and the reverse shaft.
8 Fit the reverse lever, slipper pad, pivot pin and a new circlip. If a new reverse gear pivot shaft is being used, ensure that its radial location permits engagement and clearance of the reverse slipper pad. Secure the shaft with a spring washer and nut, then recheck the engagement and clearance.
9 Remove the assembled centre plate from the vice and locate it on a suitable stand with the front of the mainshaft upwards, taking care that the reverse shaft does not slide out.
10 Fit a new front gasket to the centre plate.
11 Fit the input shaft bearing, then insert the input shaft into the casing (photo).
12 Carefully slide the casing and input shaft into position over the gear assemblies, ensuring that the centre plate dowels and selector shaft engage in their respective locations.
13 Using slave bolts and flat washers, to prevent damage to the rear face of the centre plate, draw the casing evenly onto the centre plate.
14 Place the layshaft spacer on the layshaft bearing track at the front end of the casing and temporarily fit the cover and gasket.
15 Using a dial gauge, check and record the total layshaft endfloat. **Note:** *If no layshaft endfloat can be detected, repeat Paragraph 14 using a spacer of 0.040 in (1.02 mm) and recheck.*
16 Remove the front cover and spacer, then reselect a spacer to obtain a layshaft endfloat of 0.0002 to 0.002 in (0.005 to 0.055mm).
17 Refit the front cover using the selected spacer and recheck the specified endfloat.
18 Place a small steel ball on the machined centre of the end of the input shaft so that the stylus of a dial gauge can be rested on it.
19 Check and record the combined endfloat of the input shaft and mainshaft. If it is found that side movement of the input shaft prevents an accurate reading from being obtained, it is permissible to remove the front cover and wrap about 6 turns of masking tape around the input shaft below the splines; this will stop any side movement.
20 Having determined the endfloat, select a spacer for the front cover which will give a final endfloat of 0.0002 to 0.002 in (0.005 to 0.0055 mm) (photo).
21 Remove the front cover (and tape, if used). If the front cover oil seal has not yet been renewed, this should be done now. Install it with the lips facing the gearbox, then lubricate the lips with gearbox oil.
22 Carefully fit the front cover so that the seal lips are not marked by the shaft splines.
23 Remove the slave bolts from the centre plate.
24 Fit the 5th gear spool with the longer cam towards the bottom of the gearbox, then fit the circlip.
25 Fit the 5th gear selector bracket and fork.
26 Fit a new selector shaft O-ring in the rear cover, and fit the oil ring bush.
27 Fit a new gasket to the rear of the centre plate, then engage the oil pump driveshaft in the layshaft end.
28 Fit the oil pump gears, then fit the cover.
29 Carefully fit the rear cover so that the oil pump drive engages correctly.
30 Fit the selector shaft ball, spring and screwed plug to the centre plate.
31 Fit both spool locating bosses to the gearbox casing assembly. *Note the cable clip on the front one* (photo).
32 Fit the speedometer driving gear to the mainshaft, ensuring that it engages properly with the mainshaft flats (photo).
33 Fit the circlip and sleeve, followed by the ballrace, to the mainshaft.
34 Fit the rear oil seal and lubricate the lips with gearbox oil (photo).
35 Fit the mainshaft flange, then install and tighten the nut and washer to the specified torque. Use the bracket and bolt which were used when the flange nut was removed.
36 Fit the speedometer gear and housing, securing it with the plate, washers and bolt.
37 If removed when dismantling the gearbox, refit the bellhousing.
38 Fit the clutch release bearing and lever, referring to Chapter 5
39 Install the remote control assembly, and adjust it as described in the following paragraphs.
40 Remove the remote control assembly bottom cover plate.
41 With the gear lever vertical and neutral selected, loosen the baulk plate adjusting bolts until the plate contacts the backing plate.
42 Tighten the adjusting bolts equally until they just start to move the baulk plate away from the backing plate.
43 Using a straightedge and feeler gauge, adjust the bolts equally, until a clearance of 0.050 to 0.060 in (1.27 to 1.42mm) exists between the lower face of the gear lever and the underside of the baulk plate. Tighten the locknuts, but ensure also that there is at least 0.10 in (2.54mm) clearance between the upper face of the baulk plate and the lower edge of the gear lever bush.
44 With the gear lever in the first gear plane, check that there is a clearance of 0.004 to 0.012 in (0.10 to 0.30mm) between the side of the gear lever and the edge of the baulk plate. Add or remove shims as necessary to achieve this.
45 Fit the bottom cover plate.
46 Engage 3rd gear, then position the gear lever hairpin spring bias screws so that there is a clearance of 0.020 in (0.5 mm) between the spring and crosspin on each side. Do not tighten the locknuts.
47 Apply a light load to move the gear lever to the right to take up the free play, then adjust the right-hand screw downwards until the spring leg just makes contact with the crosspin.
48 Repeat Paragraph 47, moving the gear lever to the left, then adjust the left-hand screw downwards to just contact the crosspin.
49 Move the gear lever to neutral, rock it across the gate and check that it comes to rest in the 3rd/4th plane.
50 Tighten the locknuts.
51 Adjust the reverse light switch as described in Chapter 10.
52 Top-up the gearbox with the specified grade and quantity of oil.

12 Five-speed gearchange remote control assembly - dismantling and reassembly

1 With the assembly removed from the transmission as described in Section 8, remove the two bolts and countersunk head screws securing the bias spring bridge plates.
2 Remove the bridge plates, bridge plate lines and hairpin-type bias spring.
3 Remove the bias spring adjustment bolts and locknuts.
4 Remove the two bolts and washers securing the reverse baulk plate assembly; withdraw the reverse baulk plate, springs and spacers.
5 Remove the four bolts and washers, and take off the bottom cover plate (photo).
6 Remove the reverse light switch and locknut, referring to Chapter 10 if necessary.
7 Remove the square-headed pinchbolt and take off the selector shaft elbow. Now withdraw the selector shaft.
8 Press out the selector shaft bushes from the casing.
9 Remove the circlips, and press out the pivot balls and bushes from the selector shaft elbows.
10 Reassembly is the reverse of the dismantling procedure. On completion, with the assembly attached to the gearbox, the reverse baulk plate, 1st/2nd stop gate and gear lever hairpin bias spring should each be adjusted by following the procedure given in Paragraphs 40 onwards of Section 11.

Fault diagnosis appears on page 88.

11.11 Input shaft ready for assembly into the casing

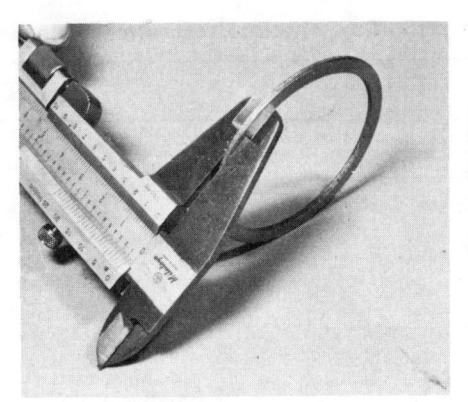

11.20 Selecting a mainshaft spacer

11.31 Cable clip on front spool locating boss

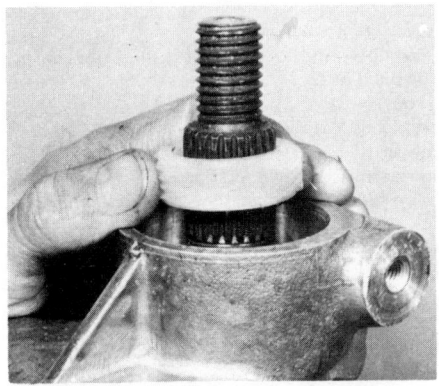

11.32 Fitting the speedometer driving gear

11.34 Rear oil seal being fitted

12.5 Bottom cover plate removed

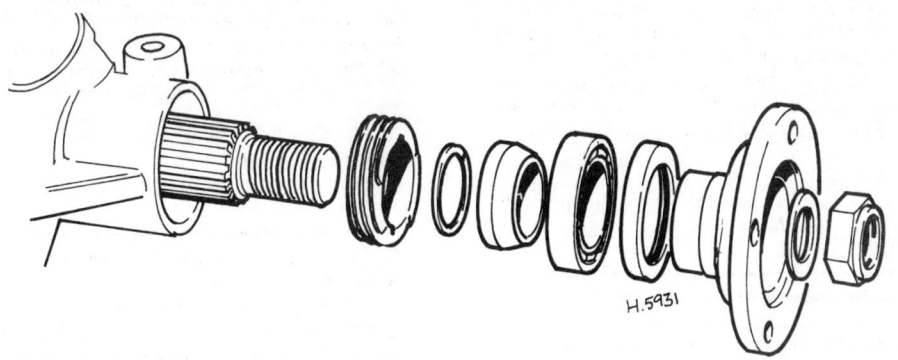

Fig. 6.16. Drive flange and associated parts

Fig. 6.17. Remove control assembly adjustment

A Gearlever/baulk plate clearance
B Gearlever bush/baulk plate clearance
C 1st/2nd gate stop shims

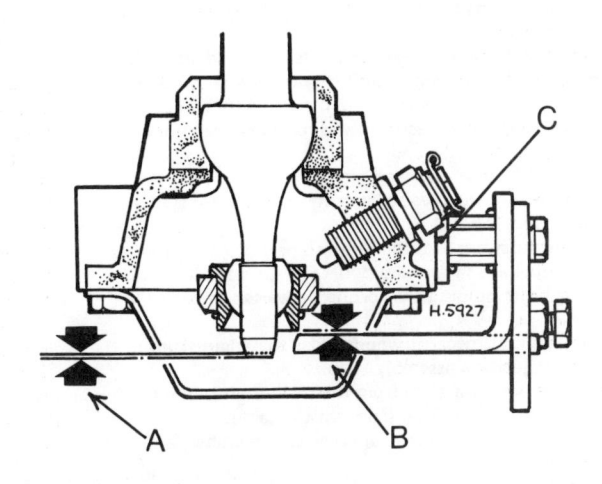

13 Fault diagnosis - manual gearbox

Symptom	Reason/s
Weak or ineffective synchromesh	Synchronising cones worn, split or damaged.
	Synchromesh dogs worn or damaged.
Jumps out of gear	Broken gearchange fork rod spring.
	Gearbox coupling dogs badly worn.
	Selector fork rod groove badly worn.
Excessive noise	Incorrect grade of oil in gearbox or oil level too low.
	Bush or needle roller bearings worn or damaged.
	Gear teeth excessively worn or damaged.

PART 2: AUTOMATIC TRANSMISSION

14 General description

The 'Borg-Warner 65' automatic transmission is available as an optional extra for certain versions of the TR7. It is a lightweight version of the earlier 'Borg-Warner 35' and due to the resighting of the hydraulic control unit to within the sump the unsightly bulge in the transmission tunnel, which was associated with former versions, is no longer necessary. The system comprises two main components:
   a) A three element hydrokinetic torque converter coupling capable of torque multiplication at an infinitely variable ratio between 2 : 1 and 1 : 1.
   b) A torque/speed responsive hydraulic epicyclic gearbox comprising a planetary gearset providing three forward ratios and one reverse ratio.

Selection of the required ratio is by means of a console-mounted lever, with the selector positions 'P', 'R', 'N', 'D', '2', '1' marked.

It is not possible to start the engine unless the selector is in the 'P' or 'N' positions. This prevents inadvertent movement of the vehicle, and is controlled by an inhibitor switch mounted on the transmission unit.

To prevent accidental engagement of '1', '2', 'R' or 'P' positions, a button on top of the selector lever must be depressed.

Note: *It is essential that a transmission oil cooler be installed where automatic transmission is fitted.*

Due to the complexity of the automatic transmission unit it is not recommended that stripping the unit is attempted. Where the unit is known to be faulty, and the fault cannot be rectified by following the procedures given in the following Sections of this Chapter, the repair should be entrusted to a Triumph dealer or automatic transmission specialist.

15 Automatic transmission - removal and installation

1 Drive the car onto a ramp, or have available adequate jacks and stands to give access to the underside.
2 Select 'N', and chock the front wheels.
3 Disconnect the battery earth.
4 Disconnect the downshift cable at the carburettor.
5 Raise the vehicle, and detach the exhaust system (see Chapter 3).
6 Disconnect the dipstick tube, and drain the transmission fluid into a container of at least 9.5 Imp pints (11.4 US pints, 5.4 litres) capacity.
7 Disconnect the selector rod at each end and remove it. Do not alter the adjustment.
8 Remove the breather hose at the gearbox.
9 Pull the connectors from the reverse lamp starter inhibitor switch.
10 Disconnect the oil cooler pipes.
11 Release the speedometer cable at the rear extension.
12 Disconnect the propeller shaft, and support it to prevent it falling.
13 Remove the four bolts holding the torque converter to the drive plate.
14 With a suitable jack and piece of wood under the engine sump, support the engine and gearbox assembly.
15 Take out the centre bolt and plate from the rear mounting.
16 Remove the steady bar. If fitted, disconnect the cable.
17 Remove the four nuts which retain the rear crossmember, and remove the crossmember.
18 Remove the heat shield bolts, shield and spacers.
19 Remove the four bolts securing the radiator lower mounting.
20 Release the cooler pipes from the clips.
21 Detach the radiator from the upper mountings and remove it forwards.
22 Remove the exhaust supporting bracket.
23 Lower the complete assembly, and take out all bellhousing and starter motor to engine bolts and nuts.
24 Take the dipstick tube out.
25 Free all wiring and earth leads from the unit.
26 Lower the gearbox rearwards.

Note: Be sure to flush the transmission oil cooler and the oil cooler pipes with clean transmission fluid before installing the torque converter or transmission.

27 To refit, position the gearbox and secure with the bellhousing to engine dowel bolts.
28 Replace the starter motor, earth leads, harness clips and dipstick tube.
29 Fit and tighten the remaining bolts.
30 Refit and secure the exhaust bracket.
31 Raise the engine/gearbox unit to the fitted position, fit the rear crossmember, and tighten the four bolts. Take away the supporting jack.
32 Offer up the rear mounting plate and steady bar, if fitted. Tighten the centre bolt.
33 Refit the steady cable (if fitted).
34 Proceed in the reverse procedure to that given in paragraphs 12 and 13.
35 Refit the heat shield, with spacers and bolts.
36 Proceed in the reverse procedure to that given in paragraphs 6 to 11.
37 Replace the radiator, and fit the four lower mounting bolts.
38 Clip the cooler pipes back on the radiator.
39 Refit the front exhaust pipe.
40 Reconnect the downshift cable.
41 Refill the gearbox.

16 Kickdown cable - adjustment

Refer to Chapter 13.

17 Front brake band - adjustment

1 Remove the transmission cover carpet and remove the access plate. If no access plate is fitted it will be necessary to raise the car for access to the transmission. Apply the handbrake and select 'P'.
2 Slacken the locknut on the adjustment screw then torque tighten the screw to 3 lbf ft (4 Nm).
3 Back the screw off 3/4 of a turn then hold it stationary in this position whilst tightening the locknut.
4 Fit the access plate and replace the carpet (where applicable).

18 Rear brake band - adjustment

1 Drive the car onto a ramp or raise it on jacks. Apply the handbrake.
2 Select 'P'; raise the ramp and slacken the adjuster locknut.
3 Torque tighten the screw to 3 lbf ft (4 Nm), then back it off ¾ of a turn.
4 Hold the screw stationary in this position and tighten the locknut.
5 Lower the car to the ground.

Chapter 6/Manual gearbox and automatic transmission

19 Front servo - removal, overhaul and installation

1 With the handbrake on and 'N' selected, remove the gearbox selector lever and rod.
2 Take out the four cover bolts and withdraw the servo assembly, spring and joint washer.
3 Remove the spring and withdraw the piston.
4 Take off all the O-rings from the piston and body, and discard them.
5 Clean all the parts in petrol and wipe them dry with a lint-free cloth.
6 Inspect the piston for scoring, corrosion or other damage and renew if necessary.
7 Inspect the cover for damage and check that the passages are unobstructed.
8 It is preferable to renew the spring if possible unless it is known to be serviceable.
9 Assemble the unit in the reverse order to dismantling, using new O-rings and a new gasket.

20 Rear servo - removal, overhaul and installation

1 Initially remove the front exhaust pipe then take out the six bolts on the rear servo cover.
2 Withdraw the servo assembly, joint washer, spring and pushrod.
3 Remove the pushrod and spring; then withdraw the piston.
4 Take off all the O-rings from the piston and body and discard them.
5 Clean all the parts in petrol and wipe them dry with a lint-free cloth.
6 Inspect the piston for scoring, corrosion or other damage and renew if necessary.
7 Inspect the cover for damage and check that the passages are unobstructed.
8 It is preferable to renew the spring if possible unless it is known to be serviceable.
9 Assemble the unit in the reverse order to dismantling using new O-rings and a new gasket.
10 Finally refit the front exhaust pipe.

21 Rear extension - removal and installation

1 Have a ramp, jacks or stands available. Raise the vehicle, chock the wheels and select 'N'.
2 Move the propeller shaft guard clear of the shaft, by removing one bolt and slackening the other.
3 Remove the flange bolts, disconnecting the propeller shaft. Support the shaft clear.
4 Remove the drive flange nut, and pull off the flange.
5 Support the transmission under the sump, using a suitable jack.
6 Remove the gearbox centre mounting bolt and plate.
7 Remove the steady bar and release the steady cable, if fitted.
8 Remove the four crossmember nuts, raise the LH captive bolts, and release and remove the crossmember.
9 Disconnect the speedometer cable by removing the clamp.
10 Remove the nut and bolt securing the exhaust pipe to the bracket. Remove the bracket and spacers.
11 Remove the eight bolts holding the extension to the main casing, and withdraw the extension.
12 To refit, reverse the dismantling procedure using a new gasket.

22 Extension rear oil seal - renewal

1 Initially carry out the procedure given in paragraphs 1, 2 and 8 of

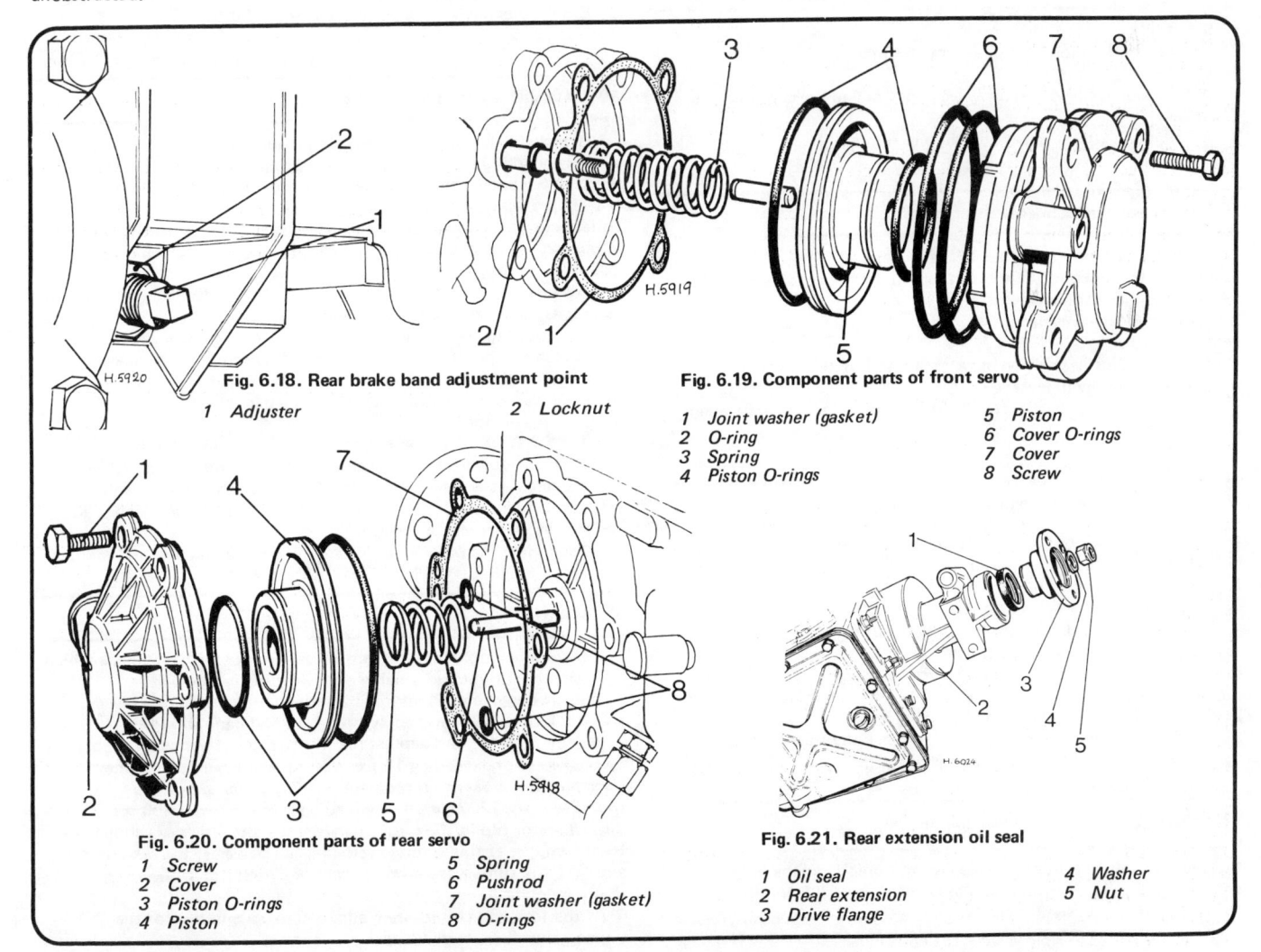

Fig. 6.18. Rear brake band adjustment point

1 Adjuster
2 Locknut

Fig. 6.19. Component parts of front servo

1 Joint washer (gasket)
2 O-ring
3 Spring
4 Piston O-rings
5 Piston
6 Cover O-rings
7 Cover
8 Screw

Fig. 6.20. Component parts of rear servo

1 Screw
2 Cover
3 Piston O-rings
4 Piston
5 Spring
6 Pushrod
7 Joint washer (gasket)
8 O-rings

Fig. 6.21. Rear extension oil seal

1 Oil seal
2 Rear extension
3 Drive flange
4 Washer
5 Nut

the previous Section.
2 Prise out the existing oil seal.
3 Using a suitable drift, carefully drive in a new oil seal. Lightly lubricate the lip of the seal.
4 Refit the parts in the reverse order to removal.

23 Governor - removal, overhaul and installation

1 Remove the rear extension as described in Section 21.
2 Take off the speedometer drive clamp tube, and withdraw the drive gear.
3 Unscrew the counterweight from the base of the governor, taking note of the spring washer.
4 Withdraw the governor from the shaft.
5 Prise off the retaining circlip and remove the weight.
6 Withdraw the stem, spring and valve.
7 Wash all the parts in petrol and dry with a lint-free cloth.
8 Check the parts for burrs and scoring and for any signs of thread damage.
9 It is best to renew the spring, if at all possible, even if apparently satisfactory.
10 When reassembling, first insert the valve into the body.
11 Next, fit the spring to the stem; then fit both parts into the body.
12 Refit the weight and a new circlip.
13 Refit the governor and replace the counterweight and spring washer.
14 Refit the speedometer drive gear and clamp tube.
15 Refit the rear extension.

24 Restrictor valve and bypass pipe - removal, cleaning and installation

1 Select 'P' and apply the handbrake.
2 Unscrew both union nuts on the bypass pipe and withdraw it.
3 Unscrew the restrictor valve.
4 Wash the pipe and restrictor valve in petrol and shake dry before replacing.
5 Installation is the reverse procedure to removal.

25 Selector rod - adjustment

1 Select 'N' and apply the handbrake.
2 Slacken the locknut on the adjuster and disconnect the rod at the selector lever (transmission end).
3 Check that the transmission is in neutral, or move the lever to neutral if necessary.
4 Alter the adjuster position as necessary to obtain neutral on the selector lever and hand lever.
5 Tighten the locknut, and finally fit the lever when the adjustment is correct.

26 Transmission sump - draining and refilling

1 Drive the vehicle onto a ramp or have available adequate jacks to provide access to the underside of the car.
2 Select 'P' and apply the handbrake.
3 Raise the ramp, or jacks.
4 Wipe around the drain plug and then remove it; drain the contents of at least 9.5 Imp pints/11.4 US pints (5.4 litres) capacity.
Note: *It is not possible to drain the torque converter completely.*
5 Refit the drain plug on completion; then add fresh fluid up to the level of the C-mark on the dipstick - no higher.
6 Give the car a warming-up run and re-check the fluid level when hot by following the procedure given in Section 29.

27 Transmission sump - removal and refitting

1 Unscrew the filler/dipstick pipe at the union on the sump sidewall. Drain the oil into a container of at least 9.5 Imp pints (11.4 US pints, 5.4 litres) capacity.
2 Remove the heat shield bolts, the shield, and the shield spacers.
3 Remove the steady bar, if fitted.
4 Remove the sump bolts, the sump and gasket.
5 To refit, reverse the removal procedure. Ensure that the sump mating surface is clean, and use a new gasket.
6 Refill the sump (see Section 26).

28 Starter inhibitor/reverse lamp switch - removal and refitting

1 Raise the vehicle, and chock the wheels if relevant. Select 'P' and apply the handbrake.
2 Take out the three bolts retaining the heat shield to the transmission sump, and remove the shield and spacers.
3 Disconnect the switch from the block connector.
4 If fitted, remove the thread protector.
5 Remove the bolt, and then the switch.
6 To refit, reverse the dismantling procedure.

29 Fluid level - checking

1 With the engine cold, initially check that the fluid level is up to the C-mark on the dipstick. Top up to this level if necessary.
2 Drive the vehicle for about 20 miles (30 km) to warm the unit properly.
3 Place the vehicle on level ground, apply the handbrake, and select 'P'.
4 Switch off the engine, wipe clean around the dipstick filler orifice, and withdraw the dipstick.
5 Wipe the dipstick with non-fluffy cloth or paper, replace it, and then withdraw it to take a reading.
6 Ensure that the level is up to the mark 'A' (hot high).
7 Do not overfill.

30 Downshift cable - pressure test

1 Run the engine until the normal operating temperature is reached.
2 Set the idling speed to 750 rpm (approximately).
3 Stop the engine and raise the car to provide access to the transmission. If jacks are used, adequately chock the wheels. Remove the plug situated in the cut-away on the lower edge of the transmission rear extension flange and fit a pressure gauge suitable for reading up to 100 lbf/in^2 (7 kgf/cm^2).
4 Lower the car to the ground and apply the footbrake and handbrake as well as having the wheels chocked.
5 Start the engine and select 'D', and check that the gauge pressure is between 60 and 75 lbf/in^2 (4.2 and 5.3 kgf/cm^2) at idle speed.
6 Increase the speed to 1000 rpm and check that the pressure increases by 15 to 20 lbf/in^2 (1.0 to 1.4 kgf/cm^2).
7 Stop the engine.
8 If the pressure increase is too small, increase the effective length of the downshift outer cable. If the pressure increase is too great, decrease the effective length of the downshift outer cable. Repeat the operation until the correct pressure increase is achieved. Switch the engine off afterwards.
9 Disconnect the pressure gauge and refit the plug on completion.

31 Stall test

This test can only be satisfactorily carried out with an engine which is in good condition and capable of developing full power.
1 Run the engine until normal operating temperature is achieved.
2 Chock the wheels and apply both foot and handbrakes.
3 Select '1' or 'R' and depress the throttle to the 'kickdown' position for a period not exceeding 10 seconds (to avoid overheating the transmission). *Note the tachometer reading which should be approximately 2200 rpm.* If the reading is lower than 1400 rpm, suspect stator slip in the torque converter; if the reading is about 1600 rpm, the engine is not developing full power; if the reading is around 2400 rpm, suspect brake band or clutch slip in the transmission unit.
4 If the test is to be repeated, allow 10 to 15 minutes for the transmission fluid to dissipate.

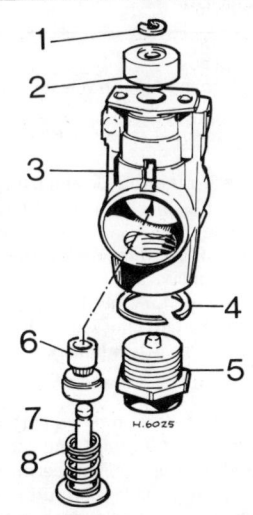

Fig. 6.22. Removing the governor

(Above) The fitted position
1 Circlip
2 Weight
3 Governor body
4 Spring washer

(Below) The component parts
5 Counterweight
6 Valve
7 Stem
8 Spring

Fig. 6.23. Restrictor valve and bypass pipe

1 Bypass pipe
2 Restrictor valve
3 Downshift cable

Fig. 6.24. Selector rod - adjustment

1 Selector rod
2 Adjuster

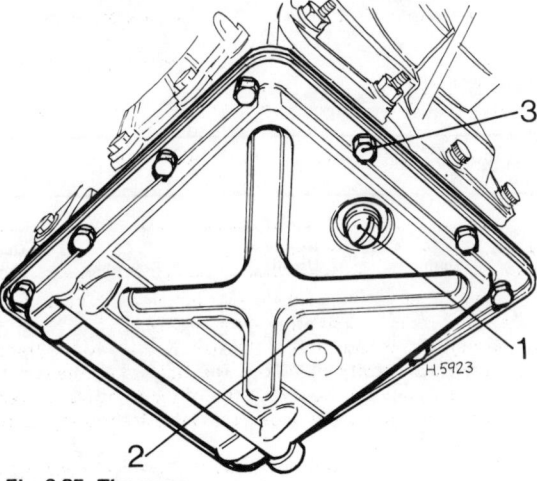

Fig. 6.25. The sump

1 Drain plug
2 Sump
3 Retaining bolts

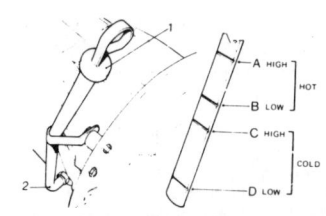

Fig. 6.26. Dipstick and filler tube

(Left)   Dipstick
(Right) Dipstick graduation
2 Filler tube

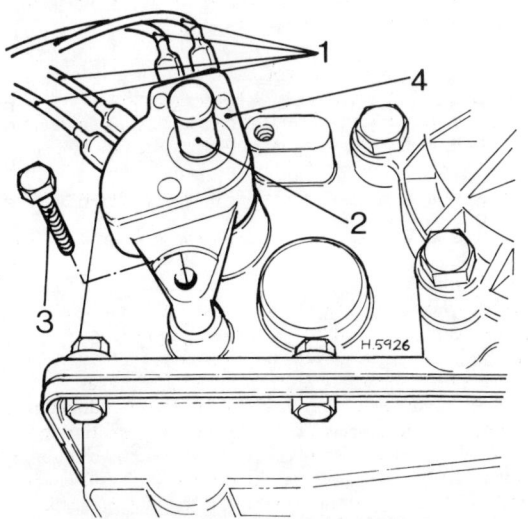

Fig. 6.27. Starter inhibitor/reverse light switch

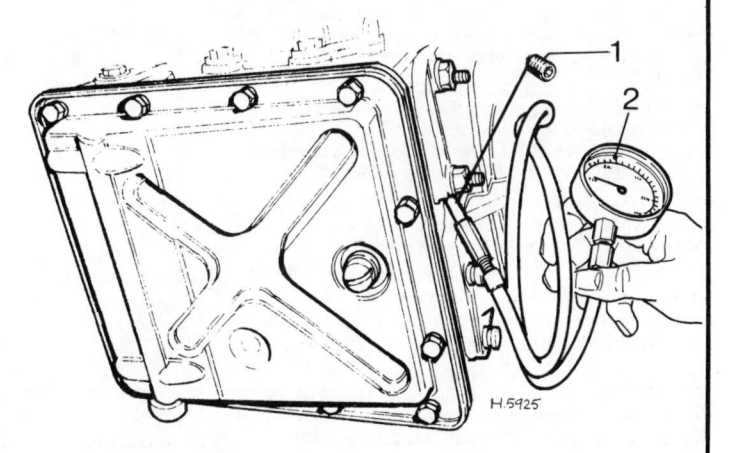

1 Electrical leads
2 Thread protector
3 Bolt
4 Switch

Fig. 6.28. Downshift cable pressure test

1 Plug
2 Pressure gauge

32 Road test to be carried out in conjunction with the fault diagnosis procedure (Section 33)

Note: *The term full throttle refers to approximately seven-eights of the available pedal travel and kickdown is equivalent to full pedal travel.*

Procedure

a Check that the starter motor operates only with the selector lever at 'P' or 'N' and that the reverse lights operate on at 'R'.
b Apply the handbrake. With the engine idling select 'N-D', 'N-2', 'N-R'. Engagement should be positive (a cushioned 'thump' under fast idling conditions is to be expected).
c With the transmission at normal running temperature, select 'D'. Release the brakes and accelerate with minimum throttle. Check the 1-2 and 2-3 shift speeds and the smoothness of the change.
d Stop the vehicle, select 'D' then re-start using 'full throttle'. Check 1-2 and 2-3 shift speeds and the smoothness of the change.
e At 40 mph (65 km/h) apply 'full throttle'. The car should accelerate in third gear and should not downshift to second.
f At a maximum speed of 56 mph (90 km/h) 'kickdown' fully. The transmission should downshift to second.
g At a maximum speed of 35 mph (56 km/h) 'kickdown' fully. The transmission should downshift to first.
h Stop the vehicle, select 'D', then re-start using 'kickdown', check the 1-2 and 2-3 shift speeds.
j At 35 mph (56 km/h), select '2' and release the throttle; check the 2-3 downshift.
k At 30 mph (50 km/h) select '1' and release the throttle; check the 2-1 downshift.
l With '1' still engaged, stop the vehicle and using 'kickdown' accelerate to over 40 mph (65 km/h). Check for 'slip', 'squawk' and loss of upshifts.
m Stop the vehicle. Select 'R', and reverse at 'full throttle' (if possible), checking for 'slip' and 'squawk'.
n Park the vehicle on a gradient. Apply the handbrake and select 'p' then release the handbrake and check that the parking pawl holds. Check that the selector lever is firmly locked in 'P'.

Converter diagnosis

Inability to start on steep hills, combined with poor acceleration from rest and low stall speeds (1400 rpm), indicates that the converter stator unidirectional clutch is slipping. This permits the stator to rotate in an opposite direction to the impeller and turbine, and torque multiplication cannot occur. Poor acceleration in third gear above 30 mph (50 km/h) and reduced maximum speed, indicates that the stator unidirectional clutch has seized. The stator will not rotate with the turbine and impeller and the 'fluid fly-wheel' effect cannot occur. This condition will also be indicated by excessive overheating of the transmission although the stall speed will be satisfactory.

33 Automatic transmission - fault diagnosis procedure

The letters indicate the suggested sequence of investigation in the road test (see Section 32)

Symptom / Diagnosis	Engagement of selected gear (Bumpy / None Delayed)	'Take-off' (None forward / None reverse / Seizure reverse / No neutral)	Upshift (Slip / Squawk / No 1–2 / No 2–3)	Quality of Upshift (Above normal speed / Below normal speed / Slip 1–2 / Slip 2–3 / Rough 1–2 / Rough 2–3 / Seizure 1–2 / Seizure 2–3)	Downshift (No 2–1 / No 3–2 / Involuntary 3–2 / Above normal speed / Below normal speed)	Quality of Downshift (Rough 3–2 / Rough 2–1 / Slip 3–2 / Slip 2–1)
Adjustment faults						
Fluid level insufficient	A A		A	A	A	
Downshift cable incorrectly assembled or adjusted	B		B	A A / B B A A	A A A A	
Manual linkage incorrectly assembled or adjusted	B b	A A / A C A	A A	C C		
Incorrect engine idling speed	A C					
Incorrect front band adjustment		A	B	D D B B A		A A
Incorrect rear band adjustment		B		A		
Hydraulic control faults						
Oil tubes incorrectly installed, missing or leaking	D C	G	G	H H / J J H E / K J F / K	H J / J K / K L K F F / F G C C	B / B C / F D	B / E E / F F	F / G C / H E / D C
Sealing ring missing or broken	G D	C F	E			
Valve block screws missing or loose	E F	B E	C			
Primary regulator valve sticking	C E F					
Secondary regulator valve sticking						
Throttle valve sticking	D			F F B B / G G G	G H D D / E / F	D D	E D
Modulator valve sticking						
Governor valve sticking, leaking or incorrectly assembled				C B C C	C C B B	C B
Orifice control valve sticking						
1–2 shift valve sticking		C	D D / C E / D F D	B B	G H C	
2–3 shift valve sticking		D				
2–3 shift valve plunger sticking						
Converter 'out' check valve sticking or missing	N					
Check valve sticking or missing	H					
Mechanical faults						
Front clutch slipping		J	D	D	J	B
Front clutch seized or plates distorted	F	B B	F			
Rear clutch slipping		K	J	E	E	J F
Rear clutch seized or plates distorted	G			E B	D	G
Front band slipping due to faulty servo or worn band				E F	E	B H
Rear band slipping due to faulty servo or worn band	L	H			D	
Uni-directional clutch slipping or incorrectly installed		E		G C		A A
Uni-directional clutch seized				H D		
Input shaft broken		G				
Front pump drive tangs on converter hub broken		H				
Front pump worn	M	J				
Converter blading and/or uni-directional clutch failed		K				

Chapter 7 Propeller shaft

Contents

General description ... 1	Universal joints - inspection, dismantling and reassembly ... 3
Propeller shaft - removal and installation ... 2	

Specifications

Type ...	Single piece, tubular
Universal joints	
Front ...	Constant velocity
Rear ...	Hookes needle bearing, or constant velocity joint velocity
Overall length, compressed	
With Hookes joint at rear ...	37.289 in (947.17 mm)
With constant velocity joint at rear ...	36.85 in (936 mm)
Torque wrench settings	lbf ft Nm
Gearbox and rear axle flange bolts/nuts ...	34 46

1 General description

Drive is transmitted from the gearbox to the rear axle by a balanced tubular propeller shaft. At the gearbox end, a constant velocity joint is used; its design also permits a small amount of longitudinal movement to allow for relative movement of the gearbox and axle. The joint at the axle end is either conventional Hookes (Hardy Spicer type) universal joint or a constant velocity joint.

Flanges are used at each end for attachment to the gearbox mainshaft flange and differential pinion.

No provision is made for lubrication during service lift.

2 Propeller shaft - removal and installation

1 Jack-up the car along one side for access to the propeller shaft flange joints. Support it on stands or blocks beneath the bodyframe sidemember. Chock the wheels which are still on the ground.
2 Index mark the front and rear flanges so that they can be installed in this original position.
3 Remove the four bolts and self-locking nuts from each flange joint; rotate the rear wheel to turn the propeller shaft for improved accessibility (photos).
4 Remove the propeller shaft from the car
5 Installation is the reverse of the removal procedure, but ensure that the index marks are aligned.

3 Universal joints - inspection, dismantling and reassembly

1 Wear in the bearings is indicated by vibration or 'clunks' in the transmission, particularly when the drive is being taken up or when going to over-run. (Backlash in the rear axle has the same effect, so check that also, if symptoms occur).
2 It is easy to check the bearings whilst the propeller shaft is still in position. Try to turn the shaft with one hand and grip on the other side of the joint with the other hand. There should be no movement between the two. If there is any, the bearings will need renewal. Check also by trying to lift the shaft and noting any movement in the joints.
3 If worn, the Hookes bearing and spider will have to be discarded and a repair kit, comprising new universal joint spider, bearings, oil seals, and retainers purchased.
4 If wear is detected in the constant velocity joint(s), a replacement propeller shaft will have to be obtained since no repair is possible.
5 To dismantle a Hookes-type rear bearing, clean away all dirt then remove the circlips which hold each set of needle roller bearings in position. If the circlip is tight, tap the face of the bearing cup inside it which may be jamming it in its groove (photo).
6 The bearing should come out if the edges of the yoke ears are tapped with a mallet. If, however, they are very tight, it should be possible to shift them by pressing them between the jaws of a vice, using two distance pieces. Two different size socket spanners are ideal, and it will be possible to force one out sufficiently far to enable it to be gripped by another suitable tool (pliers or vice again) and drawn out. Take care not to damage the yokes. An alternative method of removing the bearings is to use a socket and hammer as illustrated in the photo. If the bearings have seized up or worn so badly that the holes in the yokes are oval, then a new yoke will be needed - and if this is on the propeller shaft then that will have to be acquired too, as parts are not supplied separately (photos).
7 New bearings will be supplied with new seals and circlips. Make sure the needles are correctly in position and the cup 1/3 full of grease.
8 Fit the spider to the propeller shaft yoke.
9 Engage the spider trunnion in the bearing cup and insert the cup into the yoke.
10 Fit the opposite bearing cup to the yoke and carefully press both cups into position, ensuring that the spider trunnion engages the cups and that the needle bearings are not displaced.
11 Using two flat-faced adaptors of slightly smaller diameter than the bearing cups, press the cups into the yoke until they reach the lower land of the circlip grooves. Do not press the bearing cups below this point or damage may be caused to the cups and seals.
12 Install the circlips.

2.3a Removing the front universal joint flange bolts

2.3b Rear constant velocity type joint and flange coupling

3.5 Removing the circlips

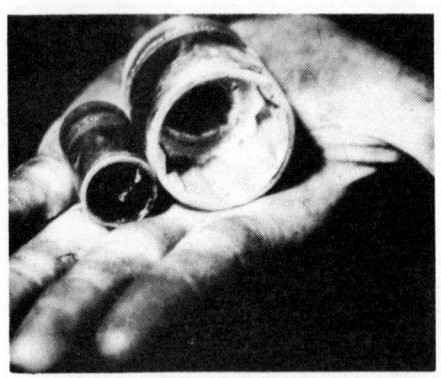

3.6a Showing the two sockets needed to press the cups out of the yoke

3.6b Using the sockets and a vice to press out the cups

3.6c Lifting out the cups with a pair of pliers

3.6d Using a hammer and socket as an alternative method of removing the cups

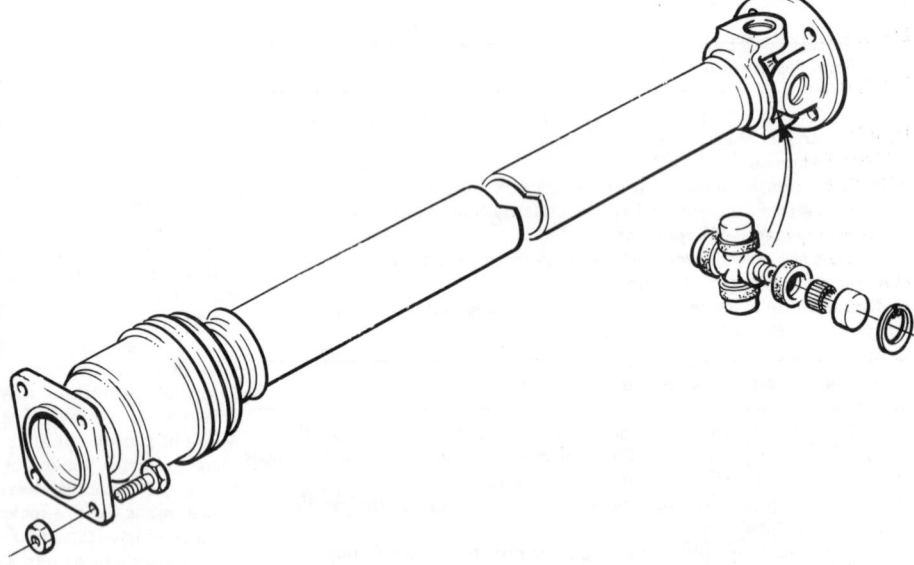

Fig. 7.1. Propeller shaft - typical, where the rear end has a Hookes needle roller bearing

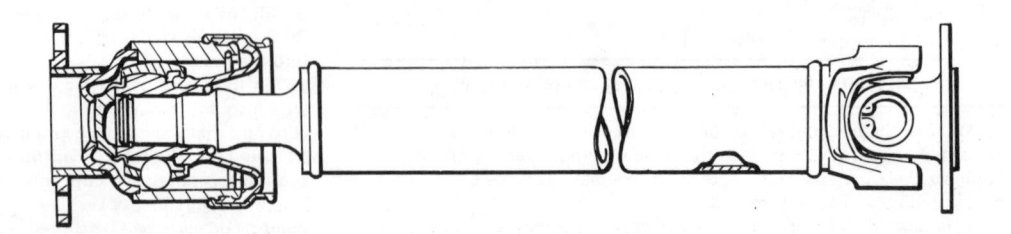

Chapter 8 Rear axle

For modifications, and information applicable to later models, see Supplement at end of manual

Contents

Differential assembly (three-quarter floating axle) - removal and refitting ... 6	Halfshaft (three-quarter floating axle) - removal and installation ... 2
Fault diagnosis - rear axle ... 9	Hub, halfshaft bearing and oil seals (three-quarter floating axle) - removal and installation ... 3
General description ... 1	Pinion oil seal (three-quarter floating axle) - removal and refitting ... 7
Halfshaft bearing and oil seals (semi-floating axle) ... 5	Rear axle (three-quarter floating and semi-floating types) - removal and refitting ... 8
Halfshaft (semi-floating axle) - removal and installation ... 4	

Specifications

Type (4-speed gearbox) ...	Three-quarter floating live axle with hypoid bevel gears and two-pinion differential.
Type (5-speed gearbox) ...	Semi-floating live axle with hypoid bevel gears and two-pinion differential.
Oil capacity (approx) ...	2¼ Imp pints (2¾ US pints/1.3 litres)
Oil type ...	SAE90EP (SAE80EP below 0°C/32°F)

Ratios
4-speed gearbox and automatic transmission ...	3.63 : 1
5-speed gearbox ...	3.9 : 1

Torque wrench settings

	lbf ft	Nm
Drain plug ...	25	34
Hub to halfshaft ...	120	160
Hypoid housing bolts ...	20	27
Hypoid housing bearing cap bolts ...	38	52

1 General description

The three-quarter floating axle has an integral crownwheel, pinion and differential assembly bolted to the front of a banjo-type axle housing, the whole unit being easily removed as an assembly.

The semi-floating axle has a differential unit mounted to the main axle casing under the differential rear cover, held in place by bearing caps. The pinion is separately mounted. The average owner is not recommended to attempt removal of the differential unit, as a 'case stretcher' is normally required.

The crownwheel and pinion run on opposed taper roller bearings, bearing preload and gear meshing being controlled by shims. Oil seals are fitted at the pinion and driveshaft extremities. Splined halfshafts transmit the drive to the wheels.

The drain plug on the three-quarter floating axle is on the rear of the housing, and that for the semi-floating axle on the front.

Operations other than those described either in this Chapter, or in Chapter 13, should in our opinion be entrusted to a Triumph dealer on account of the special tools and setting-up techniques required.

2 Halfshaft (three-quarter floating axle) - removal and installation

1 The halfshafts may be withdrawn without disturbing the differential gear. Read the whole of this Section before starting work, and remember to place a container beneath the axle casing to collect the spillage when the axle shaft is removed.
2 Jack-up the rear of the car and support it beneath the rear axle. Chock the front wheels and release the handbrake.
3 Remove the roadwheel and brake drum.
4 Detach the handbrake cable from the operating lever.
5 Detach the brake pipe(s) from the wheel cylinder. Plug the pipe(s) and fluid port(s) to prevent ingress of dirt.
6 Remove the four nuts, spring washers and bolts securing the backplate to the axle flange.
7 Temporarily install the brake drum so that the inside of the drum is facing towards you. Using a heavy soft faced hammer, or an ordinary hammer and a hardwood block, tap the drum whilst rotating it to draw out the halfshaft complete with the bearing (photo).
8 Installation is basically the reverse of the removal procedure (photo). Note the following points:

a) The halfshaft bearing should be packed with a lithium-based grease.
b) The holes in the oil seal housing plate, bearing retainer plate, backplate and oil catcher should all be aligned. Ensure that the oil drain hole in the oil seal housing aligns with the drain hole in the axle flange. **Note that the oil catcher plate trough should be below the halfshaft.**
c) Do not forget to bleed the brakes on completion. (see Chapter 8, Section 2).

3 Hub, halfshaft bearing and oil seals (three-quarter floating axle) - removal and installation

1 Remove the halfshaft as described in the previous Section. **Note:** *if only the bearing outer oil seal is to be renewed, proceed as described in paragraphs 1 to 6 of the previous Section, then refer to paragraphs 2, 3, 4 and 7 below.*
2 Remove the split pin, nut and washer, then draw off the hub using a suitable puller. Do not lose the Woodruff key.
3 Remove the oil catcher plate and backplate.
4 Remove the oil seal housing and gasket, and remove the seal from the housing.
5 Remove the inner oil seal from the axle tube.
6 Press, or drive, the bearing off the shaft. If the bearing is to be re-used, extreme care must be taken to ensure that loads are applied to the inner race only.
7 Installation is basically the reverse of the removal procedure, but the following points should be noted:
a) The inner seal should be lubricated with axle oil and pushed in with the lip towards the differential.
b) The outer seal should be lubricated with axle oil and fitted so that the lips will face the differential.
c) After packing the bearing with a lithium-based grease press it on so that there is at least 0.16 in (4.1 mm) clearance between the unshielded side towards the halfshaft collar. (see Fig. 8.3).
d) When installing the halfshaft, seal housing and new gasket, backplate and oil catcher, ensure that the drain holes align.
e) When installing the hub, tighten the securing nut to the specified torque. Either, wedge a bar between two flange studs and the ground to prevent the hub from rotating, or finally tighten the nut with the rear wheel fitted and the car lowered to the ground.
f) Do not forget to bleed the brakes on completion (see Chapter 8, Section 2).

4 Halfshaft (semi-floating axle) - removal and installation

1 Follow the procedure given in Section 2, paragraphs 1, 2 and 3.
2 Remove the four nuts and spring washers securing the bearing retainer plate to the axle tube, working through the holes drilled in the axle flange.
3 Follow the procedure given in Section 2, paragraph 7.
4 Installation is the reverse of the removal procedure. Apply a little lithium-based grease to the bearing.

5 Halfshaft bearing and oil seals (semi-floating axle) - removal and installation

1 Because of the special tools required to draw off the spacer ring, bearing and halfshaft bearing retainer oil seal, this job should be entrusted to a Triumph dealer. However, the halfshaft can be removed as described in the previous Section to minimise labour costs. (photo).
2 To remove the oil seal from the axle tube, first remove the halfshaft, as described in Section 4.
3 Prise out the seal from the axle tube.
4 Smear the replacement seal with axle oil, ensure that the axle bore is clean, then install the seal with the lip towards the differential (photo).
5 Install the halfshaft as described in Section 4.

2.7 Removing the halfshaft

2.8 Installing the halfshaft

5.1 The bearing and oil seal arrangement of the semi-floating axle

5.4 The axle tube oil seal installed

Fig. 8.1. Components of rear axle assembly - three-quarter floating axle

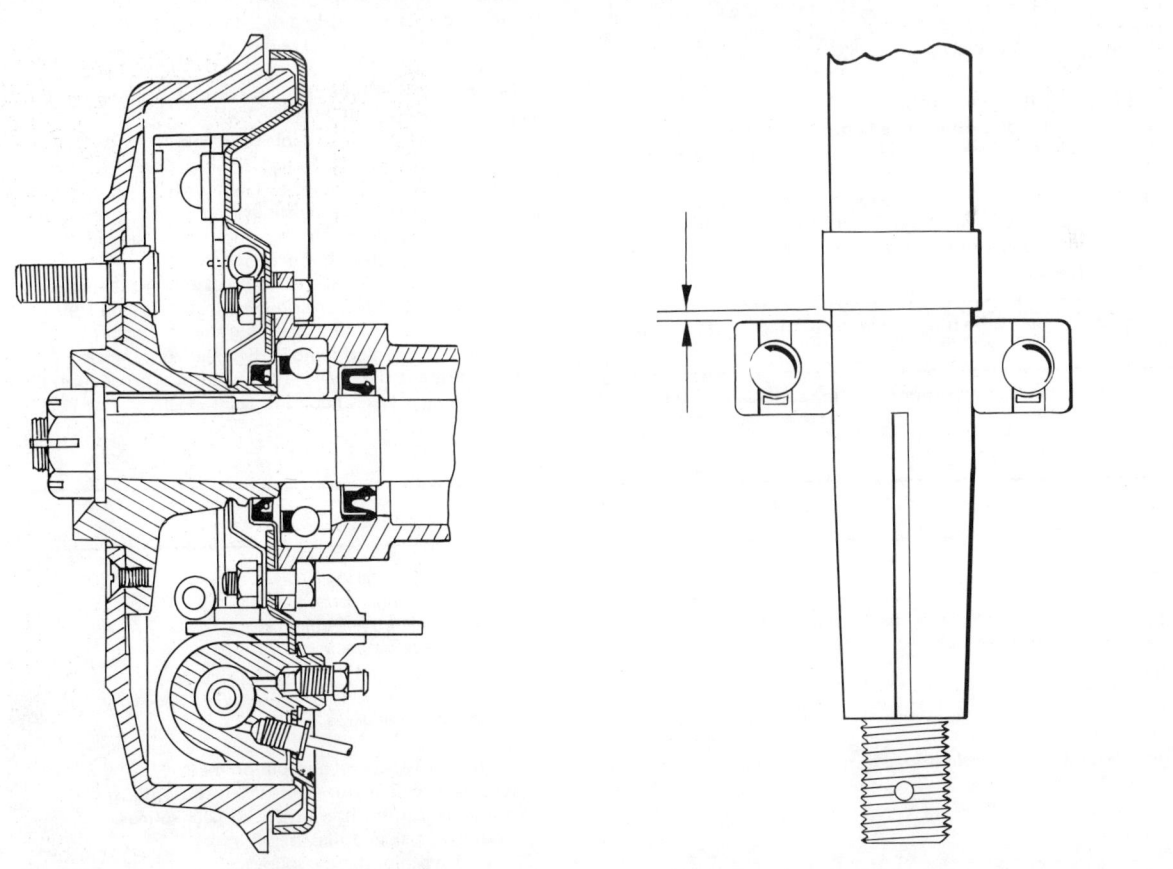

Fig. 8.2. Section view of rear hub - three-quarter floating axle

Fig. 8.3. Installed position of halfshaft bearing (see text) - three-quarter floating axle

6 Differential assembly (three-quarter floating axle) - removal and refitting

1 Remove the halfshafts as previously described.
2 Index mark the propeller shaft and axle flange to ensure correct alignment on assembly.
3 Remove the four locknuts and bolts from the drive flange, and move the propeller shaft aside.
4 Place a container beneath the differential unit to collect the oil when the unit is removed.
5 Remove the locknuts and washers securing the differential assembly to the axle housing. Carefully withdraw the differential assembly from the studs and place in an extremely clean, secure position.
6 At this point, we feel it is important to stress the fact that it is not recommended that the average owner, without access to the special tools necessary, attempts to dismantle the differential assembly any further. A new or exchange differential assembly should be purchased and fitted.
7 Clean the faces of the differential and the axle housing and fit a new joint gasket lightly smeared with a non-setting jointing compound.
8 Installation is the reverse of the removal procedure, tighten the drive flange nuts to the specified torque (see Chapter 7, Specifications) and on completion, add the correct quantity and type of axle oil.

7 Pinion oil seal (three-quarter floating axle) - removal and refitting

1 Jack-up the rear of the car and support it beneath the rear axle. Chock the front wheels for safety.
2 Index mark the propeller shaft and axle flanges to ensure correct alignment on assembly.
3 Remove the four locknuts and bolts from the drive flange, and move the propeller shaft aside.
4 Place a container beneath the pinion shaft to collect any oil spillage.
5 Remove the nut shield, then index mark the pinion shaft, nut and flange to ensure correct alignment on assembly.
6 Position two bolts in the pinion flange holes then use a bar to hold the flange while the pinion nut is loosened. Remove the nut and washer, counting the total number of turns before the nut disengages from the thread.
7 Pull off the pinion flange and prise out the oil seal.
8 Soak a new seal in engine oil for one hour, then ensure that the pinion housing is clean and push in the seal with the lip towards the differential.
9 Installation is now the reverse of the removal procedure, but ensure that all index marks are correctly aligned. The pinion nut must also be secured on the same number of turns as it was originally, or the pinion pre-load will be upset.

8 Rear axle (three-quarter floating and semi-floating types) - removal and refitting

1 Raise the rear of the car and support it using suitable stands or

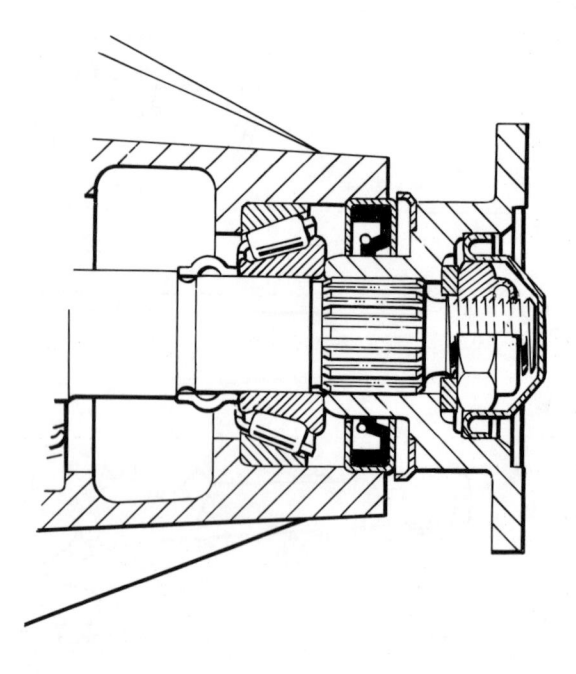

Fig. 8.4. Sectional view of pinion and oil seal

blocks beneath the bodyframe sidemembers. Chock the front wheels for safety.
2 Remove the roadwheels.
3 Index mark the propeller shaft to pinion flanges to ensure correct alignment on assembly, then detach it (four bolts and self-locking nuts).
4 Disconnect the forward end of the brake hoses (refer to Chapter 9 if necessary).
5 Detach the handbrake cables from the backplate levers (refer to Chapter 9 if necessary).
6 Detach the handbrake cables and compensator from the rear axle (refer to Chapter 9 if necessary).
7 Whilst supporting the axle weight with a jack, detach the rear dampers from the axle mountings (refer to Chapter 11 if necessary).
8 Remove the rear roadsprings and detach the radius rods from the axle (refer to Chapter 11 if necessary).
9 Remove the handbrake cable bracket from the axle.
10 Detach the rear suspension arms from the axle (refer to Chapter 11 if necessary), whilst still supporting the axle weight.
11 Lower the jack and remove the axle from the car.
12 Installation is the reverse of the removal procedure. On completion bleed the brakes (see Chapter 8, Section 2).

9 Fault diagnosis - rear axle

Symptom	Reason/s
Vibration	Worn halfshaft bearings. Loose drive flange bolts. Propeller shaft out-of-balance. Wheels require balancing.
Noise	Insufficient lubricant. Worn differential gears.
'Clunk' on acceleration or deceleration	Incorrect crownwheel and pinion mesh. Excessive backlash due to wear in differential gears. Worn halfshaft or differential side gear splines. Loose drive flange bolts. Worn drive pinion flange splines.
Oil leakage	Faulty pinion or halfshaft seals. Blocked axle housing breather.

Chapter 9 Braking system

Contents

Brake (and clutch) pedal box - removal, overhaul and installation ... 13	Front brake caliper - removal, overhaul and installation ... 5
Brake disc and disc shield - removal and installation ... 3	Front brake pads - removal and installation ... 4
Brake pipes and hoses - inspection, removal and installation ... 9	General description ... 1
Brake pressure reducing valve - description of operation ... 10	Handbrake cable assembly - adjustment ... 16
Brake pressure reducing valve - removal and installation ... 11	Handbrake cable assembly - removal and refitting ... 15
Brake servo - removal and installation ... 17	Handbrake lever assembly - removal and installation ... 14
Brake servo non-return valve and filter - removal and installation ... 18	Master cylinder - removal, overhaul and installation ... 12
Brake hydraulic system - bleeding ... 2	Rear brake backplate - removal and installation ... 8
Fault diagnosis - braking system ... 19	Rear brake shoes - removal and installation ... 6
	Rear wheel cylinder - removal, overhaul and installation ... 7

Specifications

System type ... Dual hydraulic, front discs and rear drums, with vacuum servo and rear brake pressure reducing valve. Cable operated handbrake to rear wheels.

Front brakes
Type ... Disc and caliper
Disc diameter ... 9.75 in (247.6 mm)
Disc thickness ... 0.375 in (9.5 mm)
Minimum permissible pad thickness ... 1/8 in (3 mm)
Pad material ... Ferodo 2441 - FG (DON 227 for 1975 USA models)

Rear brakes (four-speed gearbox cars, and automatic transmission models
Type ... Drum, self adjusting
Drum internal diameter * ... 7.995 to 8 in (203.20 to 203.33 mm)
Drum internal diameter, maximum ... 8.05 in (204. 47 mm)
Lining material ... DON 202 GG

Rear brakes (five-speed gearbox models) ... See Chapter 13

Servo
Boost ratio ... 2.3 : 1 (nominal)

Handbrake
Type ... Cable operated to rear wheels from floor mounted lever

Brake fluid specification ... SAE J1703d or DOT 3 (FMV SS116)

* Some models are fitted with 9 in (228.6 mm) rear brakes

Torque wrench settings

	lb f ft	Nm
Pedal mounting bracket bolt ...	21	28
Caliper attachment bolt ...	74	100
Disc to hub ...	32	43
Handbrake fulcrum compensator nut ...	21	28
Rear brake backplate bolt ...	14	19

1 General description

The brakes fitted to the front two wheels are of the rotating disc and static caliper type, with one caliper per disc, each caliper containing two piston operated friction pads, which on application of the footbrake pinch the disc rotating between them.
Application of the footbrake creates hydraulic pressure in the master cylinder and fluid from the cylinder travels via steel and flexible pipes to the cylinders in each half of the calipers, the fluid so pushing the pistons, to which are attached the friction pads, into contact with either side of each disc.

Two rubber seals are fitted to the operating cylinders. The outer seal prevents moisture and dirt from entering the cylinder. The inner seal, which is retained in a groove inside the cylinder, prevents fluid leakage and provides a running clearance for the pad irrespective of how worn it is, by moving it back a fraction when the brake pedal is released.

As the friction pad wears so the pistons move further out of the cylinders and the level of the fluid in the hydraulic reservoir drops; disc pad wear is thus taken up automatically and eliminates the need for periodic adjustments by the owner.

The rear brakes are of the self-adjusting, leading and trailing shoe type, with one brake cylinder per wheel for both shoes. A lever assembly is fitted between the two shoes of each brake unit and attached to this is a system of cables which in turn is connected to the handbrake lever. It is unusual to have to adjust the handbrake system as the efficiency of this system is largely dependent on the condition of the brake linings and the adjustment of the brake shoes. The handbrake can, however, be adjusted separately to the footbrake operated hydraulic system.

Connected to the brake pedal is a servo unit onto which is mounted the master cylinder. It increases the hydraulic line pressure whilst decreasing the driver's pedal effort. The master cylinder is a dual-type which means that the front and rear brakes operate independently of each other through a common pedal pushrod. Should one half of the system fail to operate, the brakes on the other half will still operate, although the system efficiency is reduced. A brake pressure regulator is fitted and is designed to prevent the rear wheels from locking. A separate description of this is given in the text.

Metric brake fittings are used, and it is essential that any replacements made are also metric. This also applies to the flares on the ends of metal brake pipes.

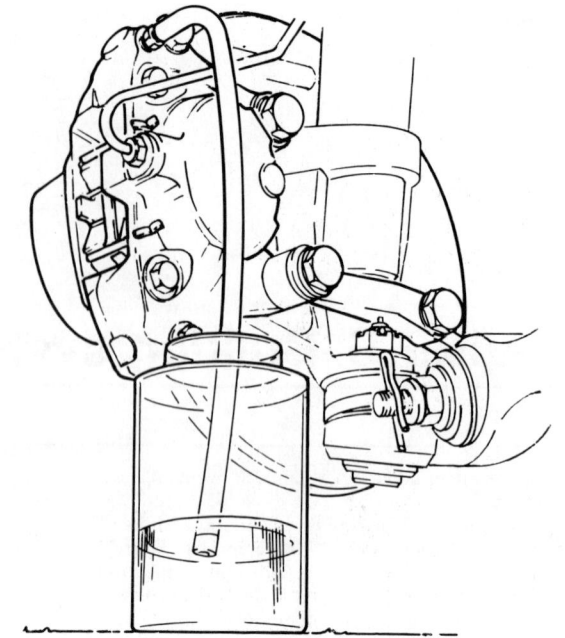

Fig. 9.1. Bleeding a front brake caliper

2 Brake hydraulic system - bleeding

1 Removal of all the air from the hydraulic system is essential to the correct working of the braking system, but before undertaking this, examine the fluid reservoir cap to ensure that the vent hole is clear; check the level of fluid and top-up if required.

2 Check all brake line unions and connections for possible seepage, and at the same time check the condition of the rubber hoses which may be perished.

3 If the condition of the wheel cylinders is in doubt check for signs of possible fluid leakage.

4 If there is any possibility of incorrect fluid having been put into the system drain all the fluid out and flush through with isopropyl-alcohol or methylated spirit. Renew all piston seals and cups since they will be affected and could possibly fail under pressure.

5 Gather together a clean jam jar, a 9 in (230 mm) length of tubing which fits tightly over the bleed nipple and a supply of the correct type of fluid. You will also require the services of an assistant.

6 Do not bleed the system with the servo in operation, i.e. with the engine running.

7 Disconnect the wires from the brake pressure failure switch, and unscrew the switch from the underside of the master cylinder.

8 To bleed the system, make sure that the car is on a level surface and release the handbrake. To gain better access to the bleed nipples it may be preferable to lift one roadwheel at a time. If this is the case, chock the roadwheels which remain on the ground.

9 Commence at the front wheel caliper furthest from the master cylinder, remove the rubber cap and wipe carefully around the bleed nipple (photo).

10 Fit one end of the rubber tube over the nipple and place the other end of the tube in a clean glass jar containing sufficient fluid to keep the end of the tube submerged during the operation.

11 Open the bleed valve with a spanner and quickly press down the brake pedal. After slowly releasing the pedal, pause for a moment to allow the fluid to recoup in the master cylinder and then depress again. This will force air from the system. Continue until no more air bubbles can be seen coming from the tube, then, with the pedal depressed, close the bleed nipple. At frequent intervals during the operation, make certain that the reservoir is kept topped-up, otherwise air will enter the system again. Always discard any fluid which is bled off since it probably contains air, dirt and moisture.

12 Now repeat the operation on the other front caliper in the same way. Do not forget to replace the rubber protective cap on the bleed nipple afterwards.

13 When the front brakes are completed, the rear ones should be bled. There is only one bleed nipple; this is on the right-hand backplate for left-hand drive cars, and on the left-hand backplate for right-hand drive cars.

14 Check the efficiency of the bleeding operation by firstly starting the engine and applying the handbrake. Apply a load of 25 lbf (111 N) to the brake pedal, when the pedal travel should not exceed 1.90 in (48.26 mm). If travel exceeds this figure, repeat the bleeding operation. If 'sponginess' is present, there is a likelihood that one of the seals in the master cylinder has failed.

15 On completion, refit the brake pressure failure switch. Tighten to the specified torque figure.

3 Brake disc and disc shield - removal and installation

1 Remove the front hub, as described in Chapter 11.

Brake disc

2 Remove the four disc-to-hub retaining bolts, and withdraw the disc.

3 Installation is the reverse of the removal procedure. Tighten the retaining bolts evenly to the specified torque.

Disc shield

4 Remove the three bolts and spring washers, and detach the disc shield from the vertical link.

5 Installation is the reverse of the removal procedure.

4 Front brake pads - removal and installation

1 Apply the handbrake, then jack-up the front of the car and remove the roadwheel.

2 Depress the retaining spring and withdraw the split pins (note that one ear of each pin needs to be straightened) (photo).

3 Noting their installed positions, lift the pads and shims out of the caliper recesses.

4 Carefully press the caliper pistons back into their bores. **Note:** This will cause the reservoir fluid level to rise, so prevent it from overflowing by loosening the caliper bleed nipple as the piston is being moved, then close it when movement is complete.

5 Installation is now the reverse of the removal procedure, but the following points should be noted:

   a) Ensure that the location in the caliper is free from dust and dirt.
   b) If the shims are corroded, obtain new ones; they should be inserted with the large cut-out uppermost.

c) Use a new pad retaining spring and split pins. Fold back one leg of each split pin.
d) Depress the pedal several times on completion to correctly locate the pads.
e) Check the reservoir fluid level on completion.

5 Front brake caliper - removal, overhaul and installation

1 Apply the handbrake, then jack-up the front of the car and remove the roadwheel.
2 Disconnect the brake union from the caliper. Seal the fluid connections to prevent ingress of dirt.
3 Remove the two bolts and spring washers retaining the steering arm (the front one is also a caliper retaining bolt), and the remaining bolt and spring washer from the caliper upper mounting. Withdraw the caliper.
4 Remove the brake pads and shims (refer to Section 4, if necessary).
5 Using a low pressure air line at the fluid port, remove the pistons. Ensure that they are not mixed-up from side-to-side; if one is seized, the whole caliper must be renewed.
6 Using a blunt screwdriver, carefully prise out the wiper seal retainers. Do not scratch any metal parts.
7 Extract the wiper dust seals and fluid seals from the bores.
8 Thoroughly clean the metal parts using clean brake fluid, methylated spirit or isopropyl alcohol. Inspect the caliper bores and pistons for wear, scoring and corrosion, renewing parts as necessary. **Do not attempt to separate the two halves of the caliper.**
9 Position new fluid seals in the caliper bores, ensuring that they are properly located. They will stand proud of the bore at the edge furthest away from the mouth of the bore.
10 Lubricate the bores with new brake fluid and squarely insert the pistons in their original positions. Leave about 5/16 in (8 mm) of each piston projecting.
11 Fit a new wiper seal into each seal retainer, and slide an assembly into each bore.
12 Press the seals and pistons fully home.
13 Installation of the caliper is now the reverse of the removal procedure. Install and lightly tighten all three bolts and washers before they are finally tightened. Bleed the brakes on completion. (see Section 2).

2.9 Front brake bleed nipple

2.13 Rear brake bleed nipple

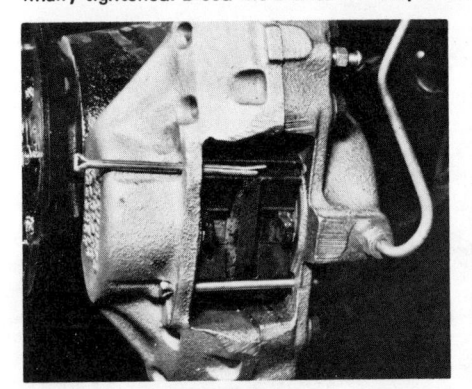

4.2 Withdrawing the split pins

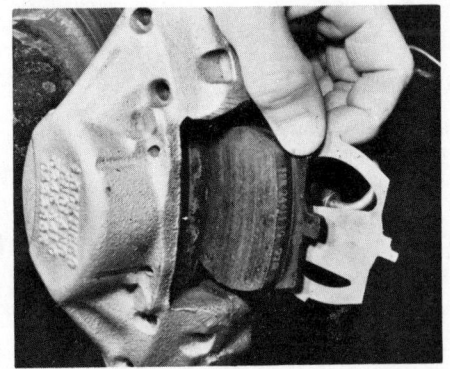

4.5 Installing a pad and shim

6.4 Layout of rear brake (9 in drum type shown)

6 Rear brake shoes - removal and installation

1 Jack-up the rear of the car and support it on stands or blocks beneath the rear axle. Chock the front wheel for safety.
2 Remove the roadwheel and release the handbrake.
3 Remove the two countersunk head screws and pull off the brake drum. Note: If the drum is difficult to remove, remove the rubber plug on the rear of the backplate and insert a small screwdriver to engage in the slotted hole in the small adjusting lever. Press downwards to contact the brake shoes.
4 Note the installed position of the shoes and springs. Remove the steady pin cups and springs; extract the pins from behind the backplate (photo).
5 Remove the shoe ends from the slots in the wheel cylinder piston heads.
6 Detach the pull-off springs and the cross-lever tension spring, to remove the brake shoes. Retain the pistons in the cylinders using a wire clip or strong rubber band.

7 Blow, or brush out, any dust etc., from the brake drum and backplate. Take care that the dust is not inhaled since it can be very harmful to the lungs. If there are signs of fluid leakage from the wheel cylinder or oil leakage from the halfshaft or bearing these points should be attended to at this stage - see Section 7 and Chapter 8. Renew brake shoes where the linings have worn down to the rivet heads or are likely to wear down to this point during the next few thousand miles of motoring.
8 When installing the shoes, insert the tension spring hook in the cross-lever and engage the other end in the web of the leading shoe. **Note:** The springs are handed for the left and right-hand brake.
9 Ease the shoe and cross-lever towards the backplate; engage the toe in the piston slot and the heel in the abutment slot.
10 Hold the cross-member and shoe against the backplate, and install the steady pin, spring and cup.
11 Hook the pull-off springs into the holes in the shoe webs. Note that they run on the backplate side of the shoes.
12 Pull the trailing shoe into position with the heel in the piston slot

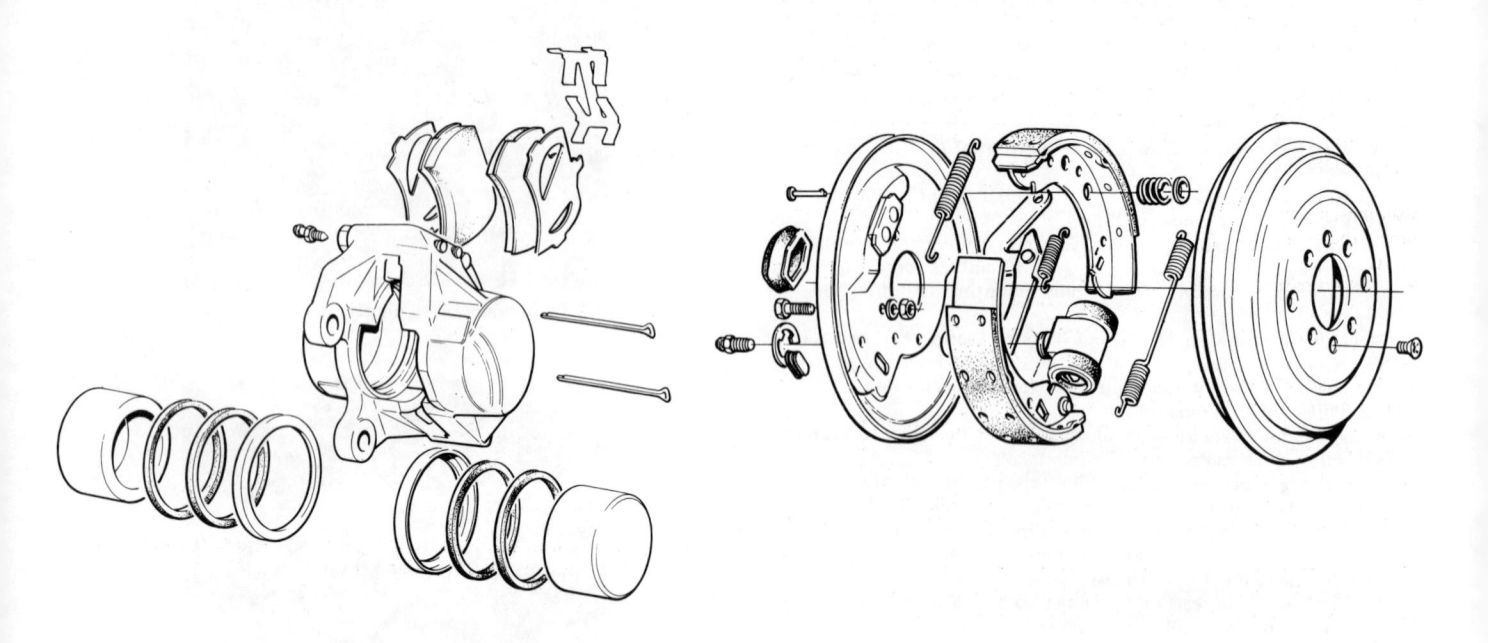

Fig. 9.2. Exploded view of brake caliper

Fig. 9.3. Exploded view of rear brake assembly

and the toe in the abutment slot. Ensure that the cross-lever cut-out engages in the adjuster plate slot.
13 Fit the second steady pin, spring and cup.
14 Provided that care is exercised to prevent the pistons from being forced out, the automatic adjuster action can be checked by gently pressing the brake pedal. As the shoes expand, the ratchet will operate; this can be cancelled by raising the ratchet plate to separate the ratchet teeth and allowing the shoes to retract under spring action.
15 Install the brake drum and roadwheel, and depress the brake pedal several times to centralize and adjust the brakes.
16 If the brake action is unsatisfactory, test run the car and apply moderately high pedal effort during several test runs at a speed of around 20 mph (33 kph).

7 Rear wheel cylinder - removal, overhaul and installation

1 Remove the brake shoes, as described in the previous Section.
2 Detach the handbrake cable from the lever behind the backplate.
3 Disconnect the pipe union and remove the bleed screw *or* the feed and transfer pipe unions from the rear of the wheel cylinder, as applicable.
4 Remove the spring clip, and take off the wheel cylinder and gasket.
5 Remove the rubber boots and take out the pistons; also remove the spring from the bore.
6 Remove the seal from each piston.
7 Thoroughly clean the metal parts using clean brake fluid, methylated spirit or isopropyl-alcohol. Inspect the pistons and cylinder bores for wear, scoring and corrosion: renewing parts as necessary.
8 Smear the cylinder bore with new brake fluid, then fit a new seal to the large groove of each piston, so that the lip of the seal faces away from the slot.
9 Locate the rubber boots into the smaller grooves of the piston, then insert the pistons into the bore with the spring between them.
10 Installation is the reverse of the removal procedure, but bleed the brakes before test-running the car.

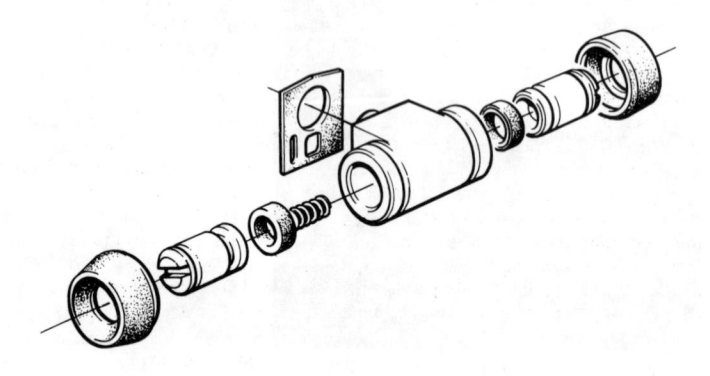

Fig. 9.4. Rear wheel cylinder - exploded view

8 Rear brake backplate - removal and installation

1 Remove the rear hub, as described in Chapter 8.
2 Refer to the procedure given in paragraphs 2 and 3 of the previous Section.
3 Remove the four nuts, spring washers and bolts securing the backplate. Withdraw the backplate and deflector plate.
4 Installation is the reverse of the removal procedures, but bleed the brakes on completion (see Section 2).

Chapter 9/Braking system

9 Brake pipes and hoses - inspection, removal and installation

1 Inspection of the braking system hydraulic pipes and flexible hoses is part of the maintenance schedule. Carefully check the rigid pipes along the rear axle, underbody and in the engine compartment, not forgetting the short runs to the front wheel calipers. Any pipes showing signs of corrosion or damage should be renewed, following which it will be necessary to bleed the system as described previously.

2 Carefully inspect the flexible hoses. There is one flexible pipe to the rear axle, and one on each suspension arm. Look for any signs of swellings, cracking and/or chafing. If any of these maladies is evident, renew the hoses straight away. Remember that your life could depend on it.

3 Where flexible hoses are to be renewed, unscrew the metal pipe union nut from its connection to the hose, and then holding the hexagon on the hose with a spanner, unscrew the attachment nut and washer.

4 The body end of the flexible hose can now be withdrawn from the chassis mounting bracket and will be quite free.

5 Disconnect the flexible hydraulic hose at the backplate by unscrewing it from the brake cylinder. **Note:** When releasing the hose from the backplate, the chassis end must always be freed first.

6 Installation is the reverse of the removal procedure, following which it will be necessary to bleed the system, as described in Section 2.

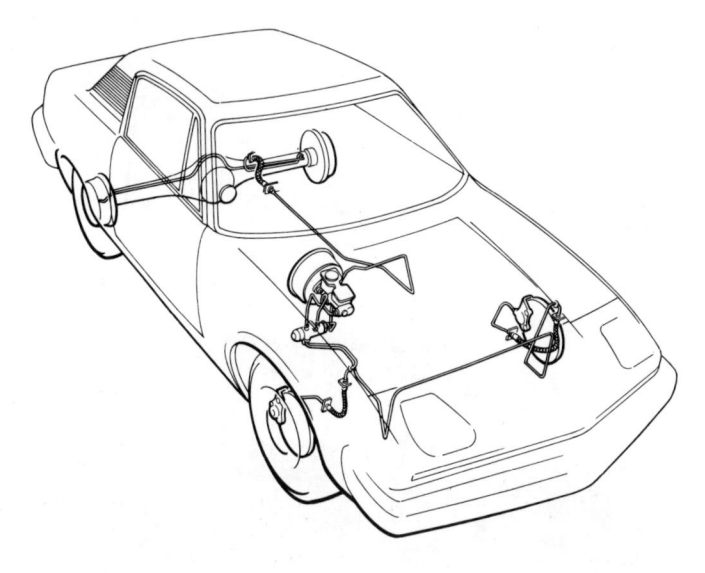

Fig. 9.5. Brake hose and pipe runs - right-hand drive (left-hand drive models are the opposite to this)

10 Brake pressure reducing valve - description of operation

See Fig. 9.6.

The reducing valve is installed in the brake circuit between the master cylinder, and the front and rear brakes. Its purpose is to limit the pressure applied to the rear brakes relative to the front brake pressure, thus minimising the possibility of rear wheel locking. In the event of a failure in the front circuit the cut-off pressure is increased and the pressure reduction ratio is alerted.

Fluid from the master cylinder primary chamber is fed into the pressure reducing valve at port 'A' and out to the front brakes via ports 'C' and 'D'. The master cylinder secondary chamber feeds into port 'B', through the internal passages in the valve plunger, past the metering valve and then to the rear brakes via port 'E'. The large spring 'S' biases the valve plunger to the left. Hydraulic pressure therefore acts on the annular area (a1 -a2) forcing the plunger to the left while the force acting on the area 'a1' and annular area (a4 - a3) tries to move the plunger to the right where it is opposed by the spring. When the net force acting to the right overcomes the pre-load provided by the spring, the plunger assembly shifts to the right and closes the metering valve 'F'. Thus, pressure at the rear outlet port 'E' falls relative to the input pressure. As pressure is increased at ports 'A' and 'B', the plunger is forced to the left, opening the metering valve 'F' and allowing a metered quantity of fluid to be fed to the rear brakes. The resultant increase in pressure acting on area 'a1' causes the plunger to again shift to the right, thus closing the metering valve. This procedure continues until there is no further increase in applied pressure from the master cylinder.

The resultant pressure at outlet 'E' is reduced after cut-off in proportion to the area 'a2' and the difference between the two annular areas (a1 - a2) and (a4 - a3). This cut-off pressure is equal to the pre-load in the spring 'S' divided by the combined areas 'a2' and (a4 - a3). Should the front brake circuit fail, there will be no pressure acting on annular area (a4 - a3) so that the net force tending to move the plunger to the right will equal the product of the input pressure and area 'a2'. Thus, as the value of the pre-load spring is unchanged, the cut-off pressure will increase considerably.

As the annular area (a4 - a3) is now redundant, the reduction ratio after cut-off changes to a value proportional to areas a1 and (a1 - a2). Should the rear brake circuit fail, the valve is inoperative and pressure is fed to the front brake system only.

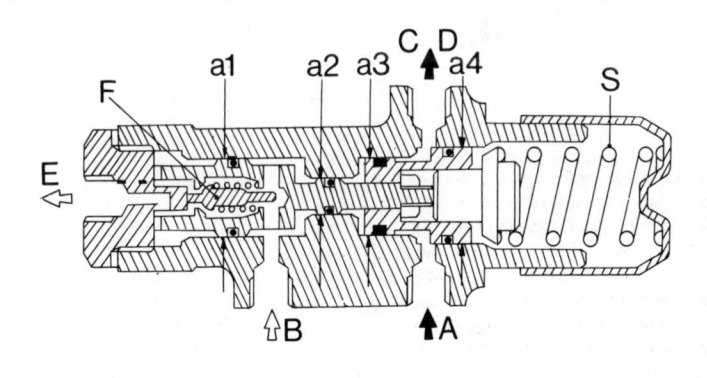

Fig. 9.6. Sectional view of brake pressure reducing valve

11 Brake pressure reducing valve - removal and installation

1 Loosen the brake pipe unions at the master cylinder and remove the two inlet pipes from the top of the reducing valve.

2 Remove the rear brake pipe from the end plug of the reducing valve.

3 Remove the two front brake pipes from the underside of the reducing valve. If necessary, detach the pipes from the clip on the inner wheel arch for improved accessibility.

4 Remove the nut, flat washer, spring washer and bolt, and detach the reducing valve and bracket from the suspension turret.

5 Installation is the reverse of the removal procedure, but bleed the brakes on completion.

12 Master cylinder - removal, overhaul and installation

1 Noting their installed position, detach the fluid lines from the master cylinder. Plug the pipes and ports to prevent ingress of dust, dirt, etc (photo).

2 Detach the leads from the brake pressure failure switch, then remove the master cylinder retaining nuts and washers. Withdraw the master cylinder from the servo.

3 Clean all dirt from the external surfaces of the master cylinder.

4 Unscrew and remove the brake pressure failure switch.

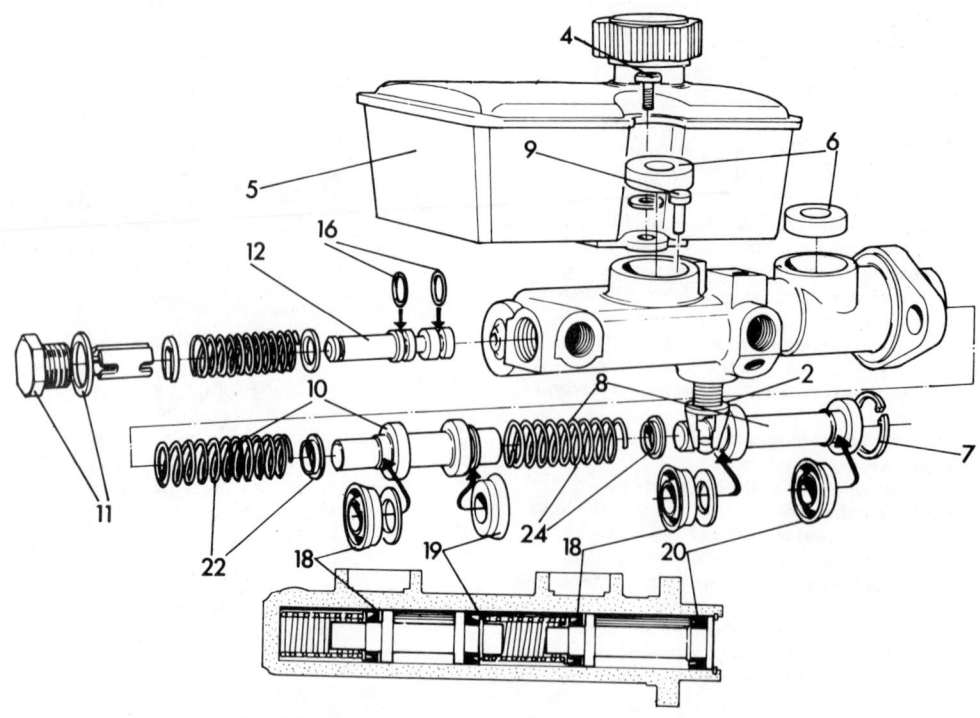

Fig. 9.7. Exploded view of the brake master cylinder

- 2 Pressure differential switch
- 4 Reservoir screws
- 5 Reservoir
- 6 Seals
- 7 Circlip
- 8 Primary piston and spring
- 9 Stop pin
- 10 Secondary piston and spring
- 11 End plug and washer
- 12 Pressure differential piston
- 16 'O' ring seals
- 18 Piston seals
- 19 Secondary piston seal
- 20 Primary piston seal
- 22 Secondary piston return spring and cup
- 24 Primary piston return spring and cup

5 Secure the master cylinder in a vice fitted with jaw protectors and unscrew and remove the two shouldered screws which secure the fluid reservoir. Withdraw the reservoir from the master cylinder body.
6 Extract the two reservoir sealing rings.
7 Extract the circlip from the end of the cylinder bore.
8 Withdraw the primary piston and return spring.
9 Using a copper or brass rod, insert it into the cylinder and depress the secondary piston so that the stop pin can be extracted from the secondary piston fluid feed port.
10 Withdraw the secondary piston and spring, either by shaking them out or by applying air pressure to the secondary outlet port.
11 Unscrew and remove the end plug and washer but do not prise off the distant piece from the end plug spigot.
12 Extract the pressure differential assembly, either by shaking it from the body or by applying air pressure to the secondary outlet port.
13 Wash all components in methylated spirit, isopropyl-alcohol or clean hydraulic fluid, and examine the surfaces of the pistons and cylinder bore for scoring or 'bright' wear areas. Where these are evident renew the complete master cylinder.
14 If the components are in good condition, discard all seals and obtain the appropriate repair kit.
15 Fit new 'O' ring seals to the pressure warning piston.
16 Install a shim washer to the primary and secondary pistons.
17 Using the fingers only, manipulate the two identical piston seals into place on the primary and secondary pistons (lip facing away from washers).
18 Of the two remaining seals contained in the repair kit, fit the thinner one to the secondary piston (lip towards primary spring seat). Fit the thicker one to the primary piston (lip towards piston seal).
19 Fit the shorter return spring and cup to the secondary piston, dip the assembly into clean hydraulic fluid and insert it into the master cylinder body. Take care not to trap the seal lip.
20 Depress the secondary piston and insert the stop pin after the head of the piston has been seen to pass the feed port.
21 Install the return spring and cup to the primary piston, dip the assembly into clean hydraulic fluid and insert it into the master cylinder body. Take care not to trap the seal lips. Refit the retaining circlip.
22 Insert the pressure differential piston into its bore and then fit a new sealing washer to the end plug and screw it in, tightening it to the specified torque.
23 Install new reservoir seals (round edge first).
24 Install the reservoir.
25 Fit the brake pressure failure switch.
26 Refitting the master cylinder is the reverse of the removal procedure, but on completion bleed the hydraulic system, as described in Section 2.

13 Brake (and clutch) pedal box - removal, overhaul and installation

1 Detach the leads from the stoplight switch, and then remove the speedometer cable from its retaining clip.
2 Remove the clevis pin from each master cylinder-to-pedal pushrod.
3 Remove the four nuts and spring washers securing the brake servo to the pedal box, and the two nuts, bolts and spring washers securing the clutch master cylinder.
4 Remove the cleat which secures the harness to the stabiliser bar, and the nut and bolt securing the stabiliser bar to the facia rail.
5 Remove the three retaining bolts and spring washers, and withdraw the pedal box.
6 Remove the stoplight switch, then remove one circlip so that the pivot rod can be taken out.
7 Remove the pedals and springs.
8 Where necessary, renew the pedal pivot bushes, and the pedal rubbers, then reassemble, using other new parts as necessary.
9 Installation of the pedal box is the reverse of the removal procedure.

14 Handbrake lever assembly - removal and installation

1 Raise one side of the car for access to the underside of the transmission tunnel. Chock the wheels which are on the ground and release the handbrake.
2 Pull back the rubber gaiter and release the cable locknut, noting its installed position (photo).
3 Unscrew the cable from the operating rod.

4 From inside the car, remove the centre console, as described in Chapter 12 (the car can be lowered to the ground if preferred).
5 Detach the handbrake warning switch connector, then remove the four bolts and spring washers securing the lever assembly to the transmission tunnel.
6 Withdraw the lever assembly, the lower plate and the gaiter.
7 Installation is the reverse of the removal procedure, but on completion check the handbrake adjustment, as described in paragraph 7 onwards, of the following Section.

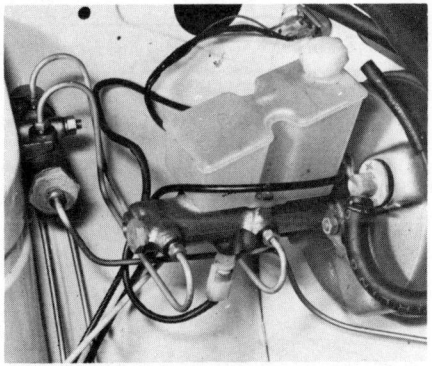

15 Handbrake cable assembly - removal and refitting

1 Proceed as described in paragraphs 1 to 3 of Section 14.
2 Slacken the nut at the forward abutment bracket, releasing the cable (see photo 14.2).
3 Remove the split pin, washer and clevis pin at each rear backplate operating lever.
4 Remove the centre nut, bolt and spring washer clamping the compensating levers together (photo).
5 Loosen the compensating levers pivot nut and bolt, until the trunnion can be freed from between them.
6 If a new cable is to be fitted, proceed as follows:

   a) Feed the cable through the guide bracket on the axle, and through the heelboard.
   b) Locate the cable in the forward abutment bracket, but do not tighten the nut.
   c) Fit the adjuster, with nut, fully into the handbrake lever connection. Tighten the nut, and fit the rubber boot.
   d) Secure the cable at the forward abutment.
   e) Position the rubber sleeves on the cable where it passes through the guide brackets.
   f) Position the compensator lever on the axle towards the left-hand wheel, offset about ½ in (12 mm) from the vertical.
   g) Pull the left-hand lever out from the backplate, and screw the cable adjuster into the fork ends using the fingers only, until slack is removed from the cable.
   h) Repeat (g) at the right-hand brake.
   j) Apply the handbrake lever 5 times using a two handed pull (about 100 lbf or 445N), to settle the cable.
   k) Repeat operations (g) and (h).
   l) Tighten the locknuts at the fork ends, and grease all pivoting and sliding parts.

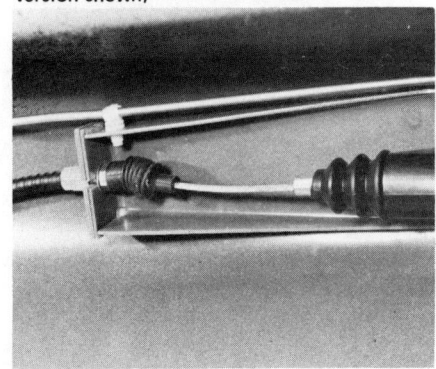

7 If the original cable is being refitted, reverse the procedure given in paragraphs 1 to 5.
8 Adjust the cable as described in Section 16.

16 Handbrake cable assembly - adjustment

1 Support the rear axle on stands.
2 Release the handbrake.
3 Disconnect the fork ends of the cable at the backplates by removing the clevis pin.
4 Use light finger pressure to push the brake levers inboard (i.e away from the backplate) to keep the levers in contact with the shoes.
5 Place the compensator lever on the axle towards the left-hand wheel, offset about ½ in (12 mm) from the vertical, and adjust the cable fork ends to permit entry of the clevis pins.
6 Fit the clevis pins, and check that the brakes are not binding.
7 Apply about 25 lbf (110 N) to the handbrake lever, when a lever travel of between five and seven notches should be produced.

12.1 Brake master cylinder location; the pressure reducing valve is attached to the suspension strut housing (right-hand drive version shown)

14.2 Handbrake cable at forward end

15.3 Handbrake operating lever on backplate (wheel and brake drum removed in this photograph)

15.4 Cable trunnion assembly

17 Brake servo - removal and installation

1 Remove the brake master cylinder, as described in Section 12.
2 Disconnect the vacuum hose from the non-return-valve then remove the servo pushrod-to-brake pedal clevis pin.
3 Remove the four retaining nuts and spring washers, and withdraw the servo.
4 Installation is the reverse of the removal procedure, but bleed the brakes on completion (see Section 2).

18 Brake servo non-return valve and filter - removal and installation

Non-return-valve
1   With the engine stopped, depress the brake pedal several times to release the servo vacuum.
2   Release the hose clip at the non-return-valve elbow on the servo, and disconnect the hose.
3   Withdraw the non-return-valve.
4   Installation is the reverse of the removal procedure; where necessary, renew the sealing rubber.

Filter
5   Remove the brake stoplight switch (refer to Chapter 10, if necessary).
6   Remove the split pin, washer and clevis pin securing the servo pad to the brake pedal.
7   Remove the pushrod rubber boot and withdraw the filter.
8   Installation is the reverse of the removal procedure.

Fig. 9.8. Handbrake lever assembly

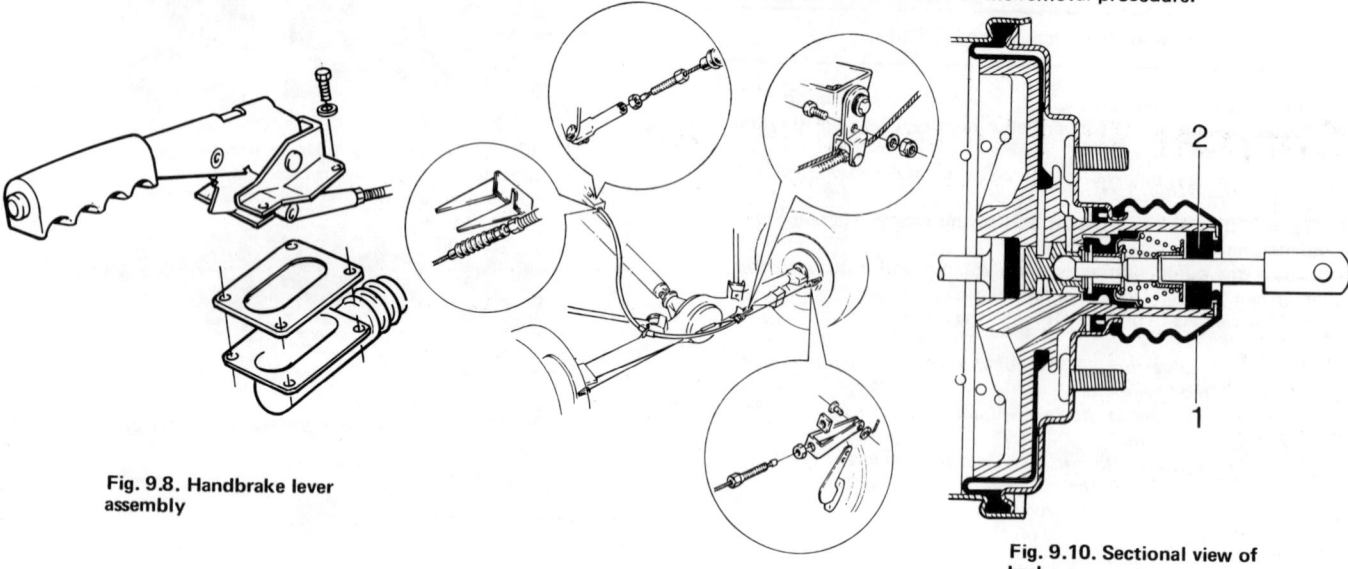

Fig. 9.9. Handbrake cable attachment points

Fig. 9.10. Sectional view of brake servo

1   Rubber boot
2   Filter

19 Fault diagnosis - braking system

Symptom	Reason/s
Pedal travels almost to floorboards before brakes operate	Brake fluid level too low.
	Caliper leaking.
	Master cylinder leaking (bubbles in master cylinder fluid).
	Brake flexible hose leaking.
	Brake line fractured.
	Brake system unions loose.
	Rear automatic adjusters seized.
Brake pedal feels springy	New linings not yet bedded in.
	Brake discs or drums badly worn or cracked.
	Master cylinder securing nuts loose.
Brake pedal feels spongy and soggy	Caliper or wheel cylinder leaking.
	Master cylinder leaking (bubbles in master cylinder reservoir).
	Brake pipe line or flexible hose leaking.
	Unions in brake system loose.
	Air in hydraulic system.
Excessive effort required to brake car	Pad or shoe linings badly worn.
	New pads or shoes recently fitted - not yet bedded-in.
	Harder linings fitted than standard causing increase in pedal pressure.
	Linings and brake drums contaminated with oil, grease or hydraulic fluid.
	Servo unit inoperative or faulty.
	One half of dual brake system inoperative.
Brakes uneven and pulling to one side	Linings and discs or drums contaminated with oil grease or hydraulic fluid.
	Tyre pressures unequal.
	Radial ply tyres fitted at one end of the car only.
	Brake caliper loose.
	Brake pads or shoes fitted incorrectly.
	Different type of linings fitted at each wheel.
	Anchorages for front suspension or rear suspension loose.
	Brake discs or drums badly worn, cracked or distorted.
Brakes tend to bind, drag or lock-on	Air in hydraulic system.
	Wheel cylinders seized.
	Handbrake cables too tight.

Chapter 10 Electrical system

For modifications, and information applicable to later models, see Supplement at end of manual

Contents

Alternator - brush renewal ... 9	Instrument illumination lamps ... 27
Alternator - general description, maintenance and precautions ... 6	Instrument panel - removal and installation ... 34
	Instrument warning lights ... 28
Alternator - removal, installation and drivebelt adjustment ... 8	Marker lamps ... 23
Alternator - testing in position in the car ... 7	Number plate lamp ... 25
Battery - charging ... 5	Panel instrument and sender units - removal and installation ... 35
Battery - electrolyte replenishment ... 4	Rear lamp assembly ... 24
Battery - maintenance and inspection ... 3	Relays and flasher units ... 29
Battery - removal and installation ... 2	Roof lamp ... 26
Cigarette lighter - removal and installation ... 33	Starter motor - general description ... 10
Fault diagnosis - electrical system ... 41	Starter motor - overhaul ... 12
Front flasher repeater lamp ... 22	Starter motor - removal and installation ... 11
Front parking/flasher lamps ... 21	Steering column switches - removal and installation ... 31
Fuses ... 30	Switches - removal and installation ... 32
General description ... 1	Windscreen washers ... 36
Headlamps - general description ... 13	Windscreen wiper arms and blades - removal and installation ... 37
Headlamp actuator - removal and installation ... 17	Windscreen wiper motor and drive assembly - dismantling, overhaul and reassembly ... 39
Headlamp actuator - overhaul ... 18	
Headlamp assembly - adjustment ... 19	Windscreen wiper motor and drive assembly - removal and installation ... 38
Headlamp assembly - overhaul ... 16	
Headlamp assembly - removal and installation ... 15	Windscreen wiper motor linkage - removal and installation ... 40
Headlamp beam - alignment ... 20	
Headlamp light unit - removal, installation and lift mechanism lubrication ... 14	

Specifications

System type ... 12 volt, negative earth

Battery
USA models ... 50 amp hr. at 20 hr. rate
Other models ... 40 amp hr. at 20 hr. rate

Alternator
Type:
    USA models ... Lucas 20 ACR or 25 ACR
    Other models ... Lucas 17 ACR

	17 ACR	20 ACR	25 ACR
Brush length, new in (mm)	0.5 (12.7)	0.5 (12.7)	0.5 (12.7)
Brush length, minimum in (mm)	0.2 (5) protrusion	0.3 (8) protrusion	0.3 (8) protrusion
Brush spring pressure oz f (kg f)	9 to 13 (255 to 370)	9 to 13 (255 to 370)	9 to 13 (255 to 370)
Stator windings	Star	Delta	Delta
Field winding resistance, ohms	$3.2 \pm 5\%$	3.5	3.0 to 3.5
Regulator type	14TR	14TR	14TR
Nominal output, hot:			
Alternator speed, rpm	6000	6000	6000
Engine speed, rpm	2540	2540	—
Control voltage	14	14	14
Current	36	66	65
Drivebelt tension in (mm)	0.75 to 1.0 (20 to 25)	0.75 to 1.0 (20 to 25)	0.75 to 1.0 (20 to 25)

Starter motor
Type	Lucas 2M100PE
Minimum commutator thickness	0.14 in (3.56 mm)
Brush length new	0.71 in (18 mm)
Brush length, minimum	0.375 in (9.5 mm) protrusion
Brush spring pressure	36 oz f (1 kg f)
Shaft endfloat, maximum	0.01 in (0.25 mm)
Solenoid pull-in winding resistance	0.25 to 0.27 ohms
Solenoid hold-in winding resistance	0.76 to 0.80 ohms

Windscreen wiper motor
Type	Lucas 16W
Running current, unloaded:	
Low speed	1.5 amp (after 60 seconds)
High speed	2.0 amp (after 60 seconds)
Armature endfloat	0.002 to 0.008 in (0.05 to 0.20 mm)
Brush length, new	0.380 in (9.7 mm)
Brush length, minimum	0.180 in (4.8 mm) or when narrow section is worn away on high speed brush
Brush spring pressure	5 to 7 oz f (140 to 200 g f)

Windscreen washer pump
Type	Lucas 103J

Headlamp actuator
Type	Lucas 15W
Running current, unloaded	1.5 amp (after 60 seconds)
Armature endfloat	0.002 to 0.008 in (0.05 to 0.20 mm)
Brush length, new	0.25 in (6.4 mm)
Brush length, minimum	0.187 in (4.75 mm)
Brush spring pressure	5 to 7 oz f (140 to 200 g f)

Relays
Type	Lucas 26RA series

Bulbs
	Wattage
Headlamps:	
LH dip	75/60 (Sealed beam)
RH dip:	
Normal	60/50 (Sealed beam)
USA	50/40 (Sealed beam)
France	45/40
Front parking lamp	4
Front indicator lamp	21
Front marker lamp	3
Rear marker lamp (where applicable)	3
Tail lamp	5
Stop lamp	21
Rear indicator lamp	21
Reverse lamp	21
Number plate lamp	5
Roof lamp	6
Instrument panel illumination	2.2
Panel warning lights	1.2
Hazard warning light	1.2

Fuses
1975 USA models	3 x 5 amp
	1 x 35 amp
Other models	3 x 50 amp
	2 x 35 amp
	1 x 15 amp
Location	At rear of glove compartment
Circuits protected	See wiring diagrams

Torque wrench settings
	lb f ft	Nm
Headlamp hinge arm to pivot bracket	30	40
Alternator mounting bolts	20	27
Alternator pulley nut	30	41
Starter motor securing bolts	34	46

Chapter 10/Electrical system

1 General description

The electrical system is of the 12-volt type and the major components comprise a 12-volt battery of which the negative terminal is earthed, a Lucas alternator which is fitted to the front left-hand side of the engine and is driven from the pulley on the front of the crankshaft, and a starter motor which is mounted on the rear left-hand side of the engine.

The battery supplies current for the ignition, lighting and other electrical circuits, and provides a reserve of electricity when the current consumed by the electrical equipment exceeds that being produced by the alternator. Normally an alternator is able to meet any demand placed upon it.

A novel feature of the electrical system is the retractable headlights. Further information on these will be found in Section 13.

When fitting electrical accessories to cars with a negative earth system, it is important, if they contain silicone diodes or transistors, that they are connected correctly, otherwise serious damage may result to the component concerned. Items such as radios, tape recorders, electronic tachometers, automatic dipping, parking lamps and anti-dazzle mirrors should all be checked for correct polarity.

It is important that the battery leads are always disconnected if the battery is to be boost charged, or if any body or mechanical repairs are to be carried out using electric arc welding equipment, otherwise, serious damage can be caused to the more delicate instruments, specially those containing semi-conductors.

2 Battery - removal and installation

1 Detach the battery leads from the terminal lugs.
2 Loosen the two nuts and swing the battery retainer downwards.
3 Lift out the battery.
4 Before installing, ensure that the retainer is in the downward position, and that the hooks are engaged in the body aperture. Check that leads and loom wires are not trapped, and that the rubber mat is correctly positioned.
5 Install the battery, tighten the retaining nuts, then fit the terminal leads. Do not hammer them on or they may jam.
6 Finally, smear the battery terminals and lead-ends with a little petroleum jelly.

3 Battery - maintenance and inspection

1 Normal weekly battery maintenance consists of checking the electrolyte level of each cell to ensure that the separators are covered ¼ in (6 mm) of electrolyte. If the level has fallen, top-up the battery using distilled water only. Do not overfill. If a battery is overfilled or any electrolyte spilled, immediately wipe away the excess as electrolyte attacks and corrodes any metal it comes into contact with very rapidly.
2 If the battery has the 'Auto-fill' device as fitted on original production of the car, a special topping-up sequence is required. The white balls in the 'Auto-fill' battery are part of the automatic topping-up device which ensures correct electrolyte level. The vent chamber should remain in position at all times except when topping-up or taking specific gravity readings. If the electrolyte level in any of the cells is below the bottom of the filling tube top-up as follows:

   a) Lift off the vent chamber cover.
   b) With the battery level, pour distilled water into the trough until all the filling tubes and trough are full.
   c) Immediately replace the cover to allow the water in the trough and tubes to flow into the cells. Each cell will automatically receive the correct amount of water.

3 As well as keeping the terminals clean and covered with petroleum jelly, the top of the battery, and especially the top of the cells, should be kept clean and dry. This helps prevent corrosion and ensures that the battery does not become partially discharged by leakage through dampness and dirt.
4 Once every three months remove the battery and inspect the battery securing bolts, the battery clamp plate, tray and battery leads for corrosion (white fluffy deposits on the metal which are brittle to touch). If any corrosion is found, clean off the deposits with ammonia or a solution of bicarbonate of soda and warm water, and paint over the clean metal with anti-rust, anti-acid paint.
5 At the same time inspect the battery case for cracks. If a crack is found, clean and plug it with one of the proprietary compounds marketed by such firms as Holts for this purpose. If leakage through the crack has been excessive then it will be necessary to refill the appropriate cell with fresh electrolyte as detailed later. Cracks are frequently caused to the top of the battery case by pouring in distilled water in the middle of winter *after* instead of *before* a run. This gives the water no chance to mix with the electrolyte and so the former freezes and splits the battery case.
6 If topping-up the battery becomes excessive and the case has been inspected for cracks that could cause leakage, but none are found, the battery is being overcharged and the alternator will have to be checked. A fairly basic check can be carried out (see Section 7) but as a general principle this sort of job is best left to a car electrical specialist.
7 With the battery on the bench at the three monthly interval check, measure the specific gravity with a hydrometer to determine the state of charge and condition of the electrolyte. There should be very little variation between the different cells, and, if a variation in excess of 0.025 is present, it will be due to either:

   a) *Loss of electrolyte from the battery at some time caused by spillage or a leak, resulting in a drop in the specific gravity of the electrolyte when the deficiency was replaced with distilled water instead of fresh electrolyte.*
   b) *An internal short circuit caused by buckling of the plates or similar malady pointing to the likelihood of total battery failure in the near future.*

8 The specific gravity of the electrolyte for fully charged conditions at the temperatures indicated, is listed in Table A. The specific gravity of a fully discharged battery at different temperatures of the electrolyte is given in Table B.

Table A

Specific gravity - battery fully charged
1.268 at 100°F or 38°C electrolyte temperature
1.272 at 90°F or 32°C electrolyte temperature
1.276 at 80°F or 27°C electrolyte temperature
1.280 at 70°F or 21°C electrolyte temperature
1.284 at 60°F or 16°C electrolyte temperature
1.288 at 50°F or 10°C electrolyte temperature
1.292 at 40°F or 4°C electrolyte temperature
1.296 at 30°F or -1.5°C electrolyte temperature

Table B

Specific gravity - battery fully discharged
1.098 at 100°F or 38°C electrolyte temperature
1.102 at 90°F or 32°C electrolyte temperature
1.106 at 80°F or 27°C electrolyte temperature
1.110 at 70°F or 21°C electrolyte temperature
1.114 at 60°F or 16°C electrolyte temperature
1.118 at 50°F or 10°C electrolyte temperature
1.122 at 40°F or 4°C electrolyte temperature
1.126 at 30°F or -1.5°C electrolyte temperature

4 Battery electrolyte replenishment

1 If the battery is in a fully charged state and one of the cells maintains a specific gravity reading which is 0.025 or more lower than the others and a check of each cell has been made with a voltage meter to check for short circuits (a four to seven second test should give a steady reading of between 1.2 and 1.8 volts), then it is likely that electrolyte has been lost from the cell with the low reading at some time.
2 Top-up the cell with a solution of 1 part sulphuric acid to 2.5 parts of water. If the cell is already fully topped-up draw some electrolyte out of it with a pipette.
3 When mixing the sulphuric acid and water **never add water to sulphuric acid** - always pour the acid slowly onto the water in a glass

5 Battery - charging

1 In winter time when heavy demand is placed upon the battery, such as when starting from cold, and much electrical equipment is continually in use, it is a good idea to occasionally have the battery fully charged from an external source at an initial charging rate of 3.5 to 4 amps.
2 Continue to charge the battery at this rate until no further rise in specific gravity is noted over a four hour period.
3 Alternatively, a trickle charge, charging at the rate of about 1.5 amps can be safely used overnight.
4 Specially rapid 'boost' charges which are claimed to restore the power of the battery in 1 to 2 hours are most dangerous as they can cause serious damage to the battery plates through overheating.
5 While charging the battery note that the temperature of the electrolyte should never exceed 100°F (37.8°C).

container. **If water is added to sulphuric acid it will explode.**
4 Continue to top-up the cell with the freshly made electrolyte and then recharge the battery and check the hydrometer readings.

6 Alternator - general description, maintenance and precautions

1 Briefly, the alternator comprises a rotor and stator. Voltage is induced in the coils of the stator as soon as the rotor revolves. This is a 3-phase alternating voltage which is then rectified by diodes to provide the necessary current for the electrical system. The level of voltage required to maintain the battery charge is controlled by a regulator unit.
2 Maintenance consists of occasionally wiping away any oil or dirt which may have accumulated on the outside of the unit.
3 No lubrication is required as the bearings are sealed for life.
4 Check the drivebelt tension at the intervals given in the 'Routine Maintenance' Section. Refer to Section 8, for the procedure.
5 Due to the need for special testing equipment and the possibility of damage being caused to the alternator diodes if incorrect testing methods are adopted, it is recommended that overhaul or major repair is entrusted to a Lucas or Triumph dealer. Alternatively, a service-exchange unit should be obtained.
6 Alternator brush renewal is dealt with in Section 9.
7 Take extreme care when connecting the battery to ensure that the polarity is correct, and never run the engine with a battery charger connected. Do not stop the engine by removing a battery lead as the alternator will almost certainly be damaged. When boost starting from another battery ensure that it is connected positive-to-positive and negative-to-negative.

7 Alternator - testing in position in the car

If the alternator is suspected as being faulty, a test can be carried out which can help in isolating any such fault. A d.c. voltmeter (range 0 - 15V) and a d.c. ammeter (suitable for the nominal output current - see Specifications) will be required.
1 Check the alternator drivebelt tension and adjust if necessary - see Section 8.
2 Disconnect the brown cable from the starter motor solenoid. Connect the ammeter between this cable and the starter motor solenoid terminal.
3 Connect the voltmeter across the battery terminals.
4 Run the engine at 2540 rpm (6000 rpm of the alternator); the ammeter reading should stabilize.
5 If the ammeter reads zero, an internal fault in the alternator is indicated.
6 If less than 10 amps is indicated, and the voltmeter shows 13.6 to 14.4 volts where it is known that the battery is in a low state of charge, the alternator is suspect and should be checked by an auto-electrical specialist. The nominal output is given in the Specifications for each type of alternator.
7 If the ammeter reads less than 10 amps and the voltmeter reads less than 13.6 volts, a fault in the alternator internal regulator is indicated. A fault in the regulator is also indicated when the voltage exceeds 14.4 volts.

8 Alternator - removal, installation and drivebelt adjustment

17 ACR alternator
1 Detach the battery earth lead, and disconnect the alternator harness plug.
2 Remove the adjustment bolt and washer, and push the alternator towards the engine so that the drivebelt can be removed.
3 Remove the nut and washer, then support the alternator and withdraw the main mounting bolt and washer. The alternator can now be lifted clear.
4 When installing tap the bush slightly rearwards, then position the alternator and insert the main mounting bolt and washer. Fit the nut and washer finger-tight.
5 Fit the drivebelt over the pulleys, then fit the adjustment bolt and washer finger-tight.
6 Adjust the drivebelt tension, as described in paragraphs 14 to 17, then connect the harness plug and battery lead.

20 ACR and 25 ACR alternators
7 Detach the battery earth lead.
8 Slide the harness plug lock inboard and down to unlock it, then remove the plug.
9 Slacken the adjustment bolt, support bracket bolt and main mounting bolt, in that order.
10 Push the alternator towards the engine so that the drivebelt can be removed.
11 Remove the adjustment bolt and washer, and the main mounting bolt nut and washer.
12 Support the alternator and withdraw the main mounting bolt and washer. The alternator can now be lifted clear.
13 Installation is basically the reverse of the removal procedure, but do not tighten the nuts and bolts until the drivebelt tension has been checked, as described in paragraphs 14 to 17.

Drivebelt tension adjustment
14 Where necessary, slacken the adjustment bolt, support bracket bolt and main mounting bolt.
15 Pull the alternator away from the engine and tighten the adjustment bolt. Check for a belt total movement of 0.75 to 1 in (20 to 25 mm) under firm thumb pressure and at the midpoint of the longest belt run. Re-adjust, if necessary. **Note:** It is permissible to apply leverage at the drive end bracket, if necessary to obtain the correct tension, but only a softwood lever, or similar item, may be used.
16 Tighten the support bracket bolt and main mounting bolt on completion.
17 If a new belt has been installed, the belt tension should be rechecked after about 150 miles (250 km) of travelling.

9 Alternator - brush removal

17 ACR alternator
1 With the alternator removed from the car, remove the end cover (two screws).
2 Make a note of the wire positions and colours on the rectifier pack, then remove them.
3 Remove the surge protection diode (one screw).
4 Remove the brush box (two screws); detach the regulator wire (one screw).
5 If the slip rings are discoloured they may be polished with fine glass paper, crocus paper or metal polish. Ensure that no residue is left afterwards.
6 Remove the brushes, noting the leaf spring which may be fitted at the side of the inner brush.
7 Reassembly of the brush box and associated parts is the reverse of the removal procedure.

20 ACR alternator
8 With the alternator removed from the car, remove the end cover (two screws, flat washers and spring washers).
9 Make a note of the wire positions and colours, then disconnect the red wire small connector, white wire small connector and red wire large connector.

Chapter 10/Electrical system

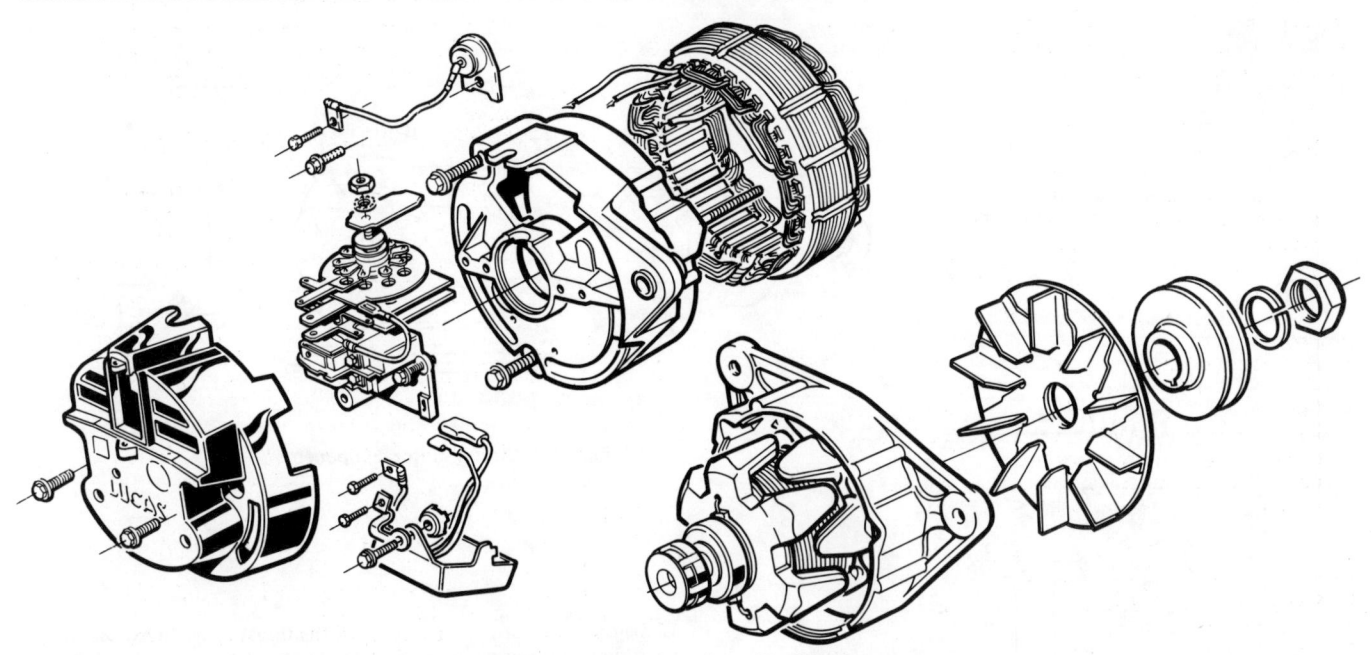

Fig. 10.1. Exploded view of the 17 ACR alternator

10 Remove the screw, flat washer and spring washer from the surge protection diode mounting.
11 Disconnect the screw and spring washer from the capacitor lug terminal to the right of the brush box.
12 Remove the screw, flat washer and spring washer to the left of the brushbox.
13 Lift out the brushbox complete with regulator and surge protection diode assembly.
14 To detach the regulator, remove the screw to release one wire eyelet, then remove the regulator retaining screw. Bend up the contact strap, then disengage the two lugs and lift the regulator off. Do not lose the spacer.
15 If the slip rings are discoloured they may be polished with fine glass paper, crocus paper or metal polish. Ensure that no residue is left afterwards.
16 Remove the brushes and fit replacements, then reassemble the parts in the reverse order to that used when dismantling.

25 ACR alternator
17 See Chapter 13.

10 Starter motor - general description

When the ignition switch is turned, current flows through the solenoid pull-in winding and starter motor, and the solenoid armature moves. At the same time a much smaller current flows through the solenoid hold-in winding directly to earth.

The movement of the solenoid armature causes the drive pinion to move and engage with the starter ring gear on the flywheel, and at the same time the main contacts close and energise the motor circuit. uit. The pull-in winding now becomes ineffective and it remains in the operated condition by the action of the hold-in winding only.

A special one-way clutch is fitted to the starter drive pinion so that when the engine commences to fire there is no possibility of it driving the starter motor.

When the ignition key is released, the solenoid is de-energised and returns to its original position. This breaks the supply to the motor and returns the drive pinion to the disengaged position.

11 Starter motor - removal and installation

Note: The procedure given below refers to vehicles with a standard exhaust system. Where a catalytic converter is installed, it may be necessary to detach this initially. This procedure is covered in Chapter 3.
1 Detach the battery earth lead and raise the car to a suitable working height for access to the front part of the exhaust system.
2 Remove the two exhaust retaining bolts, spring washers and flat washers at the gearbox bracket.
3 Remove the two exhaust retaining nuts, spring washers, flat washers and bolts at the bellhousing bracket.
4 Unhook the two rubber rings at the front silencer hanger and at the tail pipe hanger.
5 Remove the three nuts, spring washers and flat washers at the manifold flange.
6 Unclip the starter solenoid heat shield.
7 Swing the exhaust pipe under the engine to the right. Temporarily secure it.
8 Disconnect the brown and red/white leads, and the heavy cable from the starter solenoid, noting where they are fitted.

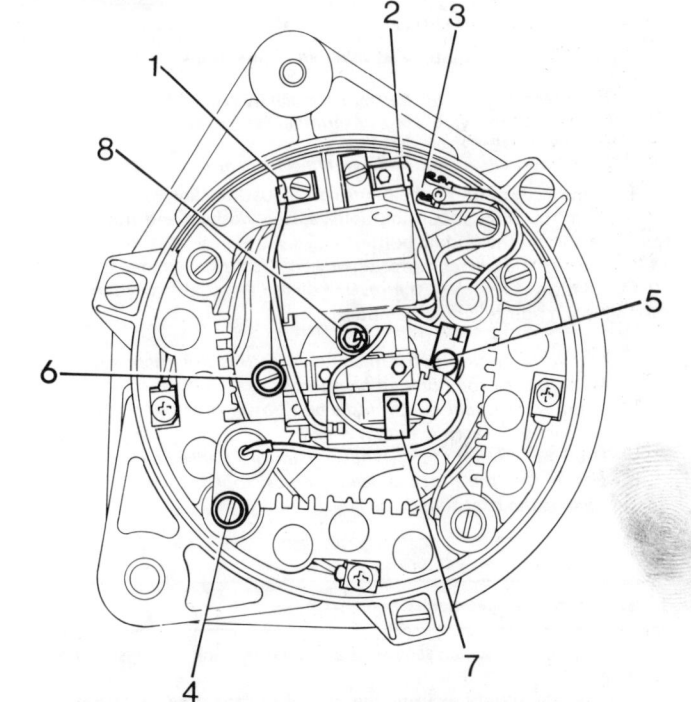

Fig. 10.2. End view of 20 ACR alternator with cover removed

1 Red wire small connector
2 White wire small connector
3 Red wire large connector
4 Surge protection diode screw
5 Capacitor lug terminal
6 Brushbox screw
7 Wire eyelet
8 Regulator retaining screw

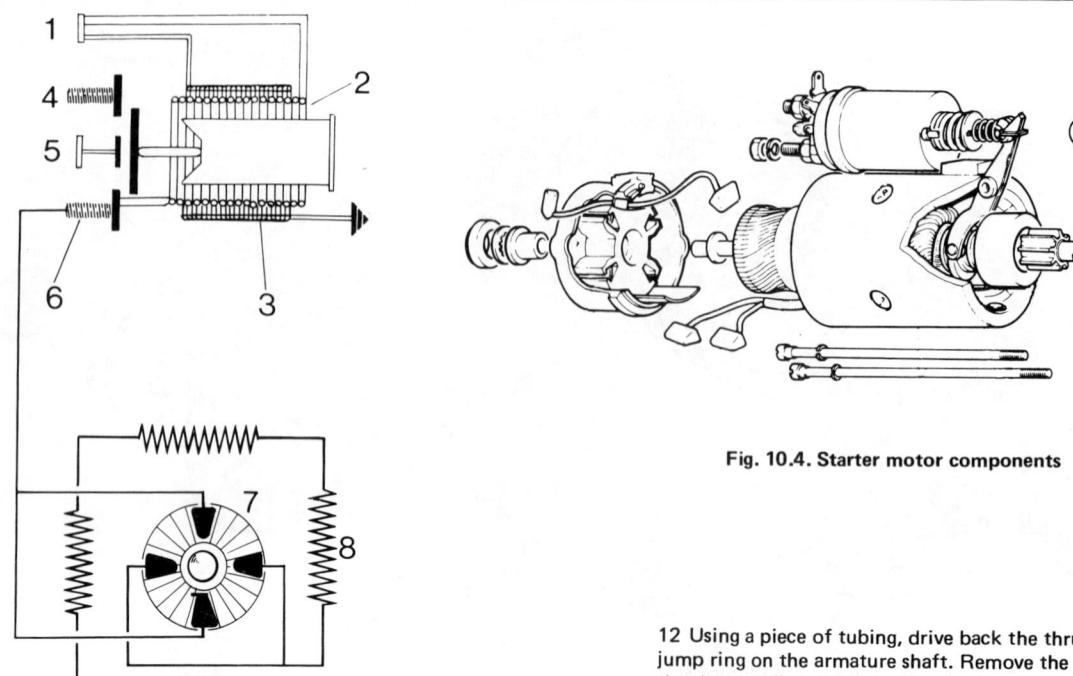

Fig. 10.3. Starter motor and solenoid connections

1 WR connector
2 Pull-in winding
3 Hold-in winding
4 Battery terminal
5 Ballast resistor/GN connector
6 Motor terminal
7 Brush gear
8 Field windings

Fig. 10.4. Starter motor components

9 Loosen one bolt on the bellhousing exhaust bracket.
10 Remove the three mounting bolts, spring washers and nuts. Note the cable clip on the middle bolt.
11 Manoeuvre the starter motor downwards from the engine.
12 Installation is basically the reverse of the removal procedure, but the following points should be noted:

 a) *Install the starter motor in an upward direction, manoeuvring it so that the top bolt can be inserted.*
 b) *Tighten the upper bolt, middle bolt, lower bolt and bellhousing exhaust bracket bolt in that order.*
 c) *Use a new gasket at the exhaust manifold flange connection.*
 d) *Ensure that the electrical leads are positioned along the inside edge of the heat shield.*

12 Starter motor - overhaul

1 Slacken the nut which secures the connecting link to the solenoid terminal 'STA'.
2 Remove the fixings securing the solenoid to the drive end bracket.
3 Lift the solenoid plunger upwards and separate it from the engagement lever. Extract the return spring, spring seat and dust excluder from the plunger body.
4 Withdraw the block from between the drive end bracket and the starter motor yoke.
5 Remove the armature end cap from the commutator end bracket.
6 Chisel off some of the claws from the armature shaft spire nut so that the nut can be withdrawn from the shaft.
7 Remove the two tie-bolts and then withdraw the commutator and cover and starter motor yoke from the drive end bracket.
8 Separate the commutator end cover from the starter motor yoke, at the same time disengaging the field coil brushes from the brush box to facilitate separation.
9 Withdraw the thrust washer from the armature shaft.
10 Remove the spire nut from the engagement lever pivot pin and then extract the pin from the drive end bracket.
11 Withdraw the armature and roller clutch drive assembly from the drive end bracket.
12 Using a piece of tubing, drive back the thrust collar to expose the jump ring on the armature shaft. Remove the jump ring and withdraw the thrust collar and the roller clutch.
13 Remove the spring ring and release the engagement lever, thrust washers and spring from the roller-clutch drive.
14 Remove the dust excluding seal from the bore of the drive end bracket.
15 Inspect all components for wear. If the armature shaft bushes require renewal, press them out or screw in a ½ in tap. Before inserting the new bushes, soak them in engine oil for 24 hours.
16 If the brushes have worn below the minimum specified length, renew them by cutting the end bracket brush leads from the terminal post. File a groove in the head of the terminal post and solder the new brush leads into the groove. Cut the field winding brush leads about ¼ in (6 mm) from the joint of the field winding. Solder the new brush leads to the ends of the old ones. Localise the heat from the field windings.
17 Check the field windings for continuity using a torch battery and test bulb. If the windings are faulty, removal of the pole shoe screws should be left to a service station having a pressure screwdriver as they are very tight.
18 Check the insulation of the armature by connecting a test bulb, torch battery and using probes placed on the armature shaft and each commutator segment in turn. If the test bulb lights at any position then the insulation has broken down, and the armature must be renewed. Discolouration of the commutator should be removed by polishing it with a piece of glass paper (not emery cloth). Do not undercut the insulation.
19 Reassembly is a reversal of dismantling but apply grease to the moving parts of the engagement lever, the outer surface of the roller clutch housing and to the lips of the drive end bracket dust seal. Install a new spire nut to the armature shaft positioning it to give the specified shaft endfloat. Measure this endfloat by inserting feeler blades between the face of the spire nut and the flange of the commutator end bush.

13 Headlamp - general description

1 The retractable headlamps are housed in light alloy box castings, which are attached to actuators and associated brackets.
2 The actuators comprise a permanent magnet motor and gearbox, with a crank arm which raises and lowers the headlamps according to the driver's requirements.
3 A hand-operated knob can be used to raise or lower the headlamp unit in the event of malfunction. This **must** be rotated in the anti-clockwise direction when viewed from the knob end, and will give a full open-closed-open cycle. When operating the headlamps by hand the actuator in-line connector **must** be detached from its mating parts, or the battery earth lead detached.

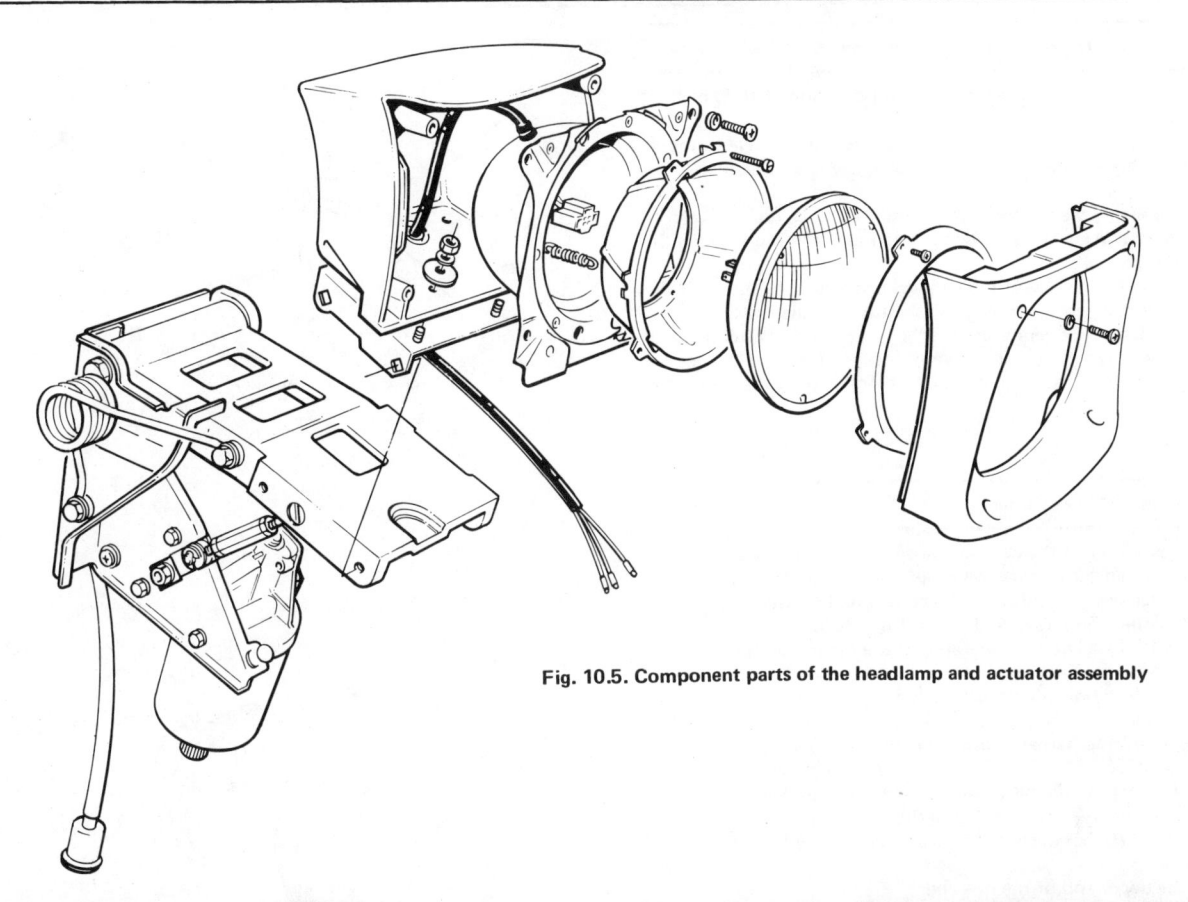

Chapter 10/Electrical system

Fig. 10.5. Component parts of the headlamp and actuator assembly

13.3a Headlamp actuator knob (arrowed)

13.3b Headlamp actuator in-line connector

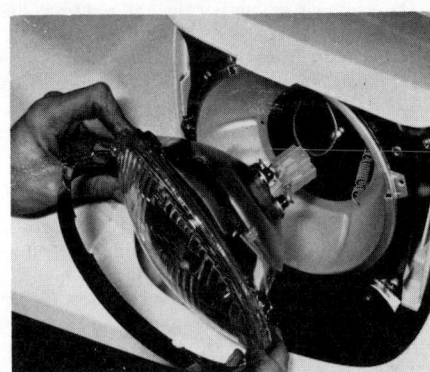

14.4 Removing the rim and light unit

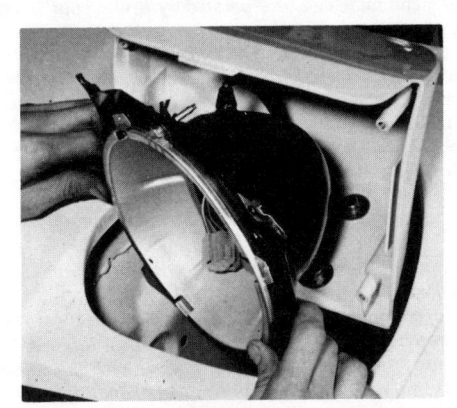

15.3 Removing a light unit

14 Headlamp light unit - removal, installation and lift mechanism lubrication

1 Raise the headlamps then detach the battery earth lead.
2 Remove the four crosshead screws and washers, and pull the rubber bezel from the box casting.
3 Manoeuvre the headlamp rubber bezel downwards into the headlamp cavity. The bezel may be manipulated, if necessary, to facilitate removal.
4 Remove the three retaining screws to release the rim and light unit (photo).
5 Pull the light unit away, detaching the connector block from the rear.
6 Lubricate the bushes at both ends of the link rod with the appropriate grease recommended by the manufacturer (part number AFU 1501).
7 Installation is the reverse of the removal procedure.

15 Headlamp assembly - removal and installation

1 Remove the rubber bezel, as described in paragraphs 1 to 3, of the previous Section.
2 Note the actuator harness and headlamp wire-runs relative to the components, then disconnect the harness plug and the three snap-connectors.
3 Whilst supporting the weight, remove the four nuts, lockwashers and flat washers, and carefully manoeuvre the assembly up through the body aperture (photo).
4 Installation is the reverse of the removal procedure. Ensure that the lead wires are connected to their correct mating colours; before connecting the battery, operate the headlamp unit over its full range of travel manually to ensure that leads are not trapped. Adjust the headlamp assembly, as described in Section 19, before fitting the rubber bezel.

16 Headlamp assembly - overhaul

Note: When working on the headlamp assembly, great care should be taken to ensure that the exterior paint surfaces are not scratched.
1 With the headlamp assembly on the bench, remove the nut, lockwasher, flat washer and screw, and take off the P-clip.
2 To assist with installation, mark around the mounting bracket sides on the light alloy member.
3 Remove the four bolts, lockwashers and flat washers, and separate the two assemblies.
4 Withdraw the lamp harness from the light alloy member.
5 Rotate the actuator hand knob to position the light alloy member against the 'up' stop on the base bracket. This restrains the spring.
6 Remove the circlip and washer from the link crank arm.
7 Remove the nut, lockwasher, flat washer and screw from the upper end of the link.
8 Detach the upper end of the link rod.
9 Detach the lower end of the link rod from the crank arm and remove the crank arm bush.
10 Remove the three screws and lockwashers, then manoeuvre the crank arm through the aperture and remove the actuator.
11 Check that the spring is still restrained (see paragraph 5) then, while retaining it with strong hand pressure, carefully remove the nut, lockwasher, small washer, large washer and bolt from the cranked end of the spring.
12 Remove the nut, lockwasher, small washer, large washer and bolt from the straight end of the spring. Remove the spring.
13 Remove the nut, two washers and bolt, and withdraw the light alloy member. Also withdraw the hinge pin.
14 Slacken the locknut and remove the rubber 'down' stop together with the locknut and washer.
15 Remove the retaining rim and light unit (three screws), then pull off the electrical connector.
16 Remove the four screws and lockwashers so that the housing and plate can be withdrawn.
17 The housing and plate can be separated by drilling out the pop rivets, where necessary.
18 Mark the outline of the painted box casting on the mounting bracket, then remove the four nuts, lockwashers and flat washers.
19 Remove the mounting bracket and take out the grommet to release the headlamp harness.
20 Reassembly of the headlamp is the reverse of the dismantling procedure. However, the following points should be noted:

  a) The hinge pin should be lubricated with Poly-Butyl-Cuprysil (PBC) grease.
  b) When installing the actuator, ensure that the harness is looped up between the actuator and the base bracket.
  c) The crank arm back should be lubricated with engine oil.
  d) The link rod ends should be positioned in the turnbuckle so that an equal number of threads is visible at each end.
  e) Before attaching the mounting bracket to the light alloy member, carry out the primary and secondary adjustment (see Section 19).

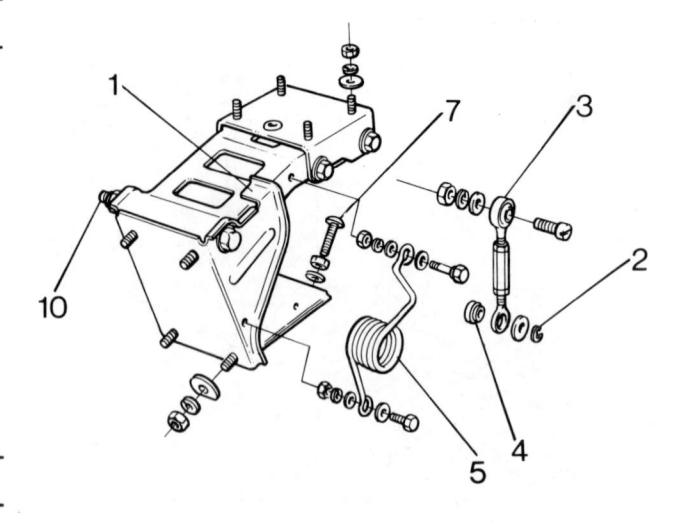

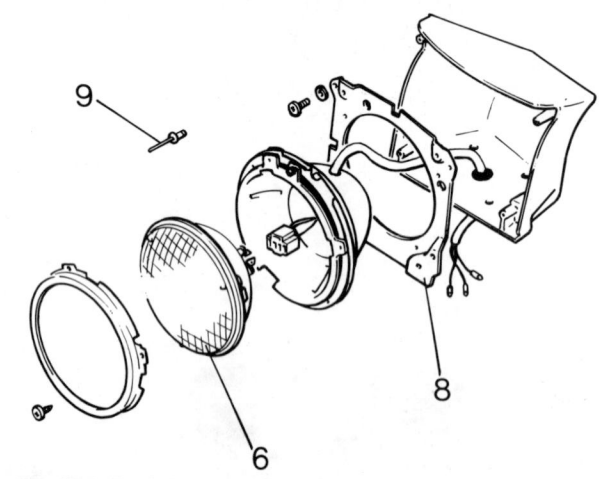

Fig. 10.6. Exploded view of headlamp assembly

1 'Lip' stop	6 Light unit
2 Circlip	7 'Down' stop
3 Link rod	8 Plate
4 Bush	9 Pop rivet
5 Spring (cranked end)	10 Hinge pin

17 Headlamp actuator - removal and installation

1 Remove the headlamp assembly, as described in Section 15.
2 Rotate the actuator hand knob to position the light alloy member against the 'up' stop on the base bracket. This retains the spring.
3 Remove the circlip and washer from the link crank arm.
4 Remove the nut, lockwasher, flat washer and screw, and take off the P-clip.
5 Remove the three screws and lockwashers, and manoeuvre the link rod from the bush.
6 Remove the bush from the crank arm, and manoeuvre the crank arm out of the aperture.
7 Installation of the actuator is the reverse of the removal procedure. However, the following points should be noted:

  a) The crank arm bush should be lubricated with engine oil.
  b) The link rod ends should be positioned in the turnbuckle so that an equal number of threads is visible at each end. Note that a left-hand thread is used at the upper end.
  c) When installing the actuator, rotate the hand knob, if necessary, to position the crank arm. Ensure that the harness is looped up between the actuator and the base bracket.
  d) Before installing the headlamp assembly, carry out the primary and secondary adjustment (see Section 19).

Chapter 10/Electrical system

18 Headlamp actuator - overhaul

Note: After any operation which involves disturbance of a joint, the unit must be sealed with an underseal-type of paint.

1 Remove the five screws and take off the cover plate and gasket.
2 Detach the two Lucas connectors, then remove the single screw and lift off the limit switch and harness.
3 Remove the shaft nut and take off the crank arm and washer.
4 Ensure that there are no burrs on the shaft, then withdraw it. Take out the dished washer.
5 Carefully prise off the hand knob from the armature shaft.
6 Remove the thrust screw (and locknut if applicable) from the housing.
7 Remove the through-bolts and slowly withdraw the cover and armature. The brushes will drop clear of the commutator, take care that they are not contaminated with grease from the worm gear.
8 Pull the armature out of the cover and remove the two thrust washers.
9 Remove the three screws to release the brush assembly; break the wire slot seal and lift the assembly from the recess.
10 Inspect all the parts for wear and damage, renewing as necessary. The motor commutator may be cleaned with a petrol moistened cloth, if necessary. Do not immerse electrical parts in cleaning fluids or the insulation may be damaged. If the armature is faulty, an attempt at repair is not recommended.
11 Reassembly is basically the reverse of the dismantling procedure, but the following points should be noted:

a) The two thrust washers should be lubricated with a molybdenum-disulphide oil.
b) The bearing in the cover should be lubricated with Shell Turbo 41 oil.
c) The self-aligning bearing in the housing should be lubricated with Shell Turbo 41 oil.
d) Ensure that the brushes are not contaminated by any grease which might be on the armature worm gear.
e) When installing the motor cover, ensure that the cover marks are aligned (see Fig. 10.8).
f) If a non-adjustable armature endfloat screw is fitted, check for an endfloat of 0.002 to 0.008 in (0.05 to 0.2 mm). If adjustment is required, add washers beneath the screw head to increase, or remove metal from the underside of the screw thread using a lathe to decrease.
g) If an adjustable armature endfloat screw is fitted, loosen the locknut and rotate the screw until it just makes contact, then screw it out ¼ turn. Tighten the locknut without altering the screw position.
h) The final gear bushes should be lubricated with Shell Turbo 41 oil.
j) When fitting the crank arm, ensure that the relationship to the final gear is as shown in Fig. 10.9.
k) The final gear and cam should be lubricated with Ragosine Listate grease.
l) Connect the red/light green wire to the red wire, and the black/light green wire to the blue wire.
m) Seal the exterior of the unit with an underseal-type of paint. Pay special attention to the joints and the wire slot seal.

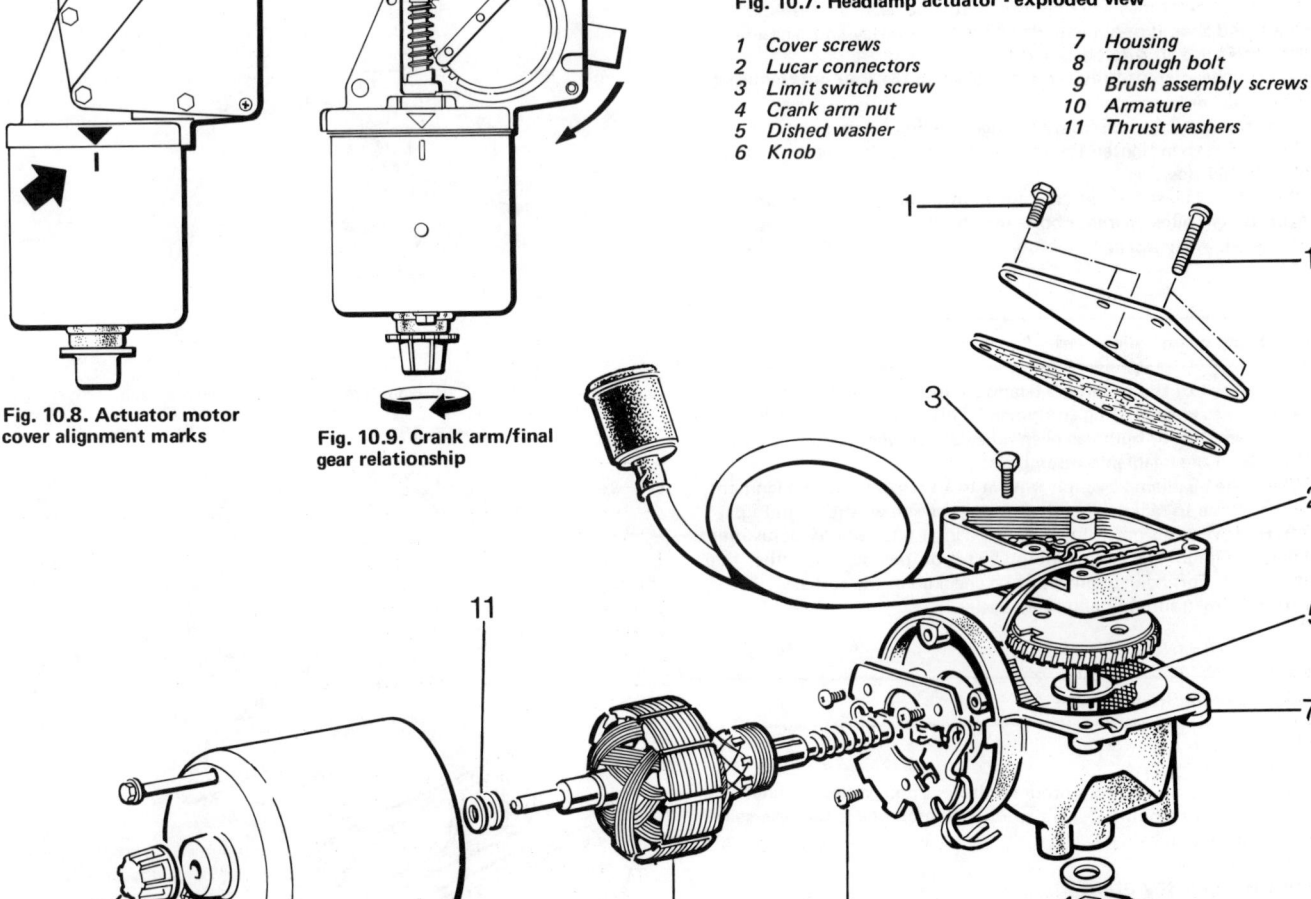

Fig. 10.7. Headlamp actuator - exploded view

1 Cover screws
2 Lucar connectors
3 Limit switch screw
4 Crank arm nut
5 Dished washer
6 Knob
7 Housing
8 Through bolt
9 Brush assembly screws
10 Armature
11 Thrust washers

Fig. 10.8. Actuator motor cover alignment marks

Fig. 10.9. Crank arm/final gear relationship

19 Headlamp assembly - adjustment

Primary adjustment
1 This adjustment may be made with the headlamp assembly removed or installed.
2 Fully raise the headlamps, then slacken the two link rod nuts (the upper one has a left-hand thread).
3 Rotate the turnbuckle by hand, until the light alloy member just contacts the metal 'up' stop on the base bracket. This provides a datum point.
4 Rotate the turnbuckle a further 4½ flats (270°) so that a pre-load will be applied to the 'up' stop.
5 Tighten the two nuts without altering the turntable position.

Secondary adjustment
6 This adjustment may be made with the headlamp assembly removed or installed, but should only be made where the primary adjustment has already been made.
7 Loosen the rubber 'down' stop locknut then screw the stop to its lowest position.
8 Fully lower the headlamps, then screw out the rubber 'down' stop until it just contacts the light-alloy member. This provides a datum point.
9 Screw out the 'down' stop a further 1 1/3 turn (480°) to provide a preload.
10 Tighten the locknut without altering the position of the 'down' stop.

Box casting adjustment
11 This adjustment may only be made with the headlamp assembly installed, and should only be made where the primary and secondary adjustments have already been made.
12 With the headlamps fully retracted, slacken the four nuts securing the base bracket to the car body.
13 Move the headlamp assembly to align the box casting with the body contour, then tighten the four nuts without altering the headlamp assembly position.
Note: If there is insufficient adjustment available, the four mounting bracket-to-light alloy member bolts may be slackened and the assembly repositioned, as necessary.

20 Headlamp beam - alignment

1 Adjustment of the headlamp beams should only be carried out by a dealer who has beam setting equipment. This should always be carried out if a beam unit or bulb has been renewed, or where adjustment has been made to the headlamp assemblies.
2 Where the headlamp beam is known to be widely out of alignment, it is permissible to make adjustment using the screws shown in Fig. 10.10. However, this must only be regarded as a temporary adjustment, and when in doubt the beams should be set on the low side rather than the high side. This adjustment must be checked at the earliest opportunity by a suitably equipped dealer.

21 Front parking/flasher lamps

Bulb renewal
1 Remove the two screws, withdraw the lens and gasket, and remove the bayonet-fitting bulb(s). Installation of the bulb(s) is the reverse of this procedure (photo).

Removal and installation
2 Remove the lamp harness from the body clip and disconnect the three snap-connectors.
3 Remove the lens and bulbs, as described in paragraph 1.
4 Remove the two nuts, spring washers and flat washers, and withdraw the lamp base and gasket.
5 Installation is the reverse of the removal procedure. Connect the wires as follows:

red to red
black to black
green/red to green (left-hand lamp)
green/white to green (right-hand lamp)

22 Front flasher repeater lamp

Bulb renewal
1 Pull off the metal bezel and remove the lens, then remove the bayonet-fitting bulb. Installation is the reverse of this procedure.

Removal and installation
2 Release the wires (green and black) from the clips and tape below the headlamp assembly (early cars) or in the engine bay behind the headlamp mounting positions, and disconnect the electrical connections.
3 Attach a cord of about 6ft 6 in (2 metres) length to the two wire ends, attach the other end to the car body crossmember to prevent it from being pulled completely through.
4 Withdraw the lamp wires from the wing cavity and detach the cord.
5 Installation is the reverse of the removal procedure. Connect the wires as follows:

green to black to black (left-hand lamp)
green to green/white (right-hand lamp)
black to black (right-hand lamp)

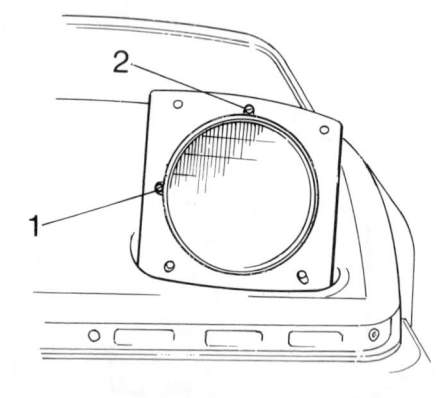

Fig. 10.10. Headlamp beam alignment screws

1 Horizontal   2 Vertical

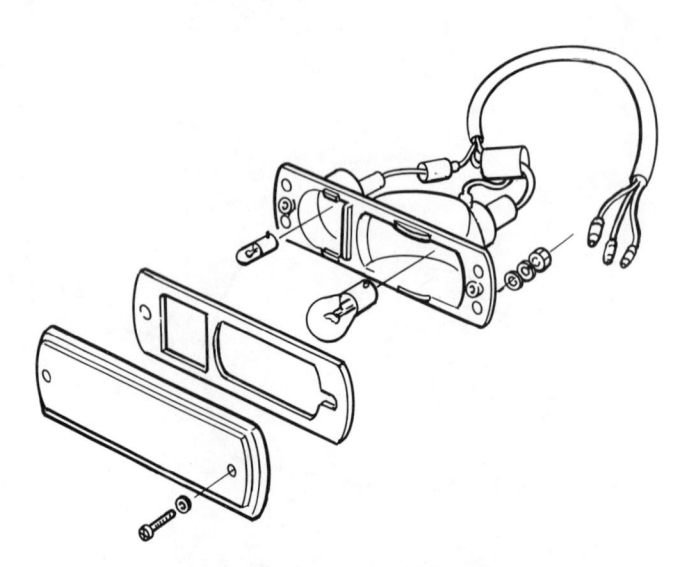

Fig. 10.11. Front parking/flasher lamp

Chapter 10/Electrical system

23 Marker lamps

Bulb renewal

1 Where applicable, remove the two nuts, spring washers and flat washers, and remove the cover. Pull out the bulb holder and remove the push-fit bulb. Installation is the reverse of this procedure.

Removal and installation - front lamp

2 Fully raise the headlamps then detach the battery ground lead.
3 Remove the bulb holder, then disconnect the two Lucas connectors.
4 Remove the two nuts, spring washers and washers, and withdraw the lamp assembly. Do not attempt to detach the lens from the base.
5 Installation is the reverse of the removal procedure; ensure that the gasket is correctly positioned.

Removal and installation - rear lamp

6 Refer to the procedure given for the front lamp, but note the cover which is also fitted.

24 Rear lamp assembly

Bulb renewal

1 From inside the luggage boot, turn back the floor mat and remove the trim panel (four screws). Rotate the bulb holders anti-clockwise then remove the bayonet-fitting bulb(s). Installation is the reverse of this procedure; note that the longer trim panel screws are the top ones (photo).

Removal and installation

2 Remove the bulbs, as described in paragraph 1, then remove the nut and spring washer to release the earth tag.
3 Remove the five remaining nuts, spring washers and flat washers.
4 Withdraw the one-piece lens and gasket, followed by the lamp base and gasket.
5 Installation is the reverse of the removal procedure. Connect the leads, as shown in Fig. 10.13.

25 Number plate lamp

Bulb renewal

1 Remove the two screws and withdraw the lens, then remove the festoon-type bulb. Installation is the reverse of this procedure.

Removal and installation

2 Remove the two screws and manoeuvre the lamp from the body panel aperture (photo).
3 Remove the festoon-type bulb and pull off the two Lucas connectors.
4 Installation is the reverse of the removal procedure.

26 Roof lamp

Bulb renewal

1 Detach the battery earth lead, then remove the lens by gently squeezing the longest sides. Prise out the festoon-type bulb, then install a replacement by following the reverse of the removal procedure (photo).

Removal and installation

2 Remove the bulb, as described in paragraph 1, then remove the two screws and withdraw the lamp base.
3 Note the wire colours and positions, then pull off the two Lucar connectors.
4 Installation is the reverse of the removal procedure. Ensure that the switch faces forwards, and that the right-hand screw makes a good earth connection to the car body.

21.1 Front parking/flasher light bulbs (typical)

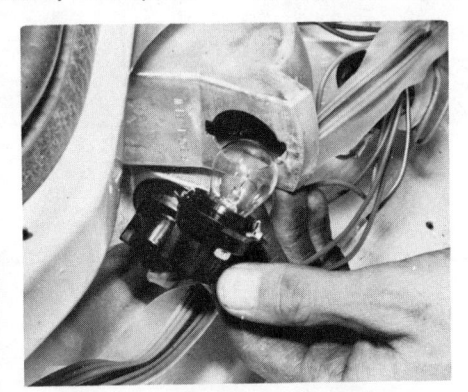

24.1 Rear lamp bulb removal (typical)

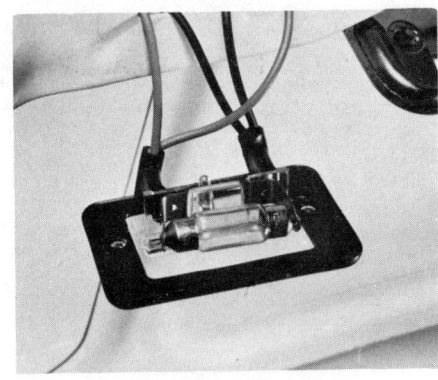

25.2 Number plate lamp removed

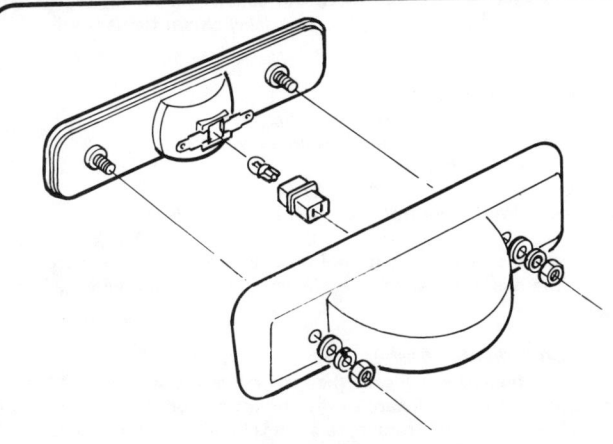

Fig. 10.12. Rear side marker lamp

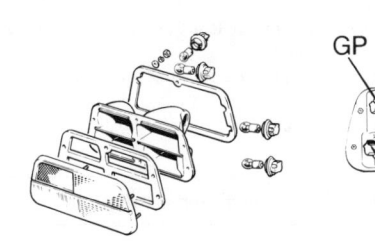

Fig. 10.13. Rear lamp lead connections (left-hand lamp shown)

GP   Green/Purple
R   Red
GN   Green/Brown
GR   Green/Red (Green/White for right-hand lamp)

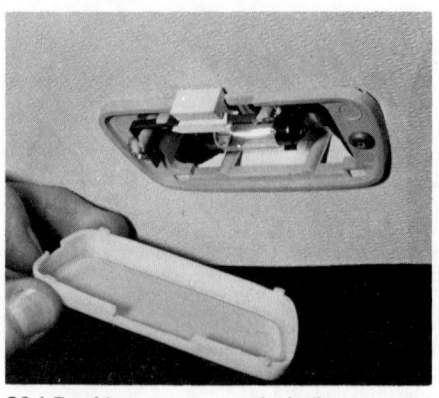

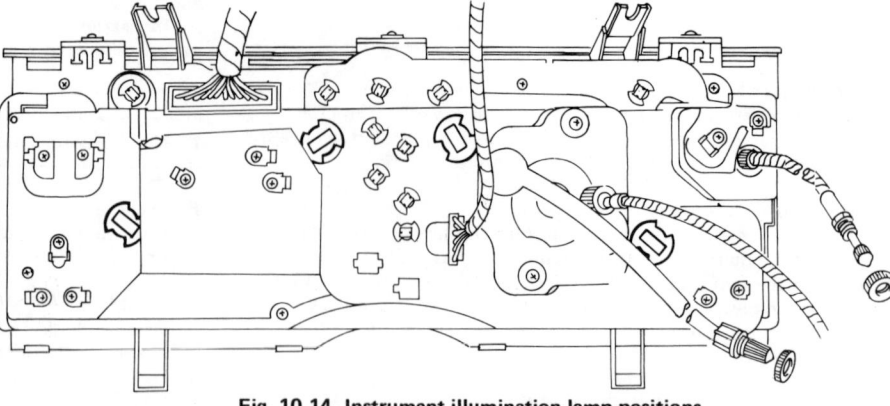

26.1 Roof lamp - access to the bulb

Fig. 10.14. Instrument illumination lamp positions

27 Instrument illumination lamps

Removal and installation

1 Isolate the battery, and remove the centre radio speaker grille and facia instrument cowl as described in Chapter 12.
2 From the concealed side of the instrument panel, rotate the bulb holder anti-clockwise and withdraw the bayonet fitting. **Note:** If satisfactory access cannot be obtained, remove the instrument panel, as described in Section 34.
3 Pull the bulb from the holder.
4 Installation is the reverse of the removal procedure.

Green dome cover - removal and installation

5 Initially proceed as described in paragraph 1.
6 Remove the three crosshead screws and washers from the top of the lens unit.
7 Slide the lens unit upwards, following its natural arc, and lift out the face panel.
8 Remove the speedometer or tachometer (see Section 35) to obtain access to the appropriate dome cover.
9 Carefully remove the dome cover, taking care not to break one of the housing claws.
10 Installation is the reverse of the removal procedure.

Fig. 10.15. Instrument panel warning light positions

28 Instrument panel warning lights - removal and installation

The procedure for removing bulbs, bulbholders and lenses is as described in the previous Section, except for paragraphs 8 and 9.

29 Relays and flasher units - removal and installation

Engine compartment items

1 Where applicable, the air-conditioning delay circuit relay and the delay circuit flasher unit are mounted on the forward engine bay panel.
2 The relay is retained by a single self-tapping screw which also secures the harness earth tag.
3 The flasher unit is secured in a clip, and has a harness plug connected to it.
4 USA models may have a horn relay mounted adjacent to the windscreen wiper motor. This relay is similar to the air-conditioning delay circuit relay.

Items behind glovebox

5 The following items are mounted behind the glovebox panel:

Starter motor interlock relay
Horn relay (except some USA models)

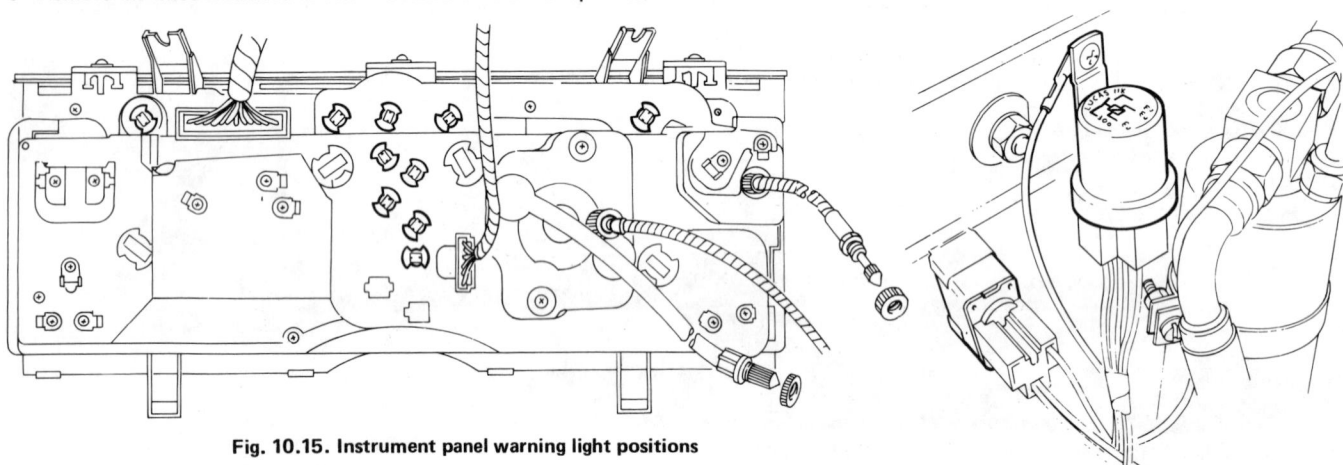

Fig. 10.16. Air-conditioning delay circuit relay and delay circuit flasher unit

Headlamp run/stop relays
Headlamp flasher relay (where applicable)
Headlamp flash control unit (where applicable)
Headlamp circuit breaker
Hazard flasher (some USA models)
Air-conditioning control relay (some USA models)

6 These items are retained by self-tapping screws. To obtain access, remove the two crosshead screws (where relevant) and withdraw the internal panel.

Direction indicator flasher unit

7 This is mounted in a clip on the facia rail at the extreme left-hand end (USA models) or forward of the choke and bonnet release controls. Detach the flasher unit from its retaining clip, or the clip from the facia rail. Detach the Lucas connectors (and the clip, where applicable).
8 Installation is the reverse of the removal procedure. Connect the

Chapter 10/Electrical system

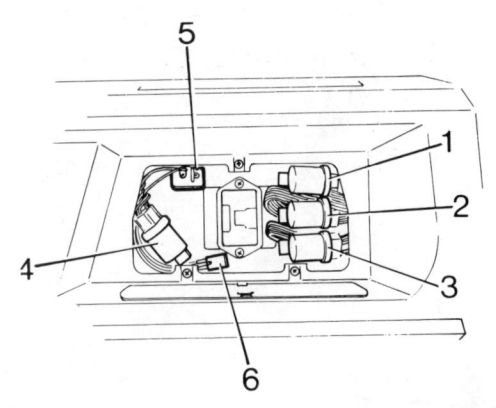

Fig. 10.17. Items behind glovebox - typical for 1975 USA models

1 Starter motor interlock relay
2 Air-conditioning control relay
3 Headlamp run/stop relay, right-hand
4 Headlamp run/stop relay, left-hand
5 Hazard flasher unit
6 Headlamp circuit breaker

light green/brown wire to the 'L' terminal, and the light green/slate wire to the 'B' terminal.

Hazard flasher unit (except some USA models)

9 Pull the flasher unit from the component mounting clip which is behind/below the glovebox. Disconnect the two Lucas connectors and remove the flasher unit from the clip.
10 Installation is the reverse of the removal procedure. Connect the light green/pink wire to the 'L' terminal and the purple wire to the 'B' terminal.

Interlock module (USA models with interlock and key warning systems)

11 The module is located at the extreme right-hand end of the facia rail in a clip. To remove it, manoeuvre it from the clip and disconnect the harness plug.
12 Installation is the reverse of the removal procedure; note that the plug is keyed and will only fit in one position.

30 Fuses

1 A list of fuse ratings is given in the Specifications Section; refer to the appropriate wiring diagram for the circuits protected.
2 The fusebox, and in-line fuses, are accessible after the internal panel has been removed from the glovebox (two crosshead screws or snap-fit retainers).
3 The in-line fuses are in bayonet type holders. The red/green and red wires connect to the front parking lamps/interior light; the white and white/black wires connect to the heated rear window.

31 Steering column switches - removal and installation

Ignition/starter switch

1 Detach the battery earth lead, then remove the steering column nacelle halves (two screws).
2 Note the run of the switch harness, then disconnect the plug.
3 Remove the two small crosshead screws and withdraw the switch.
4 Installation is the reverse of the removal procedure; note that the lock shaft and switch shaft must be correctly aligned.

Multi-function stalk switch assembly

5 Detach the battery earth lead, then remove the steering column nacelle halves (two screws).
6 Remove the steering wheel, as described in Chapter 11.
7 Note the run of the switch harness, then squeeze the projection on the harness clip to release it.
8 Disconnect the harness plugs, slacken the switch clamping screw and withdraw the complete switch assembly from the column.
9 To remove either switch drill out the two rivets and remove the

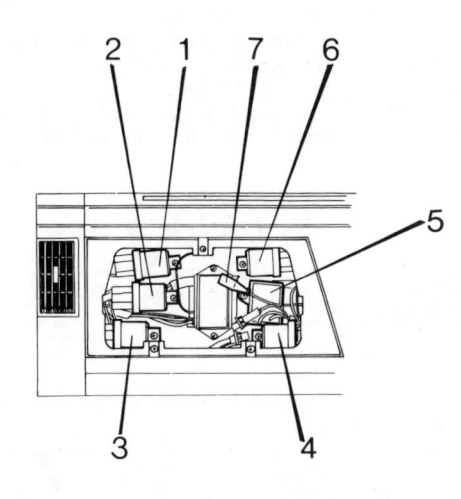

Fig. 10.18. Items behind glovebox - typical for 1976 UK models

1 Headlamp flash relay
2 Horn relay
3 Headlamp run/stop relay, left-hand
4 Headlamp run/stop relay, right-hand
5 Headlamp flash control unit
6 Starter motor interlock relay
7 Headlamp circuit breaker

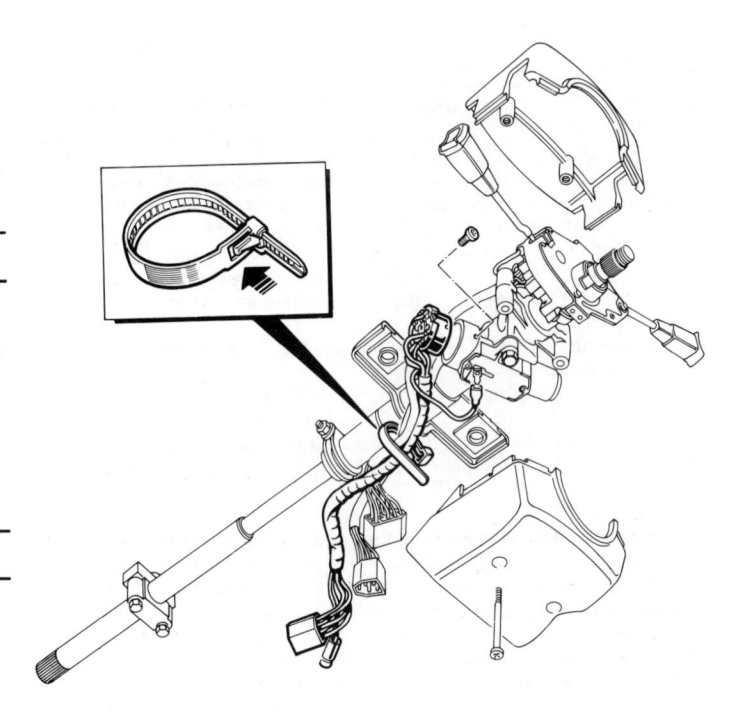

Fig. 10.19. Ignition/starter switch - typical

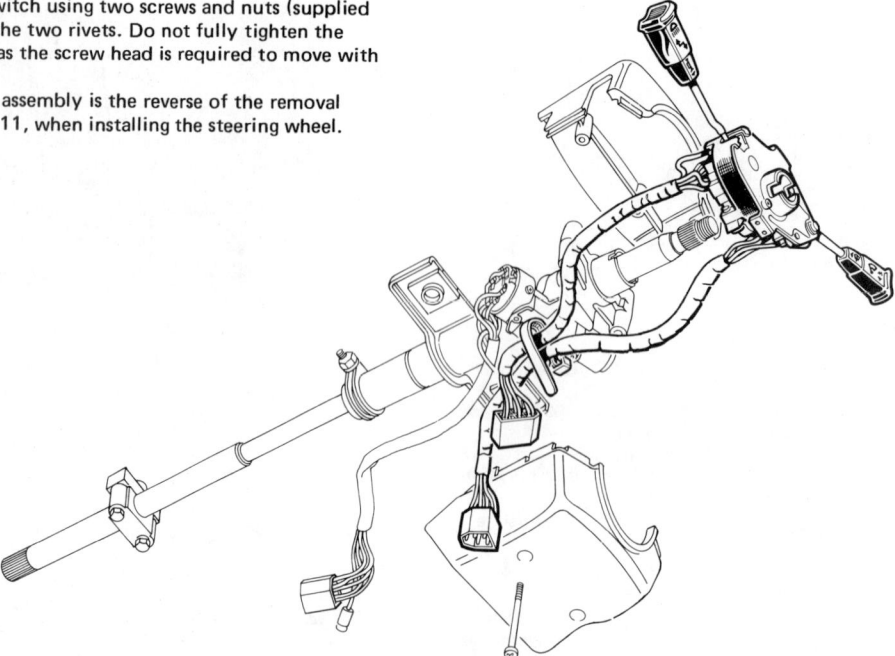

Fig. 10.20. Multi-function stalk switches - typical

single screw. **Note:** Do not remove the two screws near the steering shaft aperture.
10 Install the replacement switch using two screws and nuts (supplied with the switch) in place of the two rivets. Do not fully tighten the remaining screw and washer as the screw head is required to move with the switch arm.
11 Installation of the switch assembly is the reverse of the removal procedure. Refer to Chapter 11, when installing the steering wheel.

32 Switches - removal and installation

Master light switch, front or rear fog lamp switch, heated rear window switch and hazard switch (including hazard switch warning bulb renewal)
1 Remove two crosshead screws. If necessary, remove one crosshead screw from the instrument cowl and raise it a little. Withdraw the facia switch panel (photo).
2 Pull out the switch identification strip bulb holders and disconnect the switch harness plugs. If it is only necessary to renew the hazard switch warning bulb, simply disconnect the harness plug only and pull out the bulb.
3 Remove the facia switch panel then use a small screwdriver to prise off the two spire clips.
4 Remove the face panel and the identification strip assembly, then press in one plastic clip on the switch to release it from the panel.
5 Installation is the reverse of the removal procedure. Where applicable, the switch symbol should be at the bottom.

Panel rheostat
6 Detach the battery earth lead, then pull out the centre console tray.
7 Insert a suitable probe into the hole in the knob to depress the clip; pull off the knob.
8 Unscrew the bezel then withdraw the rheostat downwards from the centre console panel.
9 Remove the spring washer, note the lead colours and positions, then detach the two Lucas connectors.
10 Installation is the reverse of the removal procedure.

Door switch
11 With the door open, remove the single screw, withdraw the switch and detach the Lucas connector.
12 Installation is the reverse of the removal procedure.

Reverse lamp switch - 4-speed gearbox
13 Raise the car for access to the left-hand side of the gearbox extension.
14 Remove the two snap-connectors, then unscrew the switch from the base of the gearlever; collect the distance washer.
15 Installation is the reverse of the removal procedure, but take care not to overtighten the switch. No adjustment is required.

Reverse lamp switch - 5-speed gearbox
16 Unscrew the gearlever knob and lift off the gaiter.
17 Remove the four crosshead screws and washers, and lift out the draught excluder and plate.
18 From beneath the car, detach the snap-connectors from the switch on the gearbox extension.
19 From inside the car loosen the locknut and unscrew the switch.
20 When installing the switch, select reverse gear and screw in the switch until it just makes contact (this can be checked with a continuity tester or a test-lamp and battery), then screw it in another 3 flats (180°). Hold the switch in this position and tighten the locknut.
21 The remainder of the installation procedure is the reverse of that used when removing.

Gearbox interlock switch (where applicable)
22 Raise the car for access to the left-hand side of the gearbox extension.
23 Remove the two Lucas connectors from the switch nearest to the speedometer drive.
24 Loosen the locknut, unscrew the switch, then collect the locknut.
25 When installing the switch, screw it partially in and install the locknut on the selector shaft side of the bracket.
26 Select any gear, then screw in the switch until the plunger just contacts the shaft. Screw the switch in one turn further (360°) and tighten the locknut without altering the switch position.
27 Connect the Lucas connectors, then sit in the driver's seat without fastening the seatbelt, and switch on the ignition. The warning light and buzzer should now be on.
28 Select neutral; the warning light and buzzer should now be off.

Seat interlock switch
29 Remove the seat (see Chapter 12), then unhook the two front diaphragm attachment clips.
30 Note the wire run and switch position, then withdraw the wires through the diaphragm hole to allow the switch to be withdrawn.
31 Installation is the reverse of the removal procedure.

Seat belt switches
32 Refer to the procedure given in Chapter 12, Section 27.

Chapter 10/Electrical system   121

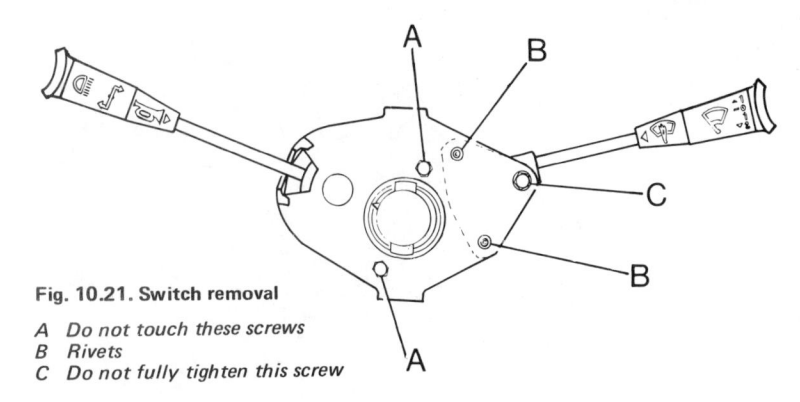

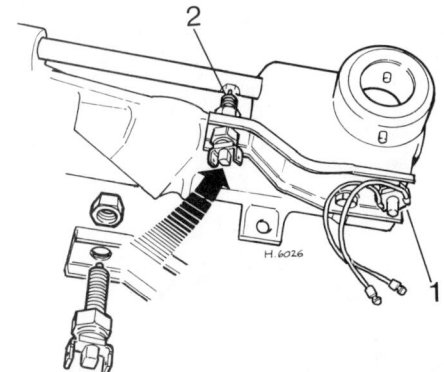

Fig. 10.21. Switch removal

A Do not touch these screws
B Rivets
C Do not fully tighten this screw

Fig. 10.22. Reverse lamp and gearbox interlock switches
(4-speed gearbox)

1 Reverse lamp switch   2 Interlock switch

32.1 Facia switch panel removal

32.40 Brake line failure switch plug

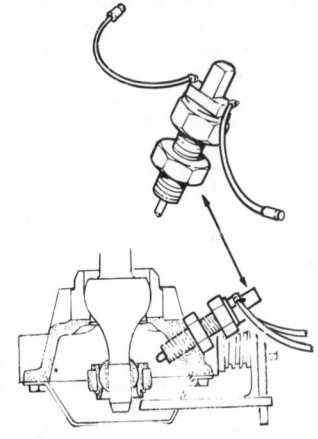

Fig. 10.23. Reverse lamp switch - 5-speed gearbox

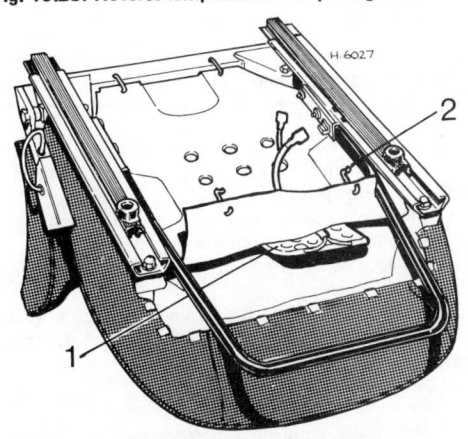

Fig. 10.24. Seat interlock switch location

1 Switch   2 Diaphragm attachment clips

Oil pressure switch

33 Detach the lead(s) from the switch on the right-hand side of the engine near the dipstick. (Note that some later vehicles may have a moulded plug which can only be fitted one way).
34 Unscrew the switch from the oil transfer adaptor.
35 Installation is the reverse of the removal procedure, but take care not to overtighten the tapered thread. On emission control cars, connect the leads as shown in Fig. 10.25.

Handbrake switch

36 Remove the console assembly, as described in Chapter 12.
37 Pull apart the touch-and-close fastener strips on the top edge of the handbrake gaiter.
38 Remove the single crosshead screw to release the switch, then detract the Lucas connector.
39 Installation is the reverse of the removal procedure.

Brake line failure switch

40 Release the harness plug claws and disconnect the plug from the switch on the lower side of the brake master cylinder (photo).
41 Unscrew the switch.
42 Installation is the reverse of the removal procedure. Note that the switch tightening torque is only 15 lb f in (1.7 Nm).

Stop light switch

43 Disconnect the two Lucas connectors and loosen the large hexagon nut on the brake pedal switch.
44 Push the pedal forward and remove the nut and washer(s) to release the switch. Do not try to rotate the switch.
45 Installation is the reverse of the removal procedure. Take care that the plastic nut is not overtightened.

Choke switch (USA manual choke models)

46 Detach the battery earth lead, then pull out the centre console tray.
47 Pull out the choke knob and unscrew the choke bezel.
48 Push the choke control downwards as far as it will go, and twist it slightly so that the small screw can be removed.
49 Slide the clip off the switch and detach the Lucar connector.
50 Installation is basically the reverse of the removal procedure. After

Chapter 10/Electrical system

installing the Lucar connector, position the switch and clip over the reduced diameter of the cable, then slide it onto the switch. Position the assembly so that the switch plunger is located in the hole in the outer cable housing, then install the screw.

Choke switch (other manual choke models)
51 Using a small screwdriver, remove the small screw securing the choke switch to the choke cable adjacent to the hand knob.
52 Proceed as described in paragraphs 49 and 50.

33 Cigarette lighter - removal and installation

1 Detach the battery earth lead, then pull out the centre console tray.
2 Withdraw the heating unit and pull the purple wire connector from the centre terminal.
3 Using a pair of long-nosed pliers on the inner well cross-piece, unscrew the inner well from the outer wall.
4 Remove the illumination ring, then disconnect the black (earth) Lucas connector and the red/white wire harness plug.
5 If the knob is to be renewed, squeeze the sides of the bulb cowl and withdraw it. Unclip the bulb from its holder and remove it from its bayonet fitting.
6 Installation is the reverse of the removal procedure; position the bulb cowl as necessary.

34 Instrument panel - removal and installation

Note: Instruments can be removed individually, if required - see Section 35.
1 Remove the centre grille (radio loud speaker grille) and facia instrument cowl, as described in Chapter 12.
2 Remove the two screws and separate the two halves of the steering column nacelle.
3 Loosen the speedometer trip reset cable and clock cable knurled nuts, then withdraw the cables from the bracket slots.
4 Depress the lever to release the catch, and pull out the speedometer cable from the instrument panel.
5 Remove the four crosshead screws and washers, and withdraw the instrument panel so that the harness plugs can be detached. Remove the panel completely.
6 If the printed circuit is to be removed, follow the procedure in paragraphs 7 to 11.
7 Remove the retaining crosshead screws and take off the contact clips.
8 Rotate all the bulb holders anti-clockwise and lift them out of the bayonet fittings.
9 Using a small screwdriver, prise up the two press clips; take care not to damage the printed circuit.
10 Release any adhesive tape and lift off the printed circuit.
11 Installation of the printed circuit is the reverse of the removal procedure.
12 Installation of the instrument panel is the reverse of the removal procedure, but ensure that the spire nut on each lower mounting is correctly positioned first of all.

35 Panel instruments and sender units - removal and installation

Battery condition indicator, temperature gauge and fuel gauge
1 Remove the centre grille (radio loud speaker grille) and facia instrument cowl, as described in Chapter 12.
2 Remove the three crosshead screws and washers, and slide the instrument panel lens upwards following its natural arc.
3 Remove the face panel, then remove the single crosshead screw and withdraw the instrument.
4 Installation is the reverse of the removal procedure. If there is any difficulty in engaging the lens tongues slacken the lower crosshead screws and reposition the instrument panel slightly.

Clock
5 Refer to the procedure given in paragraphs 1 to 4, but note that it is necessary to unscrew the knurled nut so that the cable can be withdrawn, and that the clock is retained by two screws.

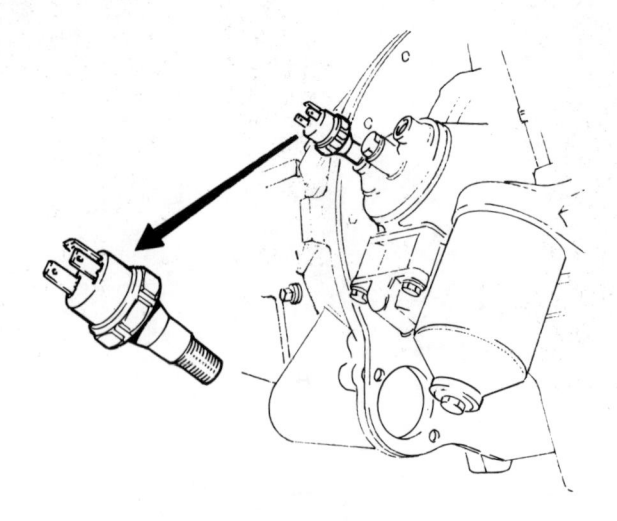

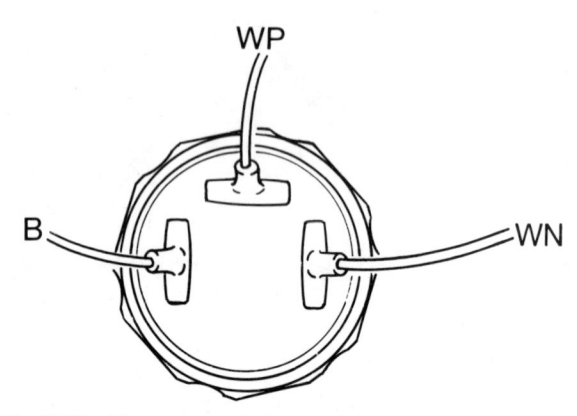

Fig. 10.25. Oil pressure switch on emission control cars

B Black
WP White/Purple
WN White/Brown

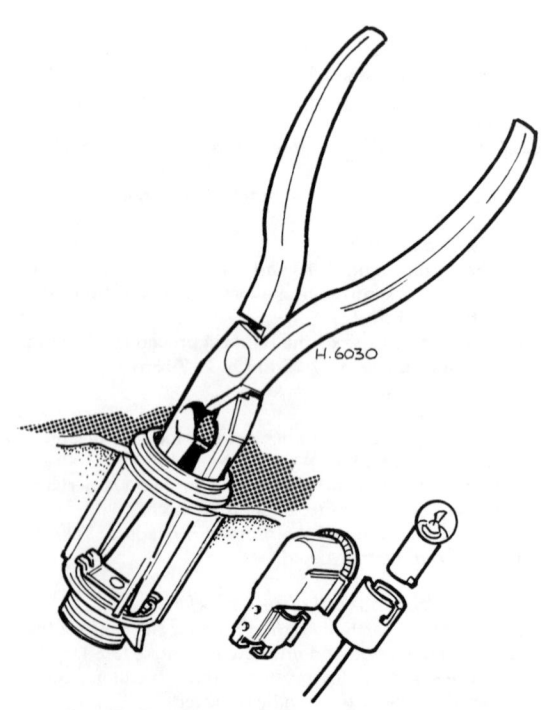

Fig. 10.26. Cigarette lighter - dismantling

123

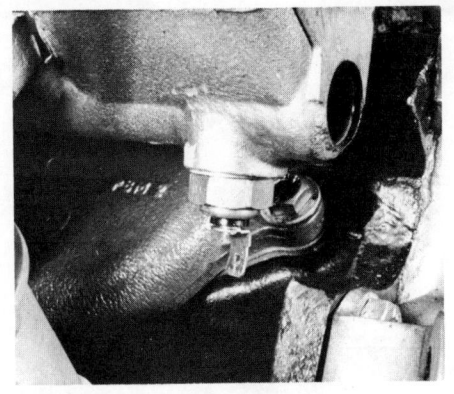

35.7 Location of temperature gauge transmitter

35.10 Tank unit

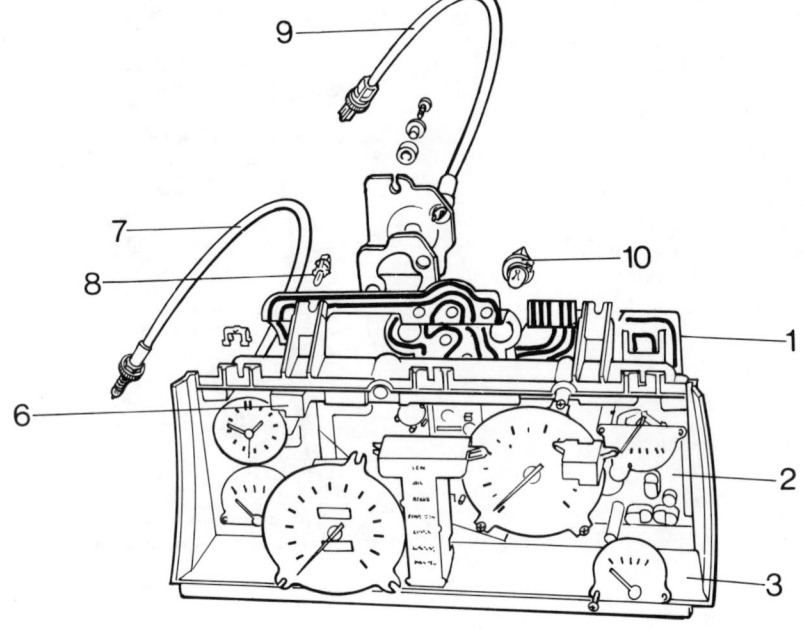

Fig. 10.27. Instrument panel assembly

1 Printed circuit
2 Housing
3 Fuel gauge
4 Face panel
5 Lens
6 Lens block
7 Clock cable
8 Warning light
9 Trip reset cable
10 Panel illumination bulb

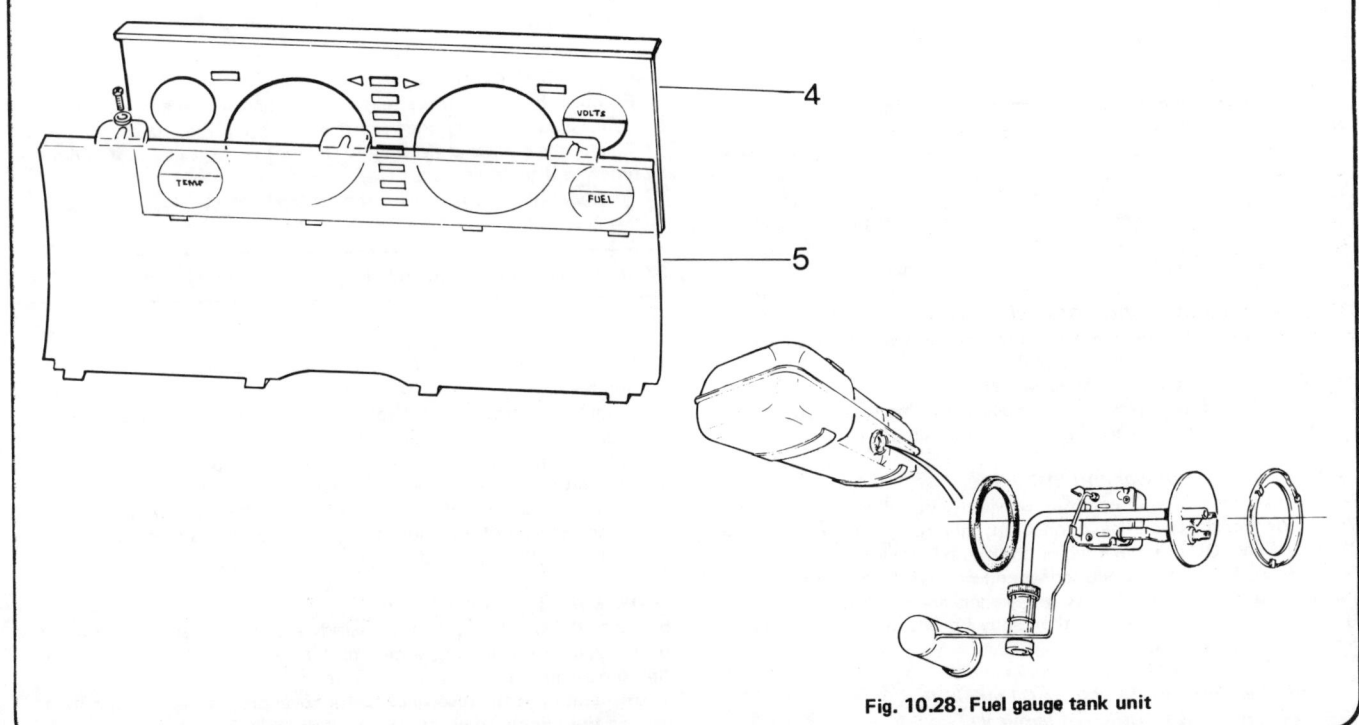

Fig. 10.28. Fuel gauge tank unit

Temperature gauge transmitter
6 Partially drain the engine coolant (refer to Chapter 2, if necessary).
7 Detach the Lucas connector, then unscrew the transmitter from the inlet manifold (photo).
8 Installation is the reverse of the removal procedure, but a new sealing washer should be used.

Fuel gauge tank unit
9 Wherever possible reduce the tank fuel level by normal usage. If this is not possible, be prepared to collect any fuel that may spill when the outlet pipe is removed.
10 Raise the rear of the car to a suitable working height, then detach the three tank unit leads (photo).
11 Pull off the fuel outlet pipe and collect any fuel that may spill out.
12 Tap the locking ring anti-clockwise to unscrew and remove it.
13 Withdraw the tank unit, being prepared for any further fuel spillage. Remove the sealing washer.
14 Installation is the reverse of the removal procedure, but a new sealing washer should be used. Connect the leads as follows:

  green/black to terminal T
  green/orange to terminal W
  black to earth

Speedometer
15 Follow the procedure given in paragraphs 1 and 2.
16 Remove the face panel. Depress the lever to release the catch, and pull out the speedometer cable from the instrument.
17 Remove three crosshead screws and withdraw the speedometer.
18 Installation is basically the reverse of the removal procedure, but to assist engagement of the trip reset cable, rotate the control to preset the alignment.

Speedometer trip reset cable
19 Remove the speedometer, as described previously.
20 Note the run of the reset cable, then slacken the knurled nut and withdraw the cable from the bracket slot.
21 Rotate the control to preset the alignment, then release the cable claws using a small screwdriver. Withdraw the cable from the instrument panel housing.
22 Installation is the reverse of the removal procedure.

Tachometer
23 Follow the procedure given in paragraphs 1 and 2.
24 Remove the face panel, then remove the three crosshead screws and withdraw the instrument.
25 Installation is the reverse of the removal procedure.

---

36 Windscreen washers

Reservoir - removal and installation
1 Pull off the washer reservoir pipe from the washer pump, then lift the reservoir from the retaining bracket.
2 Installation is the reverse of the removal procedure.

Washer jet - removal and installation
3 Pull off the pipe, then remove the nut and anti-vibration washer, taking care that they are not dropped.
4 Remove the jet and the sealing washer.
5 Installation is the reverse of this procedure. If necessary, rotate the jet using a screwdriver to provide a satisfactory aim.

Washer pump - removal and installation
6 Remove the washer reservoir, as previously described.
7 Disconnect the two Lucas connectors, then remove the pump outlet pipe ('IN' and 'OUT' are marked on the pump bracket).
8 Remove the two crosshead screws and washers, and remove the pump. Take care that the screws and washers are not dropped.
9 Installation is the reverse of the removal procedure. Connect the electrical leads as follows:

  light green/black to the positive (+) terminal
  black wire to the negative (–) terminal

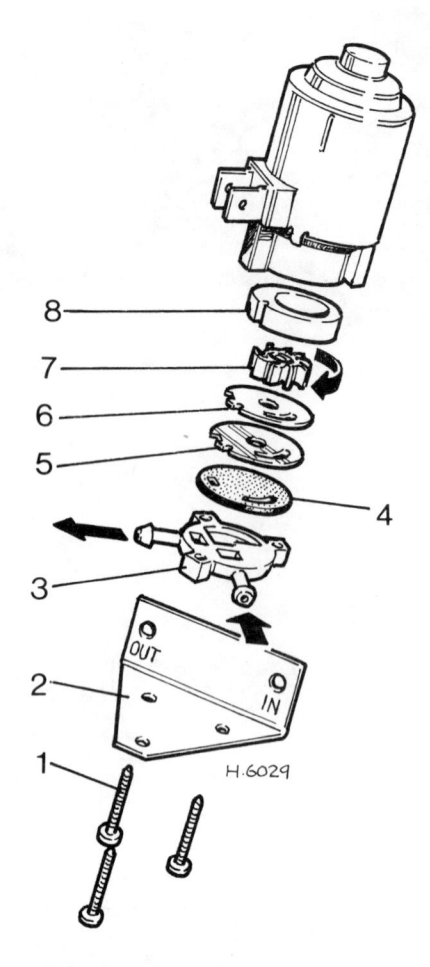

Fig. 10.29. Windscreen washer pump - exploded view

1 Screws
2 Bracket
3 Pump cover
4 Rubber disc
5 Metal disc
6 Plastic disc
7 Rotor (note direction of vanes)
8 Housing

Washer pump - overhaul
10 Remove the three screws to release the bracket. The pump can now be dismantled; the component parts are shown in Fig. 10.29.
11 Clean/renew parts as necessary. The motor is a sealed unit and must be renewed if unserviceable.
12 Reassembly is the reverse of the removal procedure.

---

37 Windscreen wiper arms and blades - removal and installation

Wiper arm - driver's side
1 Using a screwdriver, or similar tool, carefully prise up the clip from the spindle groove.
2 Remove the arm from the spindle, then remove the spindle nut (photo).
3 Take off the distance piece and wiper arm pivot plate.
4 Installation is the reverse of the removal procedure. Ensure that the tongue in the distance piece, locates through the pivot plate, rubber gasket and body slot. Position the arm on the splines to obtain the most suitable 'Park' position.

Wiper arm - passenger's side
5 Raise the wiper arm from the windscreen to the service position.
6 Using a screwdriver, or similar tool, carefully prise up the clip from the spindle groove and lift off the arm.
7 Installation is the reverse of the removal procedure. Position the arm on the spindle to obtain the most suitable 'Park' position.

Chapter 10/Electrical system

Wiper blade - driver's side

8 Lift the arm and blade away from the windscreen, then simultaneously depress the clip and withdraw the blade pin from the pivot block.
9 Installation is the reverse of the removal procedure.

Wiper blade - passenger's side

10 Raise the wiper arm from the windscreen to the service position.
11 Simultaneously lift the clip and withdraw the pin to release the blade from the arm (photo).
12 Installation is the reverse of the removal procedure.

38 Windscreen wiper motor and drive assembly - removal and installation

1 Remove the wiper arms (see Section 37). It is not necessary to remove the spindle nut, distance piece and pivot plate on the driver's side at this stage.
2 Remove the fresh air duct (see Chapter 12).
3 Where applicable, disconnect the two 35 amp Lucas connectors from the battery lead connector.
4 Remove the windscreen washer reservoir (see Section 36).
5 Remove the battery (see Section 2).
6 Disconnect the harness plug, then remove the single screw and disengage the clamp strap from the car body.
7 Remove the two crosshead screws and washers, and remove the plate adjacent to the brake master cylinder.
8 Remove the two spindle nuts, distance pieces and rubber gaskets (also remove the pivot plate on the driver's side - see Section 37, if necessary).
9 Manoeuvre the motor and drive assembly from the car.
10 Installation is the reverse of the removal procedure. If necessary, slacken the olive nuts while the motor and tubes are being installed.

39 Windscreen wiper motor and drive assembly - dismantling, overhaul and reassembly

1 With the motor and drive assembly removed, take off the gearbox cover (five screws), and lift out the slide block.
2 Remove the crankpin retaining clip and washer, then withdraw the connecting rod and second washer.
3 Take out the crosshead, rack and tube assembly.
4 Remove the final gear shaft retaining clip and washer.
5 Ensure that the shaft is not burred, then withdraw it; also remove the dished washer.
6 Remove the worm gear thrust screw and locknut.
7 Remove the through bolts and slowly withdraw the cover and armature. The brushes will drop clear of the commutator, but do not allow them to become contaminated with grease from the worm gear.
8 Pull the armature out of the cover.
9 Scribe around the limit switch box as a guide to reassembly.
10 Remove the brush assembly (three screws) and the limit switch (two screws and washers), followed by the plate.
11 Reassembly is basically the reverse of the dismantling procedure, but the following points should be noted:

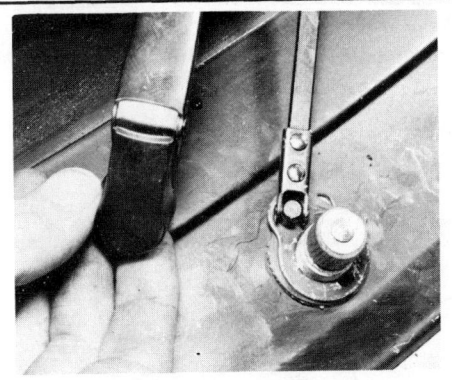

37.2 Wiper arm removal - driver's side

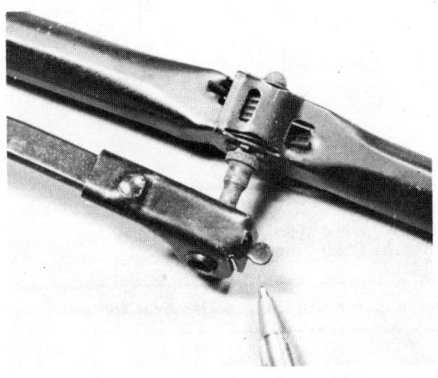

37.11 Wiper blade retaining clip - passenger's side

a) Lubricate the gearbox parts as shown in Fig. 10.31.
b) Position the plate so that the round hole will accommodate the plunger of the limit switch.
c) Lubricate the cover bearing, cover bearing felt washer and the housing self-aligning bearing with Shell Turbo 41 oil.
d) Make up small clips from paper clips or similar sized wire to retain the brushes in position prior to installing the armature.
e) Ensure that the brushes are not contaminated with grease when the armature is installed.
f) Align the cover and gearbox markings.
g) Adjust the armature by screwing in the endfloat screw until resistance is felt, then screw it out ¼ turn (90°). Hold it in this position while the locknut is tightened.
h) Lubricate the final gear bushes with Shell Turbo 41 oil.
j) Fit the dished washer with the concave side towards the final gear.

12 Installation is the reverse of the removal procedure. Position the slide block with the cam slope in the direction shown in Fig. 10.33.

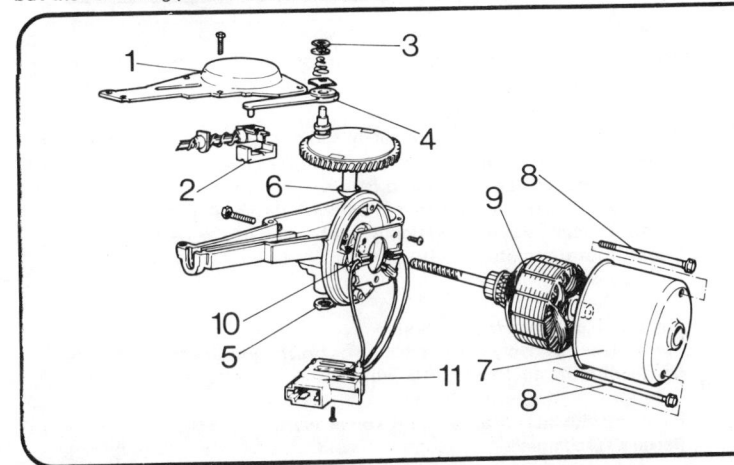

Fig. 10.30. Windscreen wiper motor parts

1 Cover
2 Slider block
3 Crankpin retaining clip
4 Connecting rod
5 Final gear shaft retaining clip
6 Dished washer
7 Cover
8 Through bolts
9 Armature
10 Brush assembly
11 Limit switch

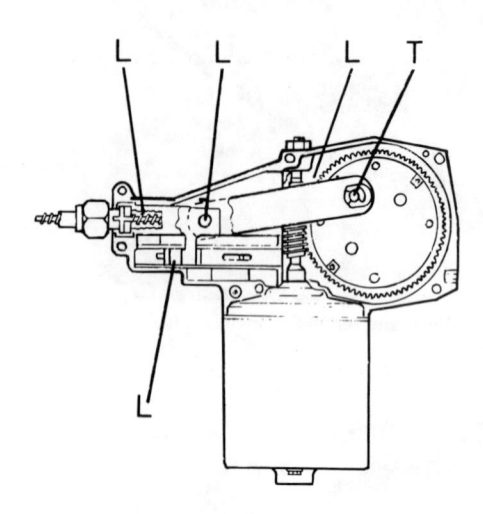

Fig. 10.31. Wiper motor lubrication

L Ragosine Listate Grease
T Shell Turbo 41 oil

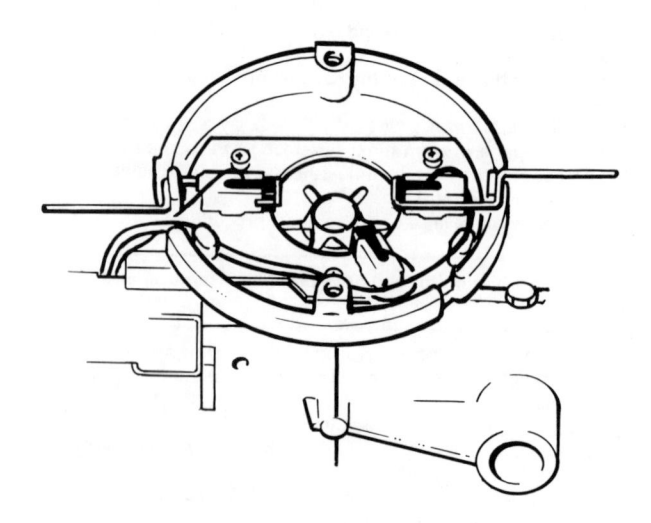

Fig. 10.32. Suitable clips for retaining the brushes

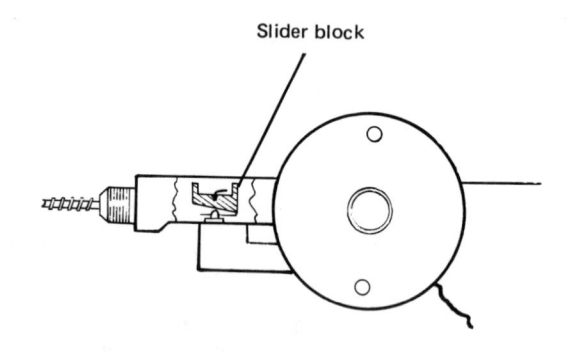

Fig. 10.33. The slider block position

40 Windscreen wiper motor linkage - removal and installation

Rack
1 Remove the motor, as described in the previous Section.
2 Withdraw the rack from the tube assembly and remove the ferrule.
3 Installation is the reverse of the removal procedure. Lubricate the rack with Ragosine Listate grease; if necessary rotate the wheelbox spindles slightly to facilitate engagement of the rack.

Wheelbox
4 Remove the motor, as described in the previous Section.
5 Scribe a line/lines to note the radial position of the tube/tubes.
6 Remove the wheelbox plate (two nuts).
7 On the driver's side, disengage and remove the short straight tube. On the passenger's side, disengage and remove the tube and far wheelbox assembly.
8 Disengage the wheelbox and remove it.
9 Installation is the reverse of the removal procedure. If a new wheelbox is to be used, scribe on a new line/lines. Lubricate the exposed section of the rack and the wheelbox with Ragosine Listate grease.

41 Fault diagnosis - electrical system

Symptom	Reason/s
Starter fails to turn engine	Battery discharged.
	Battery defective internally.
	Battery terminal leads loose or earth lead not securely attached to body.
	Loose or broken connections in starter motor circuit.
	Starter motor switch or solenoid faulty.
	Starter motor pinion jammed in mesh with flywheel gear ring.
	Starter bushes badly worn, sticking, or brush wires loose.
	Starter motor armature faulty.
	Field coils earthed.
Starter turns engine very slowly	Battery in discharged condition.
	Starter brushes badly worn, sticking or brush wires loose.
	Loose wires in starter motor circuit.
Starter spins but does not turn engine	Pinion or flywheel gear teeth broken or worn.
	Battery discharged.

Chapter 10/Electrical system

Starter motor noisy or excessively rough engagement	Pinion or flywheel gear teeth broken or worn. Starter motor retaining bolts loose.
Battery will not hold charge for more than a few days	Battery defective internally. Electrolyte level too low or electrolyte too weak due to leakage. Plate separators no longer fully effective. Battery plates severely sulphated. Fan belt slipping. Battery terminal connections loose or corroded. Alternator not charging. Short in lighting circuit causing continual battery drain. Regulator unit not working correctly.
Ignition light fails to go out, battery runs flat in a few days	Fan belt loose and slipping or broken. Alternator brushes worn, sticking, broken or dirty. Alternator brush springs weak or broken. Internal fault in alternator.

Failure of individual electrical equipment to function correctly is dealt with alphabetically, item-by-item, under the headings listed below

Horn

Horn operates all the time	Horn push either earthed or stuck down. Horn cable to horn push earthed.
Horn fails to operate	Blown fuse. Cable or cable connection loose, broken or disconnected. Horn has an internal fault.
Horns emits intermittent or unsatisfactory noise	Cable connections loose. Horn incorrectly adjusted.

Lights

Headlights do not lift up	Actuator motor faulty. Blown fuse.
Lights do not come on	If engine not running, battery discharged. Wire connections loose, disconnected or broken. Light switch shorting or otherwise faulty.
Lights come on but fade out	If engine not running, battery discharged.
Lights give very poor illumination	Lamp glasses dirty. Lamps badly out of alignment.
Lights work erratically - flashing on and off, especially over bumps	Battery terminals or earth connection loose. Lights not earthing properly Contacts in light switch faulty.

Wipers

Wiper motor fails to work	Blown fuse. Wire connections loose, disconnected or broken. Brushes badly worn. Armature worn or faulty. Field coils faulty.
Wiper motor works very slowly and takes excessive current	Commutator dirty, greasy or burnt. Armature bearings dirty or unaligned. Armature badly worn or faulty.
Wiper motor works slowly and takes little current	Brushes badly worn. Commutator dirty, greasy or burnt. Armature badly worn or faulty.
Wiper motor works but wiper blades remain static	Wiper motor gearbox parts badly worn.

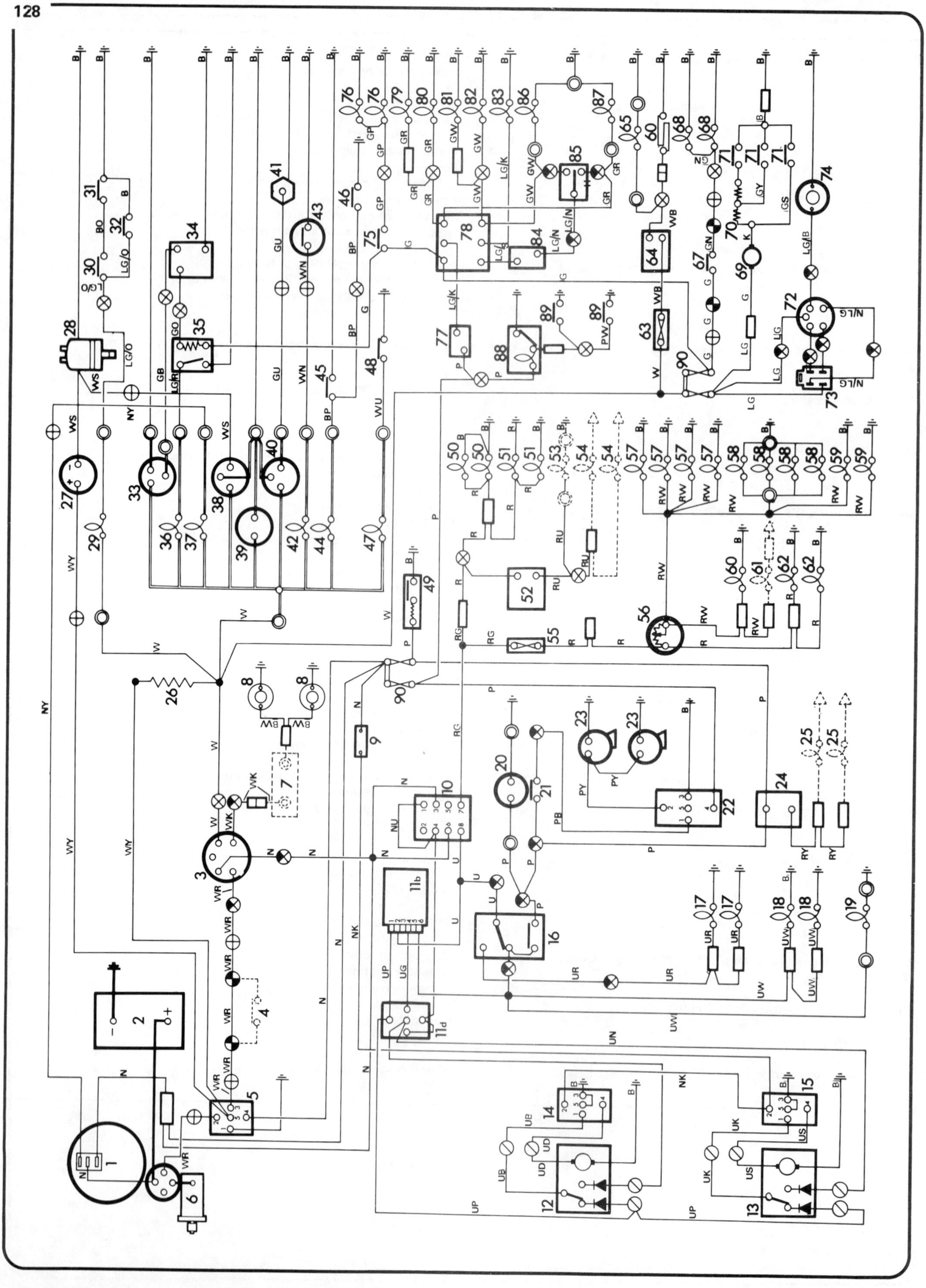

Fig. 10.34. Key to wiring diagram - UK and European models (1976 and 1977)

1	Alternator
2	Battery
3	Ignition/starter switch
4	Starter inhibitor switch (automatic transmission only)
5	Starter motor relay
6	Starter motor
7	Radio
8	Speaker
9	Headlamp circuit breaker
10	Master light switch
11a	Headlamp flash relay
11b	Headlamp flash control unit
12	Headlamp actuator passenger's side
13	Headlamp actuator driver's side
14	Headlamp run/stop relay passenger's side
15	Headlamp run/stop relay driver's side
16	Main/dip/flash switch
17	Dip beam
18	Main beam
19	Main-beam warning light
20	Clock
21	Horn push
22	Horn relay
23	Horn
24	Fog lamp switch
25	Fog lamp
26	Ballast resistor wire
27	Ignition coil
28	Ignition distributor
29	Fasten belts warning light
30	Passenger's seat switch
31	Passenger's belt switch
32	Driver's belt switch
33	Fuel indicator
34	Fuel tank unit
35	Fuel warning light delay unit
36	Fuel warning light
37	Ignition warning light
38	Tachometer
39	Battery condition indicator
40	Temperature indicator
41	Temperature transmitter
42	Oil pressure warning light
43	Oil pressure switch
44	Brake warning light
45	Brake line failure switch
46	Handbrake switch
47	Choke warning light
48	Choke switch
49	Cigarette lighter
50	Number plate illumination lamp
51	Tail lamp
52	Rear fog lamp switch
53	Rear fog lamp warning light
54	Rear fog lamp
55	In-line fuse
56	Panel rheostat
57	Heater control illumination
58	Instrument illumination
59	Facia switch panel illumination
60	Cigarette lighter illumination
61	Selector panel illumination (automatic transmission only)
62	Front parking lamp
63	In-line fuse
64	Heated rear windscreen switch
65	Heated rear windscreen warning light
66	Heated rear windscreen
67	Reverse lamp switch
68	Reverse lamp
69	Heater motor
70	Heater resistor
71	Heater switch
72	Washer/wiper switch
73	Wiper motor
74	Washer pump
75	Stop lamp switch
76	Stop lamp
77	Hazard flasher unit
78	Hazard switch
79	Left-hand front flasher lamp
80	Left-hand rear flasher lamp
81	Right-hand front flasher lamp
82	Right-hand rear flasher lamp
83	Hazard warning light
84	Turn signal flasher unit
85	Turn signal switch
86	Left-hand flasher warning light
87	Right-hand flasher warning light
88	Roof lamp
89	Door switch
90	Fuse

Colour code

B	Black	N	Brown	S	Slate
G	Green	O	Orange	U	Blue
K	Pink	P	Purple	W	White
LG	Light green	R	Red	Y	Yellow

130

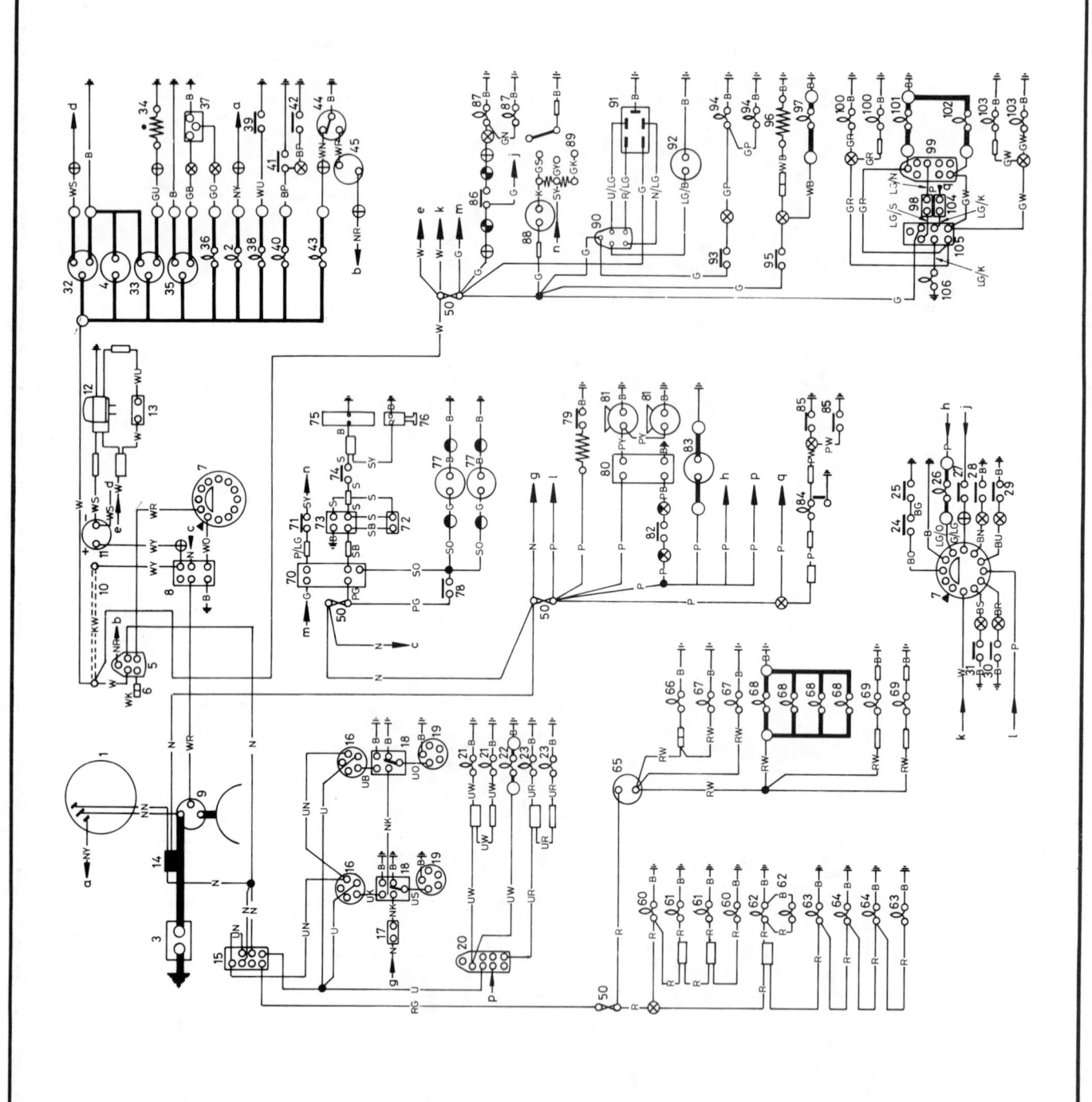

Fig. 10.35. Key to wiring diagram - USA models (1975)

1	Alternator	50	Fuse
2	Ignition warning light	51	Fuse
3	Battery	52	Fuse
4	Battery condition indicator	53	Fuse
5	Ignition/starter switch	60	Front marker lamp
6	Radio supply	61	Front parking lamp
7	Interlock module	62	Number plate illumination lamp
8	Interlock starter motor relay	63	Rear marker lamp
9	Starter motor	64	Tail lamp
10	Ballast resister wire	65	Panel rheostat
11	Ignition coil	66	Cigarette lighter illumination
12	Ignition distributor	67	Heater control illumination
13	Drive resistor	68	Instrument illumination
14	Battery lead connector	69	Facia switch panel illumination
15	Master light switch	70	Air conditioning control relay
16	Actuator - limit switch	71	Cold thermostat
17	Circuit breaker	72	Air conditioning delay circuit flasher unit
18	Headlamp run/stop relay		
19	Actuator motor	73	Air conditioning delay circuit relay
20	Main dip/flash switch	74	High pressure cut-out
21	Main beam	75	Compressor clutch
22	Main beam warning light	76	Throttle jack
23	Dip beam	77	Condenser fan motor
24	Left-hand door switch	78	Radiator switch
25	Key switch	79	Cigarette lighter
26	Fasten belts warning light	80	Horn relay
27	Interlock gearbox switch	81	Horn
28	Driver's belt switch	82	Horn push
29	Driver's seat switch	83	Clock
30	Passenger's belt switch	84	Roof lamp
31	Passenger's seat switch	85	Door switch
32	Tachometer	86	Reverse lamp switch
33	Temperature indicator	87	Reverse lamp
34	Temperature transmitter	88	Blower motor
35	Fuel indicator	89	Blower motor switch
36	Fuel warning light	90	Windscreen washer/wiper switch
37	Fuel tank unit	91	Windscreen wiper motor
		92	Windscreen washer pump
USA Federal market vehicles only		93	Stop lamp switch
38	Choke warning light	94	Stop lamp
39	Choke switch	95	Heated rear windscreen switch
		96	Heated rear windscreen
USA California market vehicles only		97	Heated rear windscreen warning light
38	Catalyst service warning light	98	Turn signal flasher unit
39	Catalyst service interval indicator	99	Turn signal switch
		100	Left-hand flasher lamp
All models		101	Left-hand turn signal warning light
40	Brake warning light	102	Right-hand turn signal warning light
41	Brake line failure switch	103	Right-hand flasher lamp
42	Handbrake switch	104	Hazard flasher unit
43	Oil pressure warning light	105	Hazard switch
44	Oil pressure switch	106	Hazard warning light
45	Anti-run on valve		

COLOUR CODE

B	Black	N	Brown	S	Slate
G	Green	O	Orange	U	Blue
K	Pink	P	Purple	W	White
LG	Light green	R	Red	Y	Yellow

Chapter 11 Suspension and steering

For modifications, and information applicable to later models, see Supplement at end of manual

Contents

Anti-roll bar - removal and installation ... 2	Rear suspension arm - removal and installation ... 9
Bump stop - removal and installation ... 10	Roadwheels and tyres ... 23
Fault diagnosis - suspension and steering ... 24	Steering column assembly - removal, overhaul and installation ... 19
Front hub - removal, dismantling, reassembly, installation and adjustment ... 5	Steering column upper universal coupling - removal and installation ... 21
Front spring and damper - removal and installation ... 3	Steering intermediate shaft - removal and installation ... 20
Front subframe - removal and installation ... 4	Steering lock/ignition switch - removal and installation ... 22
Front suspension lower link - removal, overhaul (including ball joint renewal) and installation ... 7	Steering rack gaiters - removal and installation ... 15
	Steering rack and pinion - removal and installation ... 13
Front wheel alignment - checking ... 17	Steering rack and pinion - overhaul and adjustment ... 14
General description ... 1	Steering wheel - removal and installation ... 18
Radius rod - removal and installation ... 12	Stub axle assembly - removal and installation ... 6
Rear dampers - removal and installation ... 11	Tie-rod ball joints - removal and installation ... 16
Rear spring - removal and installation ... 8	

Specifications

Front suspension

Type ... Independent McPherson strut with telescopic damper co-axial coil springs and anti-roll bar

Coil springs:
  Vehicles with manual or automatic transmission and heater, up to commission Nos ACG13000/ACW7000, and from commission Nos ACG25001/ACW30001
    Free length ... 13.74 in (349 mm)
    Rate ... 88 lbf/in (15.4 N/mm)
    Identification ... Green paint stripe
  Vehicles with manual or automatic transmission and heater, from commission Nos ACG13001/ACW7001 to commission Nos ACG25000/ACW3000
    Free length ... 13.30 in (338 mm)
    Rate. ... 88 lbf/in (15.4 N/mm)
    Identification ... One white and one blue stripe
  Vehicles with manual or automatic transmission and air conditioning, up to commission No ACG13000/ACW7000 and from commission No ACG25001/ACW30001
    Free length ... 13.85 in (352 mm)
    Rate ... 93.7 lbf/in (16.4 N/mm)
    Identification ... One red stripe
  Vehicles with manual or automatic transmission and air conditioning, from commission No ACG13001/ACW7001 to commission No ACG25000/ACW30000
    Free length ... 13.39 in (340 mm)
    Rate ... 93.7 lbf/in (16.4 N/mm)
    Identification ... One white stripe and one yellow stripe

Rear suspension

Type ... Four link system with lower trailing arms, upper trailing radius rods, coil springs, anti-roll bar and telescopic damper

Coil springs:
  Vehicles up to commission Nos ACG13000/ACW7000
    Free length ... 10.71 in (272 mm)
    Rate ... 165 lbf/in (28.9 N/mm)
    Identification ... One white stripe
  Vehicles from commission Nos ACG13001/ACW7001
    Free length ... 10.27 in (261 mm)
    Rate ... 165 lbf/in (28.9 N/mm)
    Identification ... One white stripe and one red stripe

Steering

Type ... Rack and pinion
Steering wheel turns, lock-to-lock ... 3.875
Steering gear lubricant ... General purpose, lithium based, grease

Steering/front suspension geometry

Front wheel alignment ... 0 to 1/16 in (0 to 1.6 mm) toe-in
Camber angle ... ¼° negative $\pm$ 1°
Castor angle ... 3½° positive $\pm$ 1°
King pin inclination ... 11¼° $\pm$ 1°

Chapter 11/Suspension and steering

Wheels
Type ... 5½J x 13 in

Tyres
Sizes ... 175/70 SR 13, 175/70 HR 13, 185/70 SR 13 or 185/70 HR 13 depending on vehicle. For the correct size for your vehicle consult your dealer or tyre specialist

Inflation pressures:
Front ... 24 lb/sq in (1.68 kg/sq cm)
Rear ... 28 lb/sq in (1.96 kg/sq cm)

Torque wrench settings

	lb f ft	Nm
Front suspension		
Anti-roll bar to crossmember	37	50
Anti-roll bar to lower link	59	80
Subframe attachments	59	80
Damper unit closure nut	74	100
Lower link to crossmember	59	80
Lower link to strut assembly	45	61
Strut mounting to body	21	28
Strut to mounting	44	60
Tie-rod lever to stub axle	74	100
Wheel attachment stud	74	100
Rear suspension		
Anti-roll bar to lower link:		
Bolt	37	50
Clamp nut	20	27
Damper to trailing arm:		
Lock nut	14	19
Clamp nut	20	27
Damper to body	10 to 14	14 to 19
Lower and upper link attachments	48	65
Steering		
Rack to crossmember	30	40
Steering column/clamp locating plate jam nut	37	50
Steering wheel nut	37	50
Steering shaft universal joints	21	28
Steering rack damper plug	45 to 60	61 to 80
Tie-rod inner ball joint assembly	37	50
Wheel nuts	74	100

1 General description

The front suspension is of the MacPherson strut type with telescopic dampers, co-axial coil springs and an anti-roll bar. The single lower links are pivoted at their inner ends in rubber bushes set in the subframe assembly. This subframe also carries the anti-roll bar and steering rack assembly. At the upper ends, the springs and dampers are retained in housings in the front wing valances.

The rear suspension comprises a live rear axle, coil springs, separately mounted shock absorbers and an anti-roll bar. The axle is located by twin lower trailing suspension arms and semi-trailing upper radius rods. The dampers, suspension arms and radius rods are rubber mounted at their attachment points.

The steering is a rack and pinion system with twin tie-rods connecting to the steering arms through ball joints. The rack has a spring loaded damper to ensure constant mesh between the pinion and rack teeth; the damper bears on the plain section of the rack. For lubrication, a grease plug is fitted to the top of the damper cap, which must be removed to fit a grease nipple for the routine maintenance lubrication.

The steering wheel connects through a steering shaft, a universal joint, an intermediate shaft and a second universal joint to the steering gear pinion.

The steering and suspension is maintenance-free except for periodic inspection and lubrication of the steering rack.

2 Anti-roll bar - removal and installation

Front
1 Apply the handbrake then raise the front of the car, and support it on blocks or axle stands beneath the subframe or body frame members. For convenience, also remove the roadwheels.
2 Remove two bolts, two nyloc nuts, and a saddle bracket from each side (photo), together with a distance piece washer if fitted. On later models, remove two additional bolts and nuts inboard from the saddle brackets.
3 Remove the spring pin, nyloc nut, flat and dished washer, and bush at each end of the anti-roll bar (photo).
4 Withdraw the anti-roll bar from the car, then take off the inner bush and dished washer from each end.
5 If the rubbers have deteriorated, they should be cut away with a sharp knife. When installing replacements, ensure that the anti-roll bar is clean, then smear it with a proprietary rubber grease or glycerine and slide on the bushes.
6 Installation is basically the reverse of the removal procedure. Attach the outer ends of the anti-roll bar first, followed by the saddle brackets.

Rear
7 Raise the rear of the car and support it on blocks or axle stands beneath the rear axle or suspension arms. Chock the front wheels for safety.
8 Remove the two nuts and bolts at each side, and detach the bar from the suspension. Collect and retain any shims which might be fitted (photo).
9 Installation is the reverse of the removal procedure. If a new anti-roll bar is being installed, shims should be used to just eliminate any endplay.

3 Front spring and damper - removal and installation

1 Apply the handbrake, then raise the front of the car at the appropriate side. Support it beneath the subframe or bodyframe member.

Chapter 11/Suspension and steering

2.2 Front anti-roll bar saddle bracket

2.3 Front anti-roll bar spring pin

2.8 Rear anti-roll bar attachment

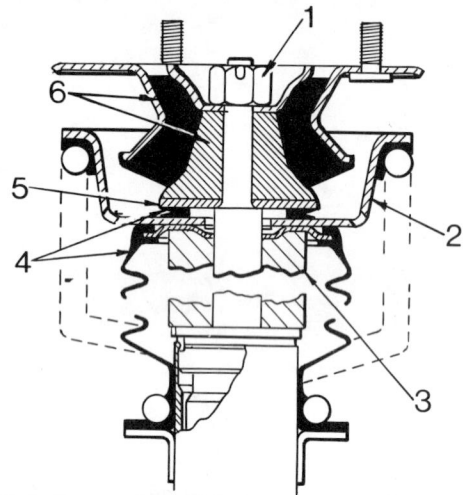

Fig. 11.1. Front spring and damper - top mounting

1. Nut
2. Spring pan
3. Bump stop rubber
4. Seal and thrust collar
5. Large washer
6. Rubber mounting and dished washer

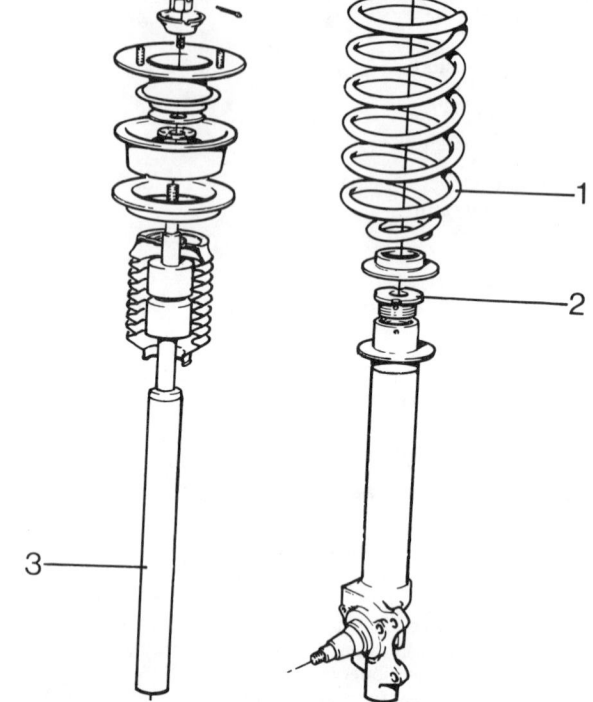

Fig. 11.2. Front damper and spring

1. Spring
2. Closure nut
3. Damper cartridge

3.2 Steering arm retaining bolts

2   Remove the roadwheel, then remove the two bolts to detach the steering arm from the stub axle assembly (photo).
3   Loosen the locknut which secures the brake hose union to the damper tube.
4   Remove the one remaining bolt securing the brake caliper, then support the caliper to prevent strain on the brake pipe.
5   Detach the ball joint from the stub axle assembly, as described in Section 7, paragraphs 4 and 5.
6   Remove the three nyloc nuts at the top damper mounting, then pull the strut clear of the car.
7   Using proprietary spring compressors, compress the spring evenly, and remove the split pin and slotted nut from the top of the damper piston rod.
8   Remove the spring pan, top mounting and swivel assembly, then withdraw the spring.
9   If a suitable tool is available, the closure nut can be unscrewed from the top of the damper (tool number RTR 359 is designed for this application). The damper cartridge can then be removed.
10 Installation of a new damper cartridge is straightforward, the closure nut being tightened to the specified torque.
11 Installation of the spring and damper is basically the reverse of the removal procedure, but ensure that the following points are observed:
  a) With the damper piston rod fully extended, the lower insulating ring, rubber gaiter, spring, upper insulating ring and spring pan should be installed.
  b) At the upper end, fit the parts as shown in Fig.11.2. Lightly coat the large washer with a general purpose molybdenum disulphide grease, and fit it with the ground face downwards.
  c) At the upper mounting, ensure that the spring turret is clean, and apply a proprietary sealant such as Plasti-seal to the mounting flange.

Chapter 11/Suspension and steering

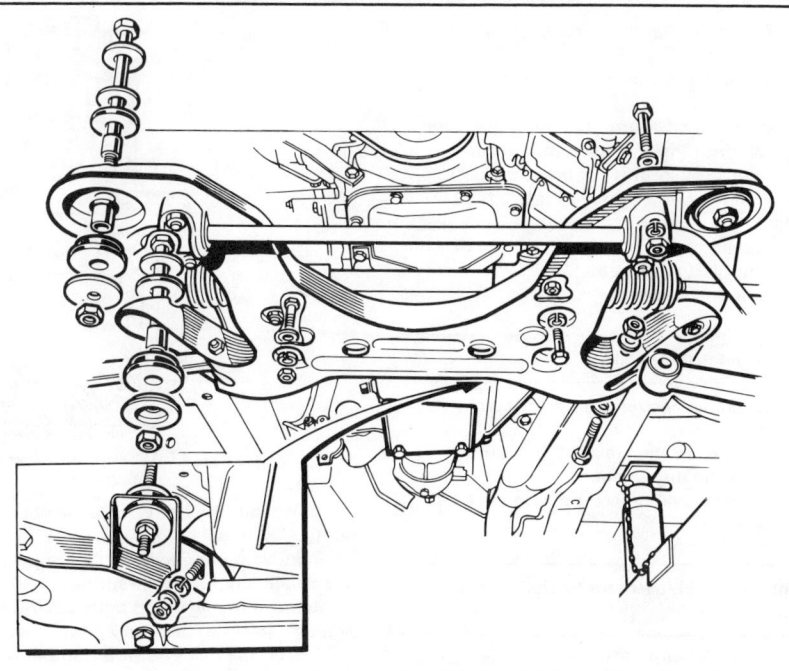

Fig. 11.3. Subframe attachment points (left-hand drive version shown)

d) Do not tighten the brake caliper upper mounting bolt until the steering arm bolts are tightened (the rear steering arm bolt also secures the brake caliper).

4 Front subframe - removal and installation

1 Remove the engine stabilizer, referring to Chapter 1, if necessary (applicable to early models only).
2 Remove the nut, spring washer and flat washer securing the engine mounting bracket to the subframe.
3 Jack-up the car, and support it on the bodyframe sidemembers so that the weight of the body is just taken off the front suspension.
4 Support the engine with a jack placed beneath the sump coupling plate.
5 Detach the steering rack from the subframe (refer to Section 13, if necessary).
6 Detach the anti-roll bar mountings from the subframe (refer to Section 2, if necessary).
7 Remove the two nuts, spring washers and flat washers securing the lower links to the subframe.
8 Adjust the jacks, if necessary, to relieve any load from the bolts, then drive the bolts out to release the lower links.
9 Remove the nuts, retainers, mounting rubbers and sleeves from the two rear subframe mounting points.
10 Support the subframe and remove the nuts, retainers, mounting rubbers and sleeves from the two front subframe mounting points.
11 Lower the subframe from the car.
12 Installation is the reverse of the removal procedure. Ensure that the four mounting rubbers and washers are correctly positioned between the subframe and body, and ensure that the nuts are torque tightened before any other parts are attached.
13 On completion, adjust the engine tie-rod, as described in Chapter 1.

5 Front hub - removal, dismantling, reassembly, installation and adjustment

1 Proceed as described in Section 3, paragraphs 1 to 4, but ensure that the brake caliper is supported clear of the brake disc.

2 Prise off the hub grease cap and wipe away the surplus grease from the end of the stub axle.
3 Remove the split pin, retaining cap, nut and washer.
4 Withdraw the hub and disc, complete with bearings and oil seal.
5 Remove the outer bearing; remove the inner oil seal and inner bearing.
6 Drive out the bearing inner races from the hub.
7 Clean all the parts carefully in petrol or paraffin (gasoline or kerosene), and check for obvious wear, scoring, signs of overheating (a bluish colour), etc. Brown grease stains are of no significance. Ensure that the bearings run smoothly when assembled to their outer tracks. Replace bearings, where necessary, as complete assemblies, and on no account mix up parts of bearings from one side of the car to the other.
8 Press in the bearing outer tracks, then lubricate the bearing races with a general purpose grease, working it well in with the fingers.
9 Partially pack the hub with grease, then install the bearing races.
10 Install a new seal with the lip facing the inner bearing. Lubricate

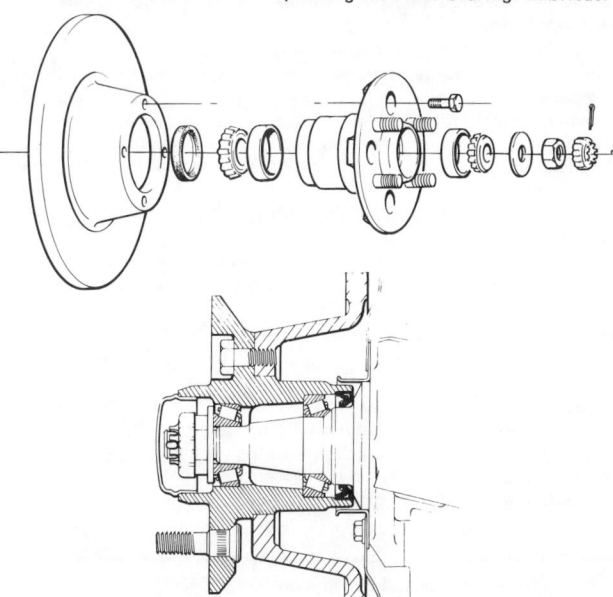

Fig. 11.4. Front hub sectional view and components

the seal, and push on the hub. Take care that the outer bearing is not dislodged.
11 Fit the washer to the stub axle and screw on the nut finger-tight only.
12 Install the brake pipe, caliper and steering arm. Do not tighten the brake caliper upper mounting bolt until the steering arm bolts are tightened (the rear steering arm bolt also secures the brake caliper).
13 Adjust the hub by tightening the nut to a torque of 5 lb f ft (0.69 kg fm) then back it off one flat.
14 Install the retaining cap and retain it with a new split pin. Install the hub grease cap (there is no need to fill it with grease).
15 Install the roadwheel and lower the car to the ground.

6 Stub axle assembly - removal and installation

1 Remove the front spring and damper cartridge, as described in Section 3.
2 Remove the front hub, as described in the previous Section.
3 Remove the three bolts securing the disc shield.
4 Installation is the reverse of the removal procedure, reference being made to Sections 3 and 5, as necessary.

7 Front suspension lower link - removal, overhaul (including ball joint renewal) and installation

Note: It is necessary to remove the lower link if the suspension ball joint is to be renewed.
1 Apply the handbrake, then raise the front of the car at the appropriate side. Support it beneath the subframe or the bodyframe member.
2 Remove the roadwheel, then detach the end of the anti-roll bar from the lower link (refer to Section 2, if necessary).
3 Remove the two bolts securing the steering arm to the stub axle assembly; push the steering arm to one side.
4 Remove the split pin and unscrew the castellated nut from the ball joint; remove the flat washer.
5 Pull back the rubber boot from the ball joint then use a proprietary ball joint separator or split wedges to separate the ball joint from the stub axle. If neither tool is available, the joint can usually be separated by driving in the taper of a cold chisel between the ball joint casing and the stub axle flange at the base of the taper.
6 Remove the bolt and nyloc nut which secure the lower link to the subframe, and withdraw the link.
7 To remove the ball joint from the lower link, remove the retaining circlip then carefully drive or press it out of the link.
8 Ensure that the ball joint bore in the link is clean and free from burrs, then lightly lubricate it and squarely press in the replacement. Take care that load is applied to the housing only, and not to the spring loaded cap.
9 Retain the ball joint with a new circlip.
10 Fill the rubber boot with a general purpose grease and secure it with the rubber retaining ring.
11 Press out the old rubber bush and sleeve, then lubricate a new one with a proprietary rubber grease or glycerine, and press it in.
12 To install the lower link, first locate the anti-roll bar into the lower mounting hole.
13 Install the link in the pivot in the subframe, but do not tighten the nyloc nut at this stage.
14 Ensure that the ball joint taper is clean and very lightly greased, then install it in the stub axle. Secure it with the flat washer, castellated nut and a new split pin.
15 Raise the lower link with a jack to locate the outer rubber bush, dished washer, flat washer and nyloc nut onto the end of the anti-roll bar.
16 Tighten the nyloc nut to the stop and install the spring pin.
17 Lower the lower link jack, and install the steering arm.
18 Fit the roadwheel and lower the car to the ground.
19 With the weight of the car on the suspension, tighten the nyloc nut at the subframe pivot to the specified torque.

8 Rear spring - removal and installation

1 Jack-up the rear of the car and support it on blocks or an axle

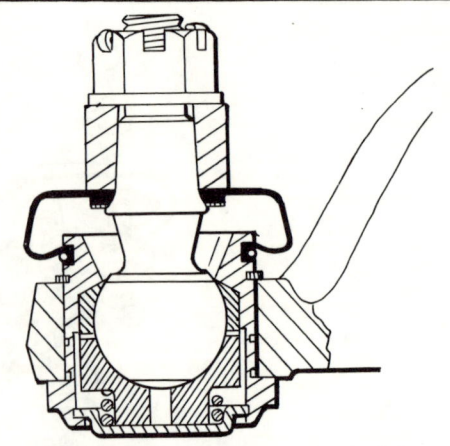

Fig. 11.5. Sectional view of suspension balljoint

stand beneath the bodyframe member. Chock the front wheels for safety.
2 Remove the roadwheel, then support the suspension arm on a jack to partially compress the spring.
3 Remove the nuts and bolts securing the anti-roll bar to the suspension arm (refer to Section 2, if necessary).
4 Detach the rear end of the suspension arm from the axle (one nut and bolt).
5 Carefully lower the jack, then remove the spring and, its upper and lower insulating rubbers.
6 Installation is the reverse of the removal procedure, but the following points should be noted:
   a) Ensure that the upper and lower spring insulating rubbers are correctly positioned.
   b) Do not finally tighten the suspension arm bush nut and bolt until the weight of the car is on the wheels.

9 Rear suspension arm - removal and installation

1 Remove the rear spring, as described in the previous Section.
2 Remove the nut and bolt securing the suspension arm to the bodyframe, and detach the arm from the body.
3 If the bushes have deteriorated, they should be pressed out and replacements pressed in. Ensure that the front bush is positioned as shown in Fig. 11.7.
4 Installation is the reverse of the removal procedure. Do not fully tighten either of the suspension arm nuts and bolts until the weight of the car is on the wheels.

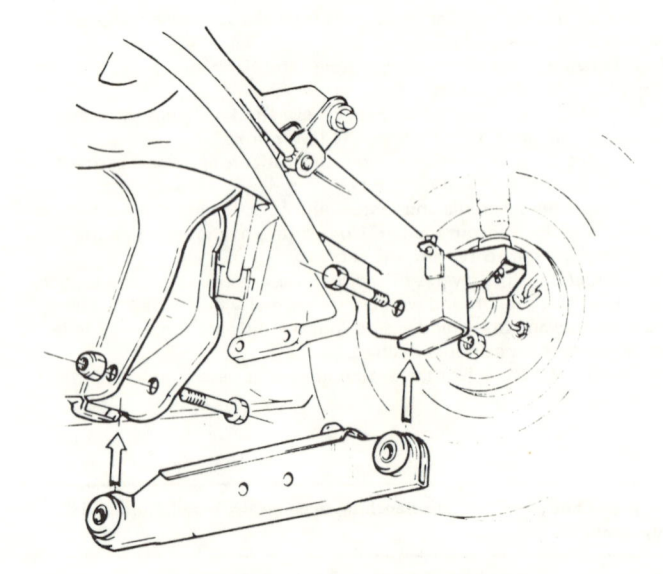

Fig. 11.6. Rear suspension arm removal

Chapter 11/Suspension and steering

10 Bump stop - removal and installation

1 The bump stops are a press fit in the bodyframe sidemembers and can be readily prised out if renewal becomes necessary.
2 Replacements are simply pressed into position.

11 Rear dampers - removal and installation

1 Jack-up the rear of the car, and support it on blocks or on axle stands beneath the bodyframe sidemember. Chock the front wheels for safety.
2 Remove the roadwheel.
3 *Left-hand damper:* Remove the damper access plate (three screws and washers) from inside the luggage compartment.
4 *Right-hand damper:* Remove the fuel filler cap and filler assembly, referring to Chapter 3 if necessary.
5 Remove the locknut, nut, flat washer and rubber bush securing the damper at the upper end.
6 Similarly detach the lower damper mounting from the axle bracket.
7 From each end of the damper, remove the rubber bush and plain washer.
8 Installation is basically the reverse of the removal procedure, but it is preferable to attach the upper mounting before the lower one. On the left-hand damper apply a proprietary sealant such as Plasti-seal to the damper cover plate.

12 Radius rod - removal and installation

1 Jack-up the rear of the car and support it on blocks or an axle stand beneath the bodyframe sidemember. Chock the front wheels for safety.
2 Remove the retaining nut and bolt from each end of the radius rod, then withdraw the rod downwards.
3 If the rubber bushes have deteriorated they should be pressed out and replacements installed. Installation is made easier if a little glycerine or a proprietary rubber lubricant is used.
4 Installation of the radius rod is a straightforward reversal of the removal procedure.

13 Steering rack and pinion - removal and installation

1 Apply the handbrake then raise the front of the car, and support it on blocks or axle stands beneath the bodyframe members. For convenience also remove the roadwheels.
2 With the wheels in the straight-ahead position, index mark the pinion shaft and lower steering coupling to assist with alignment when installing. (This is not necessary if the steering rack is to be dismantled or a different steering rack installed).
3 Remove the nuts from the tie-rod ball joints on the steering arms.
4 Pull back the rubber boot on each ball joint then use a proprietary ball joint separator or split wedges to separate the ball joint from the steering arm. If neither tool is available the joints can usually be separated by driving in the taper of a cold chisel between the ball joint casing and the steering arm at the base of the taper.
5 Remove the pinch bolt securing the steering coupling to the pinion splined end.
6 Remove the two bolts, spring washers and flat washers securing the rack to the subframe at the pinion end.
7 Remove the two nyloc nuts and flat washers at the opposite end of the rack, and withdraw the bolts.
8 Disconnect the lower coupling from the pinion shaft and pull out the rack from the driver's side of the car.
9 When installing, check that the wheels are in the straight-ahead position and push in the rack from the driver's side.
10 Engage the lower coupling and pinion shaft so that the index markings are aligned. If there are no markings, remove the centre plug from the thrust pad and locate the dimple in the rack shaft by passing a stiff wire through the plug aperture. Move the pinion shaft as necessary to get the dimple in the centre of the aperture. Fit a bolt in place of the centre plug and lightly tighten it to hold the rack in the centred position.

10.1 Rear bump stop

13.6 Steering rack retaining bolts ...

13.7 ... and nuts

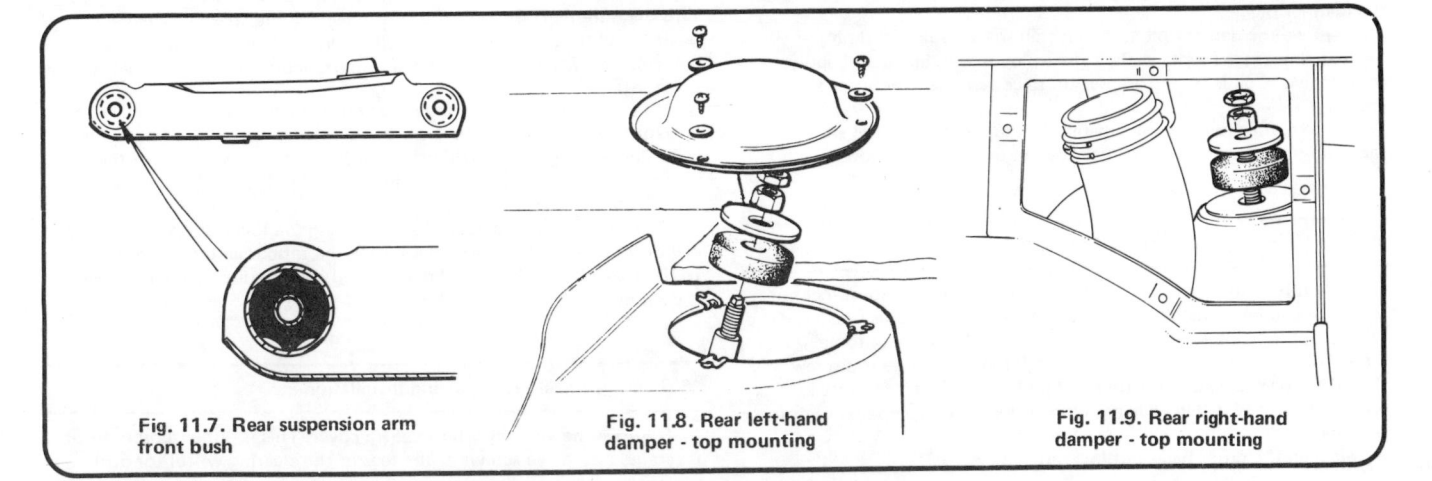

Fig. 11.7. Rear suspension arm front bush

Fig. 11.8. Rear left-hand damper - top mounting

Fig. 11.9. Rear right-hand damper - top mounting

11 The remainder of the installation procedure is the reverse of that used for removal. Where applicable do not forget to install the damper centre plug; on completion the front wheel alignment should be checked - see Section 17.

14 Steering rack and pinion - overhaul and adjustment

Note: The steering rack and pinion is easily dismantled, reassembled and adjusted. However, in cases where there is a great deal of wear, it is recommended that spare part availability is checked before dismantling begins as it may prove more worthwhile to obtain a service exchange item.

1 With the steering rack removed from the car, remove the damper hexagon plug, spring, shim(s) and damper.
2 Remove the pinion shaft seal then unscrew the pinion retaining plug using a suitable tool.
3 Withdraw the pinion and bearing, then remove the circlip securing the bearing. The bearing can now be pressed or pulled off.
4 Remove the gaiter clips, and slide the gaiters clear of the rack.
5 Slide the pinion end of the rack housing towards its tie-rod inner ball joint.
6 Using protective jaw clamps grip the exposed end of the rack. Now unscrew both tie-rod inner ball joint assemblies and withdraw the rack shaft.
7 The rack housing bush can now be driven out with a suitable drift.
8 Clean all the parts in petrol or paraffin (gasoline or kerosene), dry them carefully and inspect for wear and damage. Renew parts as necessary.
9 Reassembly is basically the reverse of the removal procedure, but the following points should be noted:
   a) With the plain portion of the rack shaft held in a vice with protective jaw clamps, fit the inner ball joint, tighten the cap nut to the specified torque and fold over the washer tabs.
   b) With the rack positioned in the housing, install the pinion-end ball joint in a similar manner.
   c) Smear the pinion shaft with general purpose grease before installing it.
   d) The ends of the rack shafts and the tie-rod inner ball joints should be packed with general purpose grease.
   e) After installing the rack damper, adjust it as described in the following paragraphs.

Adjustment

10 With the centre plug of the damper plug removed, locate the dimple in the rack shaft by passing a stiff wire through the plug aperture. Move the pinion shaft, as necessary, to get the dimple in the centre of the aperture; the rack is now in the mid-position. Mark the pinion shaft and housing for reference.
11 Remove the damper plug, spring and shims then gently screw in the damper plug on its own, until the plunger grips the rack, eliminating all side play.
12 Using feeler gauges, measure the gap between the pinion housing and the underside of the damper plug flange. To the measured gap, add the required rack side movement of 0.001 to 0.007 in (0.03 to 0.18 mm). This gives the total amount of shim thickness required beneath the damper plug.
13 Remove the damper plug, then install spring, shims and plug again.
14 Install a grease nipple in the centre plug screw thread and apply four or five strokes from a grease gun filled with general purpose grease.
15 Remove the grease nipple, install the plug, then refit the steering rack and pinion to the car, as described in the previous Section.

15 Steering rack gaiter - removal and installation

1 Carefully mark the position of the tie-rod outer ball joint locknut so that it can be re-installed in the same position and thus prevent upsetting the front wheel alignment. If new ball joints are to be installed, note the distance from the beginning of the tie-rod screw thread to the centre of the outer ball joint cap. (This is because a replacement may differ slightly in dimension when compared with the original one).
2 Slacken the outer ball joint locknut.
3 Detach the outer ball joint, as described in paragraphs 3 and 4, of Section 13.
4 Unscrew the ball joint and locknut from the tie-rod.
5 Remove the gaiter retaining clips and slide the gaiter off the tie-rod.
6 When installing, lubricate the inner ball joint with a general purpose grease and slide on the gaiter.
7 With the rack centralized, install the inboard clip on the gaiter.
8 Install the outboard clip on the gaiter whilst checking that the steering can be moved from lock-to-lock without the gaiter being strained.
9 The remainder of the installation procedure is the reverse of that used when removing.
10 On completion, if there is any doubt about the ball joint(s) not being installed in their original positions, the front wheel alignment must be checked - see Section 17.

16 Tie-rod ball joints - removal and installation

Outer
1 Refer to the procedure given in paragraphs 1 to 4, of the previous Section.
2 Install the replacement ball joint on the tie-rod. If the original position is not known, screw it on until the distance between centres of the ball joints is approximately 13.33 in. (338 mm).
3 The remainder of the installation procedure is the reverse of that used when removing.
4 On completion, if there is any doubt about the ball joint(s) not being installed in their original positions, the front wheel alignment must be checked - see Section 17.

Inner
5 Remove the outer ball joint and detach the gaiter, by reference to the procedure in Section 15.
6 Wipe the inner ball joints to remove excess dirt and grease.
7 Fold back the locking tabs, then unscrew the ball socket halves. To prevent stress being applied to the rack pinion, it is recommended that an assistant holds the opposite ball joint with a spanner.
8 Installation is the reverse of the removal procedure. In view of the difficulty in checking the dimension between centres of the inner and outer ball joints - which should be 13.33 in. (338 mm) - it is strongly recommended that the front wheel alignment is checked on completion. For further information refer to the following Section.

17 Front wheel alignment - checking

1 In order to minimize tyre wear, and retain the correct steering and roadholding characteristics, it is essential that the front wheels are correctly aligned. Ideally, the alignment should be checked using special gauges: It is, therefore, recommended that the job is done by a Triumph dealer. However, it is possible to do the check with a reasonable amount of accuracy if care is taken.
2 The front wheels are correctly aligned when they are turning inwards at the front by 0 to 1/16 in. (0 to 1.6 mm); this is the toe-in. This measurement is made with the wheels in the straight-ahead position, with the steering rack in the mid-position of its travel, the ball centres of the tie-rods equal and the steering wheel centre spot horizontal.
3 It is important that the measurement is taken on a centre line drawn horizontally and parallel to the ground through the centre-line of the hub. The exact point should be in the centre of the sidewall of the tyre and not on the wheel rim which could be distorted and so give inaccurate readings.
4 The adjustment is effected by loosening the locknut on each tie-rod outer ball joint and also slackening the rubber gaiter clip holding it to the tie-rod, both tie-rods being turned equally until the adjustment is correct.

18 Steering wheel - removal and installation

1 Remove the steering wheel centre cover. This is either a press fit or is retained by three screws at the rear of the steering wheel (photo).

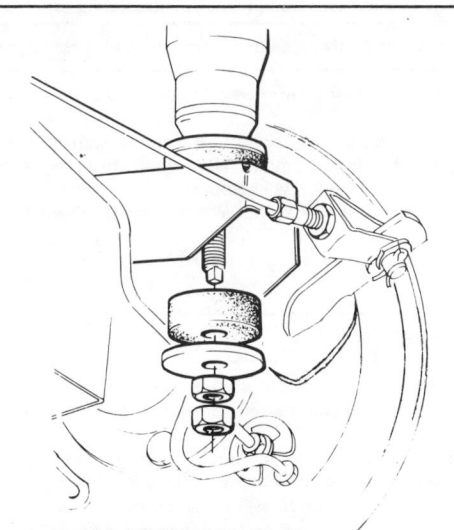

Chapter 11/Suspension and steering

Fig. 11.10. Rear right-hand damper - lower mounting

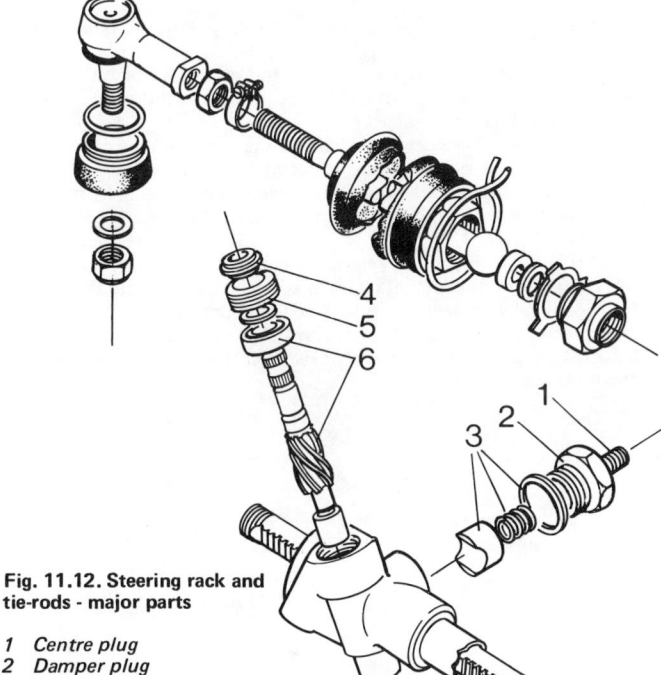

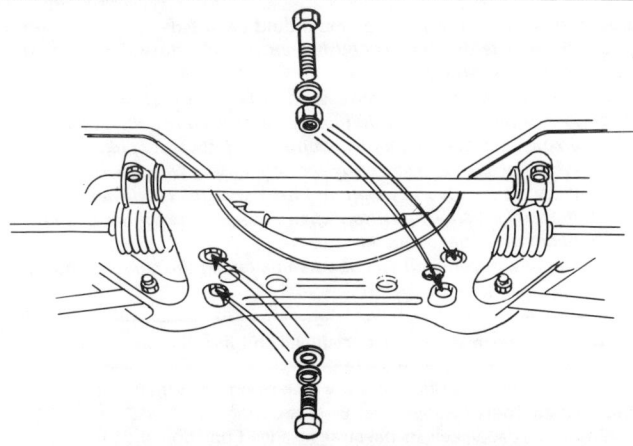

Fig. 11.11. Steering rack attachment points (right-hand drive version shown)

Fig. 11.12. Steering rack and tie-rods - major parts

1. Centre plug
2. Damper plug
3. Spring, shim(s) and damper
4. Rubber seal
5. Pinion retaining plug
6. Bearing and pinion

18.1 Steering wheel centre cover removed

21.1 Steering column upper universal coupling

2 With the roadwheels in the straight-ahead position, remove the steering wheel retaining nut and washer.
3 Index mark the steering wheel hub and the top of the steering mast to assist with alignment when installing.
4 Using a suitable extractor, pull off the steering wheel. If one is not available, apply blows from the ball of the hand; do not drive or tap the wheel off in any other way.
5 Installation is the reverse of the removal procedure, but ensure that the arrow on the direction indicator cancelling collar aligns with the centre of the direction indicator stalk.

19 Steering column assembly - removal, overhaul and installation

1 Disconnect the battery earth lead then remove the pinchbolt from the steering mast universal coupling.
2 Remove the wiring loom cleat from the steering column, then disconnect the three or four plug-in connectors.

3 Remove the two screws and take off the steering column nacelle halves.
4 Either remove the two shear-head screws securing the column housing to the body using a hammer and a small chisel, or centre-punch and drill them, then use a screw extractor.
5 With the roadwheels in the straight-ahead position withdraw the steering column assembly. Note the position of the flat and wavy washers.
6 Remove the steering wheel using the procedure given in the previous Section.
7 Loosen the clamping screw and withdraw the switch stalk assembly.
8 Remove the two shear-head screws securing the steering lock, as described for the column housing screws, in paragraph 4.
9 Withdraw the column housing from the steering mast, then remove the nut and bolt and take off the clamp.
10 Using a suitable drift, drive out the top and bottom column housing bushes.
11 Reassembly and installation is basically the reverse of the foregoing

procedure, but the following points should be noted:
a) When installing the new bushes, ensure that slots align with the lugs in the housing.
b) Use new shear-head screws when installing the steering lock.
c) When fitting the direction indicator cancelling collar, align the arrow with the centre of the direction indicator stalk.
d) When installing the steering column, ensure that the flat and wavy washers are in position. Check that the roadwheels are in the straight-ahead position when engaging the splines of the upper universal coupling.
e) Use new shear-head screws when installing the column housing.

20 Steering intermediate shaft - removal and installation

1 Remove the pinchbolt from the rack pinion coupling, and from the intermediate shaft-to-upper universal coupling.
2 With the roadwheels in the straight-ahead position slide the intermediate shaft upwards to disengage the lower end.
3 Now slide the shaft downwards to remove it from the upper coupling.
4 Installation is the reverse of the removal procedure, but ensure that the roadwheels are still in the straight-ahead position.

21 Steering column upper universal coupling - removal and installation

1 Slacken the pinchbolt on the steering mast side of the upper universal coupling then turn the steering wheel so that it can be removed (photo).
2 Remove the pinchbolt on the intermediate shaft side of the coupling.
3 With the roadwheels in the straight-ahead position, slide the coupling downwards to clear the steering mast. Note the position of the two washers on the steering mast to ensure correct assembly.
4 Remove the coupling from the intermediate shaft.
5 Installation is the reverse of the removal procedure, but ensure that the steering wheel and roadwheels are in the straight-ahead position. Do not forget the two washers on the steering mast.

22 Steering lock/ignition switch - removal and installation

1 Withdraw the key from the switch and remove the nacelle halves (two screws).
2 Either remove the two shear-head screws securing the steering lock/ignition switch using a hammer and a small chisel, or centre-punch and drill them, then use a screw extractor.
3 Remove the plug-in connector to the ignition switch then remove the steering lock.
4 Installation is the reverse of the removal procedure.

23 Roadwheels and tyres

1 Whenever the roadwheels are removed it is a good idea to clean the insides to remove accumulations of mud and in the case of the front ones, disc pad dust.
2 Check the condition of the wheel for rust and repaint if necessary.
3 Examine the wheel stud holes. If these are tending to become elongated or the dished recesses in which the nuts seat have worn or become overcompressed, then the wheel will have to be renewed.
4 With a roadwheel removed, pick out any embedded flints from the tread and check for splits in the sidewalls or damage to the tyre carcass generally.
5 Where the depth of tread pattern is 1 mm or less, the tyre must be renewed.
6 Rotation of the roadwheels to even out wear is a worthwhile idea if the wheels have been balanced off the car. Include the spare wheel in the rotational pattern.
7 If the wheels have been balanced on the car then they cannot be moved round the car as the balance of wheel, tyre and hub will be upset. In fact their exact stud fitting positions must be marked before removing a roadwheel so that it can be returned to its original 'in-balance' state.
8 It is recommended that wheels are re-balanced halfway through the life of the tyres to compensate for the loss of tread rubber due to wear.
9 Finally, always keep the tyres (including the spare) inflated to the recommended pressures and always refit the dust caps on the tyre valves. Tyre pressures are best checked first thing in the morning when the tyres are cold.

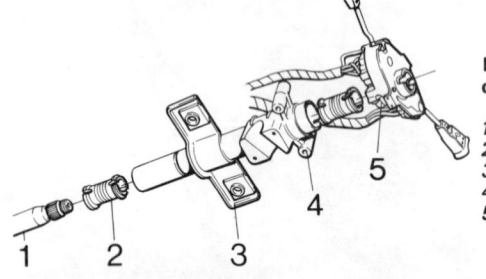

Fig. 11.13. Steering column assembly

1 Steering mast
2 Bottom bush
3 Column housing
4 Top bush
5 Switch assembly

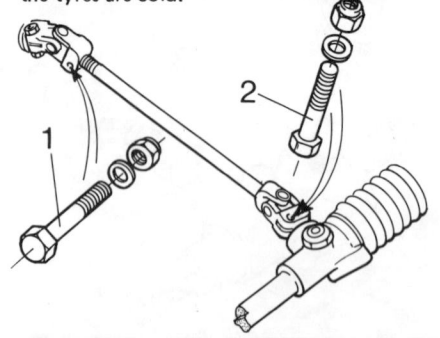

Fig. 11.14. Intermediate shaft coupling bolts

1 Pinch bolt - shaft to upper coupling
2 Pinch bolt - lower coupling to rack pinion

24 Fault diagnosis - suspension and steering

Symptom	Reason/s
Steering feels vague, car wanders and 'floats' at speed	Tyre pressures uneven. Dampers worn. Steering gear ball joints badly worn. Steering mechanism free play excessive. Front rear suspension pick-up points out of alignment.
Stiff and heavy steering	Tyre pressures too low. No grease in steering gear. Front wheel toe-in incorrect. Steering gear incorrectly adjusted too tightly. Steering column badly misaligned.
Wheel wobble and vibration	Wheel nuts loose. Front wheels and tyres out of balance. Steering ball joints badly worn. Hub bearings badly worn. Steering gear free play excessive. Front springs weak or broken.

Chapter 12 Bodywork and fittings

For modifications, and information applicable to later models, see Supplement at end of manual

Contents

Air-conditioning ... 36	Fan motor - removal and installation ... 35
Air hoses and swivelling vents - removal and installation ... 33	Front grille - removal and installation ... 7
Air vent grille - removal and installation ... 8	General description ... 1
Bonnet - removal and installation ... 9	Glovebox lid and lock - removal and installation ... 22
Bonnet catches and controls ... 10	Heated rear window - removal and installation ... 30
Bumpers - removal and installation ... 6	Heater unit - removal and installation ... 34
Centre grille (radio loudspeaker grille) and facia instrument cowl - removal and installation ... 21	Luggage compartment lid - removal and installation ... 11
Console assembly - removal and installation ... 23	Luggage compartment lock and striker ... 12
Control cowl - removal and installation ... 24	Maintenance - bodywork and underframe ... 2
Demister ducts - removal and installation ... 32	Maintenance - upholstery and carpets ... 3
Doors - removal and installation ... 14	Major body damage - repair ... 5
Door glass and regulator - removal and installation ... 19	Minor body damage - repair ... 4
Door lock and outside handle - removal and installation ... 16	Rear quarter trim and parcel tray - removal and installation ... 25
Door lock remote control - removal and installation ... 15	Seats and seat runners - removal and installation ... 26
Door lock striker - removal, installation and adjustment ... 17	Seatbelts - removal and installation ... 27
Door quarter light - removal and installation ... 18	Sliding roof - removal and installation ... 31
Door rattles - tracing and rectification ... 13	Windscreen - removal and installation ... 29
Facia and glovebox cowl - removal and installation ... 20	Windscreen lower finisher strip - removal and installation ... 28

1 General description

The body and underframe are a unitary welded steel construction; it is provided in a closed top, 2-door, 2-seat form only.

The doors are forward hinged, with anti-burst locks and flush door handles. A laminated glass windscreen and heated rear window are fitted as standard equipment.

2 Maintenance - bodywork and underframe

1 The general condition of a car's bodywork is the one thing that significantly affects its value. Maintenance is easy but needs to be regular and particular. Neglect, particularly after minor damage, can lead quickly to further deterioration and costly repair bills. It is important also to keep watch on those parts of the car not immediately visible, for instance the underside, inside all the wheel arches and the lower part of the engine compartment.
2 The basic maintenance routine for the bodywork is washing - preferably with a lot of water, from a hose. This will remove all the loose solids which may have stuck to the car. It is important to flush these off in such a way as to prevent grit from scratching the finish.
   The wheel arches and underbody need washing in the same way to remove any accumulated mud which will retain moisture and tend to encourage rust. Paradoxically enough, the best time to clean the underbody and wheel arches is in wet weather when the mud is thoroughly wet and soft. In very wet weather the underbody is usually cleaned of large accumulations automatically, and this is a good time for inspection.
3 Periodically it is a good idea to have the whole of the underside of the car steam cleaned, engine compartment included, so that a thorough inspection can be carried out to see what minor repairs and renovations are necessary. Steam cleaning is available at many garages and is necessary for removal of accumulation of oily grime which sometimes is allowed to cake thick in certain areas near the engine, gearbox and back axle. If steam facilities are not available, there are one or two excellent grease solvents available which can be brush applied. The dirt can then be simply hosed off.
4 After washing paintwork, wipe off with a chamois leather to give an unspotted clear finish. A coat of clear protective wax polish will give added protection against chemical pollutants in the air. If the paintwork sheen has dulled or oxidised, use a cleaner/polisher combination to restore the brilliance of the shine. This requires a little effort, but is usually caused because regular washing has been neglected. Always check that the door and ventilator opening drain holes and pipes are completely clear so that water can drain out. Bright work should be treated the same way as paintwork. Windscreens and windows can be kept clear of the smeary film which often appears, if a little ammonia is added to the water. If they are scratched, a good rub with a proprietary metal polish will often clear them. Never use any form of wax or other body or chromium polish on glass.

3 Maintenance - upholstery and carpets

1 Mats and carpets should be brushed or vacuum-cleaned regularly to keep them free of grit. If they are badly stained remove them from the car for scrubbing or sponging and make quite sure they are dry before replacement. Seats and interior trim panels can be kept clean by a wipe over with a damp cloth. If they do become stained (which can be more apparent on light coloured upholstery) use a little liquid detergent and a soft nail brush to scour the grime out of the grain of the material. Do not forget to keep the head lining clean in the same way as the upholstery. When using liquid cleaners inside the car do not over-wet

This sequence of photographs deals with the repair of the dent and paintwork damage shown in this photo. The procedure will be similar for the repair of a hole. It should be noted that the procedures given here are simplified – more explicit instructions will be found in the text

In the case of a dent the first job – after removing surrounding trim – is to hammer out the dent where access is possible. This will minimise filling. Here, the large dent having been hammered out, the damaged area is being made slightly concave

Now all paint must be removed from the damaged area, by rubbing with coarse abrasive paper. Alternatively, a wire brush or abrasive pad can be used in a power drill. Where the repair area meets good paintwork, the edge of the paintwork should be 'feathered', using a finer grade of abrasive paper

In the case of a hole caused by rusting, all damaged sheet-metal should be cut away before proceeding to this stage. Here, the damaged area is being treated with rust remover and inhibitor before being filled

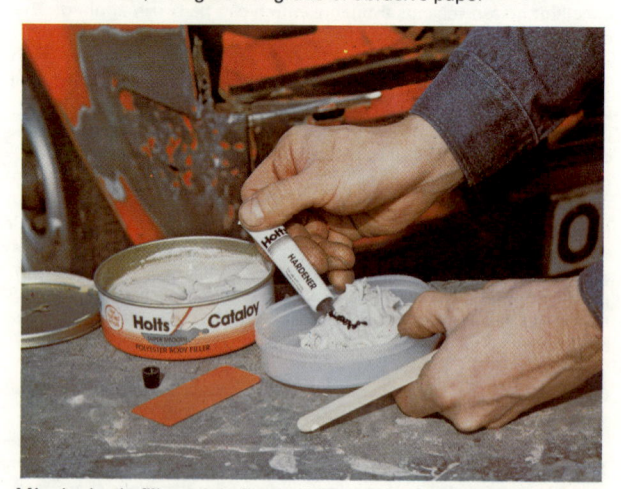

Mix the body filler according to its manufacturer's instructions. In the case of corrosion damage, it will be necessary to block off any large holes before filling – this can be done with aluminium or plastic mesh, or aluminium tape. Make sure the area is absolutely clean before ...

... applying the filler. Filler should be applied with a flexible applicator, as shown, for best results; the wooden spatula being used for confined areas. Apply thin layers of filler at 20-minute intervals, until the surface of the filler is slightly proud of the surrounding bodywork

Initial shaping can be done with a Surform plane or Dreadnought file. Then, using progressively finer grades of wet-and-dry paper, wrapped around a sanding block, and copious amounts of clean water, rub down the filler until really smooth and flat. Again, feather the edges of adjoining paintwork

The whole repair area can now be sprayed or brush-painted with primer. If spraying, ensure adjoining areas are protected from over-spray. Note that at least one inch of the surrounding sound paintwork should be coated with primer. Primer has a 'thick' consistency, so will find small imperfections

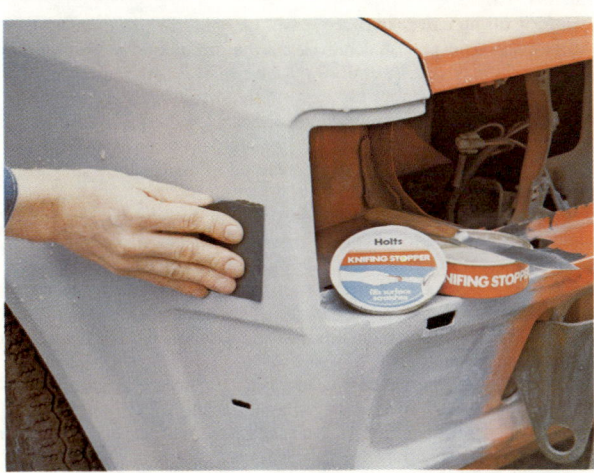

Again, using plenty of water, rub down the primer with a fine grade wet-and-dry paper (400 grade is probably best) until it is really smooth and well blended into the surrounding paintwork. Any remaining imperfections can now be filled by carefully applied knifing stopper paste

When the stopper has hardened, rub down the repair area again before applying the final coat of primer. Before rubbing down this last coat of primer, ensure the repair area is blemish-free – use more stopper if necessary. To ensure that the surface of the primer is really smooth use some finishing compound

The top coat can now be applied. When working out of doors, pick a dry, warm and wind-free day. Ensure surrounding areas are protected from over-spray. Agitate the aerosol thoroughly, then spray the centre of the repair area, working outwards with a circular motion. Apply the paint as several thin coats

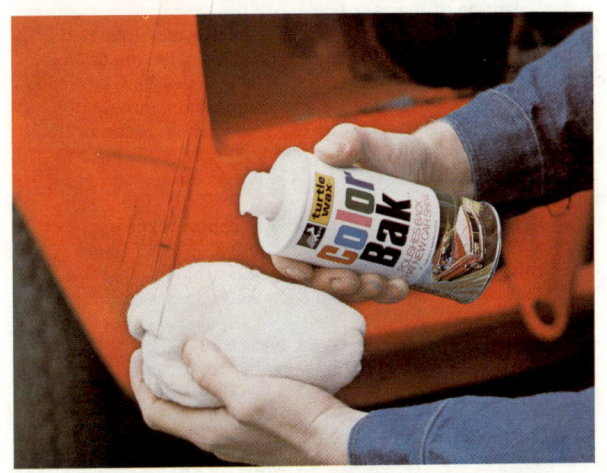

After a period of about two weeks, which the paint needs to harden fully, the surface of the repaired area can be 'cut' with a mild cutting compound prior to wax polishing. When carrying out bodywork repairs, remember that the quality of the finished job is proportional to the time and effort expended

the surfaces being cleaned. Excessive dampness could get into the seams and padded interior causing stains, offensive odours or even rot. If the inside of the car gets wet accidentally it is worthwhile taking some trouble to dry it out properly particularly where carpets are involved. Do not leave oil or electric heaters inside the car for this purpose.

4 Minor bodywork damage - repair

The photograph sequences on pages 142 and 143 illustrates the operations detailed in the following Sections.

Repair of minor scratches in the car's bodywork

If the scratch is very superficial, and does not penetrate to the metal of the bodywork, repair is very simple. Lightly rub the area of the scratch with a paintwork renovator (eg: T-Cut), or a very fine cutting paste, to remove loose paint from the scratch and to clear the surrounding bodywork of wax polish. Rinse the area with clean water.

Apply touch-up paint to the scratch using a thin paintbrush, continue to apply thin layers of paint until the surface of the paint in the scratch is level with the surrounding paintwork. Allow the new paint at least two weeks to harden; then, blend it into the surrounding paintwork by rubbing the paintwork, in the scratch area with a paintwork renovator (eg: T-Cut), or a very fine cutting paste. Finally apply wax polish.

An alternative to painting over the scratch is to use Holts "Scratch-Patch". Use the same preparation for the affected area; then simply pick a patch of a suitable size to cover the scratch completely. Hold the patch against the scratch and burnish its backing paper; the patch will adhere to the paintwork, freeing itself from the backing paper at the same time. Polish the affected area to blend the patch into the surrounding paintwork. Where the scratch has penetrated right through to the metal of the bodywork, causing the metal to rust, a different repair technique is required. Remove any loose rust from the bottom of the scratch with a penknife, then apply rust inhibiting paint (eg. Kurust) to prevent the formation of rust in the future. Using a rubber or nylon applicator fill the scratch with bodystopper paste. If required, this paste can be mixed with cellulose thinners to provide a very thin paste which is ideal for filling narrow scratches. Before the stopper-paste in the scratch hardens, wrap a piece of smooth cotton rag around the top of a finger. Dip the finger in cellulose thinners and then quickly sweep it across the surface of the stopper-paste in the scratch; this will ensure that the surface of the stopper-paste is slightly hollowed. The scratch can now be painted over as described earlier in this Section.

Repair of dents in the car's bodywork

When deep denting of the car's bodywork has taken place, the first task is to pull the dent out, until the affected bodywork almost attains its original shape. There is little point in trying to restore the original shape completely, as the metal in the damaged area will have stretched on impact and cannot be reshaped fully to its original contour. It is better to bring the level of the dent up to a point which is about 1/8 inch (3 mm) below the level of the surrounding bodywork. In cases where the dent is very shallow anyway, it is not worth trying to pull it out at all.

If the underside of the dent is accessible, it can be hammered out gently from behind, using a mallet with a wooden or plastic head. Whilst doing this, hold a suitable block of wood firmly against the impact from the hammer blows and thus prevent a large area of bodywork from being 'belled-out'.

Should the dent be in a section of the bodywork which has a double skin or some other factor making it inaccessible from behind, a different technique is called for. Drill several small holes through the metal inside the dent area - particularly in the deeper sections. Then screw long self-tapping screws into the holes just sufficiently for them to gain a good purchase in the metal. Now the dent can be pulled out by pulling on the protruding heads of the screws with a pair of pliers.

The next stage of the repair is the removal of the paint from the damaged area, and from an inch or so of the surrounding 'sound' bodywork. This is accomplished most easily by using a wire brush or abrasive pad on a power drill, although it can be done just as effectively by hand using sheets of abrasive paper. To complete the preparations for filling, score the surface of the bare metal with a screwdriver or the tang of a file, or alternatively, drill small holes in the affected area. This will provide a really good 'key' for the filler paste.

To complete the repair see the Section on filling and respraying.

Repair of rust holes or gashes in the car's bodywork

Remove all paint from the affected area and from an inch or so of the surrounding 'sound' bodywork, using an abrasive pad or a wire brush on a power drill. If these are not available a few sheets of abrasive paper will do the job just as effectively. With the paint removed you will be able to gauge the severity of the corrosion and therefore decide whether to replace the whole panel (if this is possible) or to repair the affected area. Replacement body panels are not as expensive as most people think and it is often quicker and more satisfactory to fit a new panel than to attempt to repair large areas of corrosion.

Remove all fittings from the affected area except those which will act as a guide to the original shape of the damaged bodywork (eg. headlamp shells etc.). Then, using tin snips or a hacksaw blade, remove all loose metal and any other metal badly affected by corrosion. Hammer the edges of the hole inwards in order to create a slight depression for the filler paste.

Wire brush the affected area to remove the powdery rust from the surface of the remaining metal. Paint the affected area with rust inhibiting paint (eg. Kurust); if the back of the rusted area is accessible treat this also.

Before filling can take place it will be necessary to block the hole in some way. This can be achieved by the use of one of the following materials: Zinc gauze, Aluminium tape or Polyurethane foam.

Zinc gauze is probably the best material to use for a large hole. Cut a piece to the approximate size and shape of the hole to be filled, then position it in the hole so that its edges are below the level of the surrounding bodywork. It can be retained in position by several blobs of filler paste around its periphery.

Aluminium tape should be used for small or very narrow holes. Pull a piece off the roll and trim it to the approximate size and shape required, then pull off the backing paper (if used) and stick the tape over the hole; it can be overlapped if the thickness of one piece is insufficient. Burnish down the edges of the tape with the handle of a screwdriver or similar, to ensure that the tape is securely attached to the metal underneath.

Polyurethane foam is best used where the hole is situated in a section of bodywork of complex shape, backed by a small box section (eg. where the sill panel meets the rear wheel arch - most cars). The unusual mixing procedure for this foam is as follows: Put equal amounts of fluid from each of the two cans provided in the kit, into one container. Stir until the mixture begins to thicken, then quickly pour this mixture into the hole, and hold a piece of cardboard over the larger apertures. Almost immediately the polyurethane will begin to expand, gushing frantically out of any small holes left unblocked. When the foam hardens it can be cut back to just below the level of the surrounding bodywork with a hacksaw blade.

Bodywork repairs - filling and respraying

Before using this Section, see the Sections on dent, deep scratch, rust hole, and gash repairs.

Many types of bodyfiller are available, but generally speaking those proprietary kits which contain a tin of filler paste and a tube of resin hardener (eg. Holts Cataloy) are best for this type of repair. A wide, flexible plastic or nylon applicator will be found invaluable for imparting a smooth and well contoured finish to the surface of the filler.

Mix up a little filler on a clean piece of card or board - use the hardener sparingly (follow the maker's instructions on the packet) otherwise the filler will set very rapidly.

Using the applicator, apply the filler paste to the prepared area; draw the applicator across the surface of the filler to achieve the correct contour and to level the filler surface. As soon as a contour that approximates the correct one is achieved, stop working the paste - if you carry on too long the paste will become sticky and begin to 'pick-up' on the applicator. Continue to add thin layers of filler paste at twenty-minute intervals until the level of the filler is just 'proud' of the surrounding bodywork.

Once the filler has hardened, excess can be removed using a Surform plane or Dreadnought file. From then on, progressively finer grades of abrasive paper should be used, starting with a 40 grade production paper and finishing with 400 grade 'wet-and-dry' paper. Always wrap the abrasive paper around a flat rubber, cork, or wooden block -

otherwise the surface of the filler will not be completely flat. During the smoothing of the filler surface the 'wet-and-dry' paper should be periodically rinsed in water. This will ensure that a very smooth finish is imparted to the filler at the final stage.

At this stage the 'dent' should be surrounded by a ring of bare metal, which in turn should be encircled by the finely 'feathered' edge of the good paintwork. Rinse the repair area with clean water until all of the dust produced by the rubbing-down operation is gone.

Spray the whole repair area with a light coat of grey primer - this will show up any imperfections in the surface of the filler. Repair these imperfections with fresh filler paste or bodystopper, and once more smooth the surface with abrasive paper. If bodystopper is used, it can be mixed with cellulose thinners to form a really thin paste which is ideal for filling small holes. Repeat this spray and repair procedure until you are satisfied that the surface of the filler, and the feathered edge of the paintwork are perfect. Clean the repair area with clean water and allow to dry fully.

The repair area is now ready for spraying. Paint spraying must be carried out in a warm, dry, windless and dust free atmosphere. This condition can be created artificially if you have access to a large indoor working area, but if you are forced to work in the open, you will have to pick your day very carefully. If you are working indoors, dousing the floor in the work area with water will 'lay' the dust which would otherwise be in the atmosphere. If the repair area is confined to one body panel, mask off the surrounding panels; this will help to minimise the effects of a slight mis-match in paint colours. Bodywork fittings (eg. chrome strips, door handles etc.,) will also need to be masked off. Use genuine masking tape and several thicknesses of newspaper for the masking operation.

Before commencing to spray, agitate the aerosol can thoroughly, then spray a test area (an old tin, or similar) until the technique is mastered. Cover the repair area with a thick coat of primer; the thickness should be built up using several thin layers of paint rather than one thick one. Using 400 grade 'wet-and-dry' paper, rub down the surface of the primer until it is really smooth. While doing this, the work area should be thoroughly doused with water, and the 'wet-and-dry' paper periodically rinsed in water. Allow to dry before spraying on more paint.

Spray on the top coat, again building up the thickness by using several thin layers of paint. Start spraying in the centre of the repair area and then, using a circular motion, work outwards until the whole repair area and about 2 inches of the surrounding original paintwork is covered. Remove all masking material 10 to 15 minutes after spraying on the final coat of paint.

Allow the new paint at least 2 weeks to harden fully; then, using a paintwork renovator (eg. T-Cut) or a very fine cutting paste, blend the edges of the new paint into the existing paintwork. Finally, apply wax polish.

5 Major body damage - repair

Where serious damage has occurred or large areas need renewal due to neglect, it means certainly that completely new sections or panels will need welding in and this is best left to professionals. If the damage is due to impact it will also be necessary to completely check the alignment of the body shell structure. Due to the principle of construction the strength and shape of the whole can be affected by damage to a part. In such instances the services of a Leyland agent with specialist checking jigs are essential. If a body is left misaligned it is first of all dangerous as the car will not handle properly and secondly uneven stresses will be imposed on the steering, engine and transmission, causing abnormal wear or complete failure. Tyre wear may also be excessive.

6 Bumpers - removal and installation

Front bumper
1 Disconnect the six snap-connectors for the lamp leads.
2 Remove the outer nuts, spring washers and flat washers.
3 Whilst supporting the bumper, remove the two inner nuts, spring washers and flat washers. Lift away the bumper.
4 Installation is the reverse of the removal procedure. Ensure that the bumper is correctly aligned before the nuts are tightened.

Rear bumper
5 Remove the nuts, spring washers and flat washers from each side bracket.
6 Remove the inner nuts, spring washers and flat washers. Where applicable, remove the handling brackets.
7 Whilst supporting the bumper, remove the outer nuts, spring washers and flat washers. Lift away the bumper.
8 Installation is the reverse of the removal procedure. Ensure that the bumper is correctly aligned before the nuts are tightened.

7 Front grille - removal and installation

1 On USA models, remove the side marker lamps, as described in Chapter 10. On UK models, remove the screws and washers, and take off the blanking plate.
2 Press out the centre pieces of the three plastic rivets and remove the grille.
3 When installing, apply a mastic sealing compound such as Seelastik to the area shown in Fig. 12.3. Install the plastic rivets using a suitable tool.

8 Air vent grille - removal and installation

1 Remove the rear quarter trim pad (Section 25).
2 Remove the four nuts, spring washers and flat washers, and two spacers.
3 Lift away the grille, complete with T-bolts.
4 Installation is the reverse of the removal procedures. Ensure that the long spacer is used for the front fixing and the short one used for the rear fixing along the lower edge.

9 Bonnet - removal and installation

1 Remove the windscreen washer pump tube and attach a length of string to it to facilitate installation. Pull the tube through the holes in the inner wheel arch (photo).
2 Mark around the hinges to facilitate installation then support the bonnet in the open position. Help from an assistant is necessary at this stage.
3 *Short stay:* Detach the two nuts and one bolt from the stay bracket attachment.
4 *Extending stay:* Remove the two screws and spring washers securing the stay to the bonnet.
5 Remove the bolts and spring washers securing the hinges to the adjuster plates. Remove the plates and lift off the bonnet.
6 Installation is the reverse of the removal procedure. Ensure that the bonnet is correctly aligned before the hinge bolts are finally tightened.

10 Bonnet catches and controls

Catch - removal, installation and adjustment
1 Support the bonnet in the open position then remove the two bolts, spring washers and flat washers to release the catch.
2 Installation is the reverse of the removal procedure.
3 If necessary, adjust the catch to give a positive locking action and to eliminate free-movement, as described in the following paragraphs.
4 Pull back the spring and loosen the locknut on the adjuster bolt.
5 Rotate the screw in, or out, as necessary, then tighten the locknut and check the operation. Adjust further as necessary (photo).

Bonnet lock - removal and installation
6 Remove the cable trunnion, then slacken the pinch bolt and remove the cable from the lock.
7 Remove the four bolts, spring washers and flat washers, and detach the lock.
8 Installation is the reverse of the removal procedure.

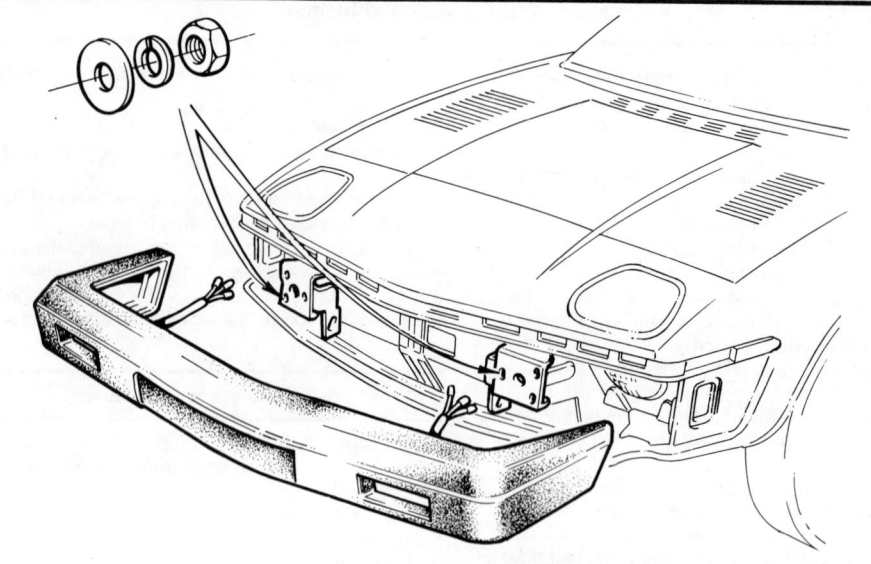

Fig. 12.1. Front bumper attachments

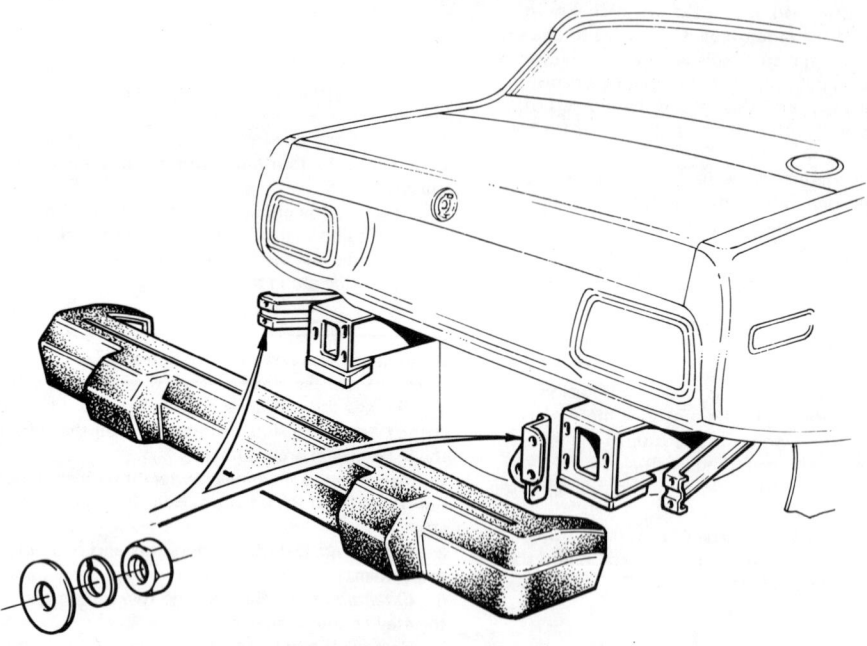

Fig. 12.2. Rear bumper attachments

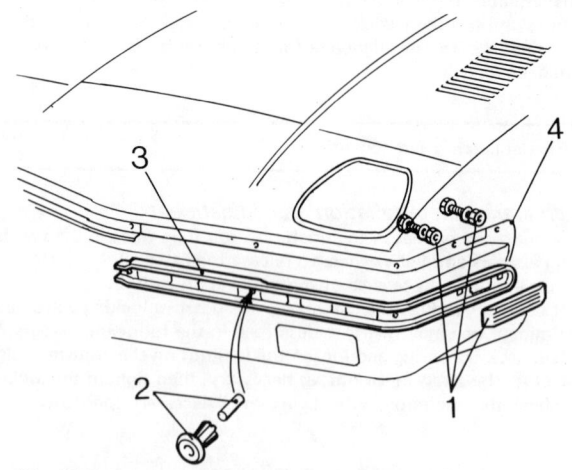

Fig. 12.3. Front grille attachments - typical

1 Screws and blanking plate
2 Rivet
3 Grille
4 Apply sealing compound at this point

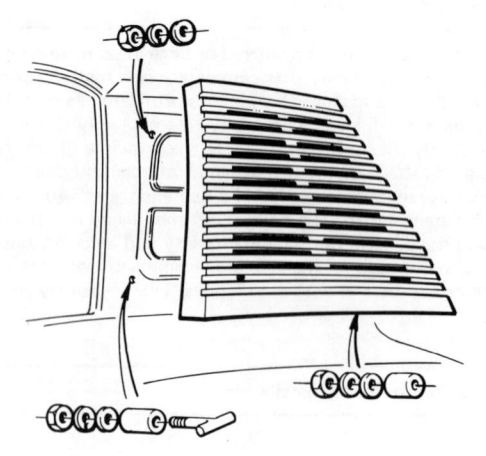

Fig. 12.4. Air vent grille attachments

Chapter 12/Bodywork and fittings 147

9.1 Windscreen washer tube in bonnet

10.5 Adjusting the bonnet adjuster bolt

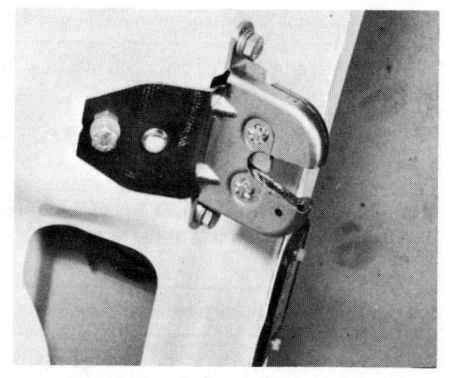

12.1 Luggage compartment lock

Release cable - removal and installation

9 Initially proceed as described in paragraph 6. **Note:** Do not close the bonnet with the release cable detached.
10 Unscrew the nut from the cable securing bracket beneath the facia, then withdraw the cable through the grommet in the bulkhead. Do not lose the nut and shakeproof washer.
11 Installation is the reverse of the removal procedure, adjustment being made at the cable end for satisfactory operation.

11 Luggage compartment lid - removal and installation

1 Mark around the hinges on the lid to facilitate installation.
2 Disconnect the number plate leads from the connector in the luggage compartment at the right-hand side.
3 With help from an assistant, to support the lid, remove the bolt(s) and washer(s) from the stay bracket.
4 Remove the hinge bolts, spring washers and flat washers, and lift off the bonnet.
5 If necessary, the hinges can be removed by removing the bolts, spring washers and flat washers.
6 Installation is the reverse of the removal procedures. Ensure that the lid is correctly aligned before the bolts are finally tightened.

12 Luggage compartment lock and striker

Lock - removal and installation

1 Remove the three bolts, spring washers and flat washers and lift away the catch (photo).
2 Pull the spring plate to one side, then withdraw the lock and gasket.
3 Installation is the reverse of the removal procedure.

Lock striker - removal and installation

4 Carefully pull off the body weatherstrip from the area of the lock striker.
5 Mark the position of the striker bolts on the rear panel, then remove them together with the spring and flat washers. Remove the striker.
6 Installation is the reverse of the removal procedure. If necessary, adjust the striker position for satisfactory operation.

13 Door rattles - tracing and rectification

1 The most common cause of door rattles is a misaligned, loose or worn striker plate, but other causes may be:
   a) Loose door handles, window winder handles or door hinges
   b) Loose worn or misaligned door lock components
   c) Loose or worn remote control mechanism
or a combination of these.
2 If the striker catch is worn, renew and adjust, as described later in this Chapter.
3 Should the hinges be badly worn then it may become necessary for new ones to be fitted.

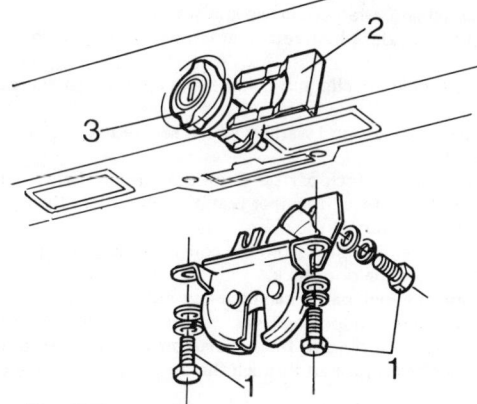

Fig. 12.5. Luggage compartment lock parts

1 Retaining bolts
2 Spring plate
3 Lock

14 Doors - removal and installation

1 Detach the battery earth lead.
2 Unscrew the knob from the lock plunger rod.
3 Take out the two screws and remove the armrest.
4 Remove the screw, and take off the handle and bezel.
5 Using a suitable small lever, prise out the nine trim pad retaining clips. Where applicable, detach the radio loudspeaker inline connectors.
6 Lift the trim pad up over the plunger rod and away from the door.
7 Mark around the hinge positions on the door to facilitate installation.
8 With help from an assistant to support the door, remove the six nuts, spring washers and flat washers, and the adjuster plates. Lift away the door.
9 Installation is the reverse of the removal procedure. If necessary, adjust the position of the hinges for satisfactory alignment.

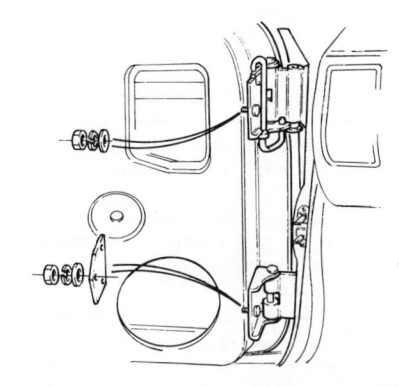

Fig. 12.6. Door hinge attachment points

15 Door lock remote control - removal and installation

1 Remove the trim pad, as described in paragraphs 2 to 6, of Section 14.
2 Remove the retaining clip and detach the lock control rod.
3 Remove the retaining screw and take off the escutcheon.
4 Remove the retaining screw and push the control assembly rearwards to disengage it from the door (photo).
5 Withdraw the remote control assembly through the door aperture.
6 Installation is the reverse of the removal procedure.

16 Door lock and outside handle - removal and installation

1 Remove the trim pad, as described in paragraphs 2 to 6, of Section 14.
2 Remove the bolt, spring washer and flat washer securing the rear glass channel. Detach the insert from the channel, then carefully withdraw the channel and weather curtain (photo).
3 Remove the four door lock retaining screws and lift off the disc latch.
4 Remove the retaining clip and unhook the remote control rod from the lock.
5 Remove the two nuts and washers which secure the clamp bracket to the outside handle.
6 Taking care that the lock parts and linkages are not bent, manoeuvre the outside handle and seal, together with the door lock, through the handle apertures and out of the door.
7 Remove the retaining clips from the control rods, and separate the outside handle from the door lock.
8 If necessary, the lock barrel can be taken out by removing the circlip, wave washers and locking lever.
9 Installation is the reverse of the removal procedure. Ensure that the lock plunger rod is pushed through the aperture before any of the parts are secured.

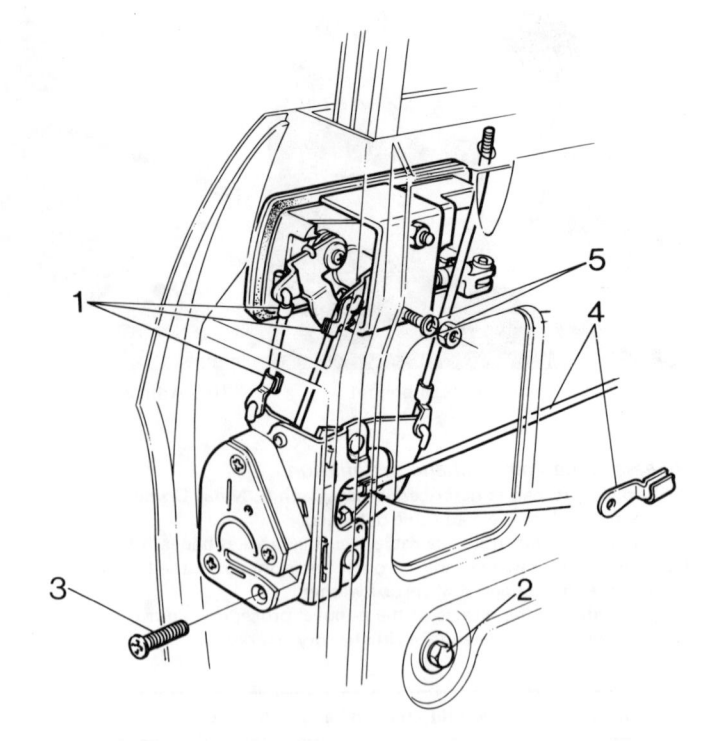

Fig. 12.7. Door lock attachment points

1 Retaining clips
2 Rear glass channel bolt
3 Lock assembly screws
4 Remote control rod and retaining clip
5 Clamp bracket nut and washer

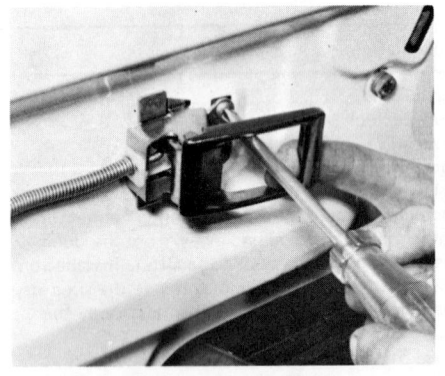

15.4 Removing the control assembly

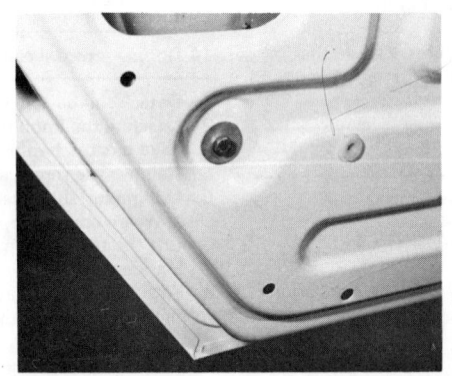

16.2 Rear glass channel bolt

16.3 The door lock retaining screws

17 Door lock striker - removal, installation and adjustment

1 Remove the rear quarter trim pad, as described in Section 25.
2 Remove the two screws, and lift off the striker complete with outer striker plate and seal.
3 Installation is the reverse of the removal procedure.
4 If problems arise with the door failing to close satisfactorily, loosen the screws and reposition the striker plate, as necessary.

18 Door quarter-light - removal and installation

1 Fully lower the door glass then remove the door trim pad, as described in paragraphs 2 to 6, of Section 14.
2 Drill out the rivet which secures the top of the channel to the door.
3 Remove the two bolts, spring washers and flat washers securing the front glass channel, then pull the top of the channel away from the quarter-light weatherstrip (photos).
4 Carefully ease the quarter-light and weatherstrip rearwards and upwards, and away from the door.
5 Installation is the reverse of the removal procedure. If a suitable rivet cannot be used for the top glass channel, a small nut and bolt or self-tapping screw can be substituted.

19 Door glass and regulator - removal and installation

1 Remove the door lock remote control, as described in Section 15.
2 Fully raise the glass using the handle, then turn the handle back slightly so that the regulator mounting bolts are not under a load. Take off the handle again.

Chapter 12/Bodywork and fittings

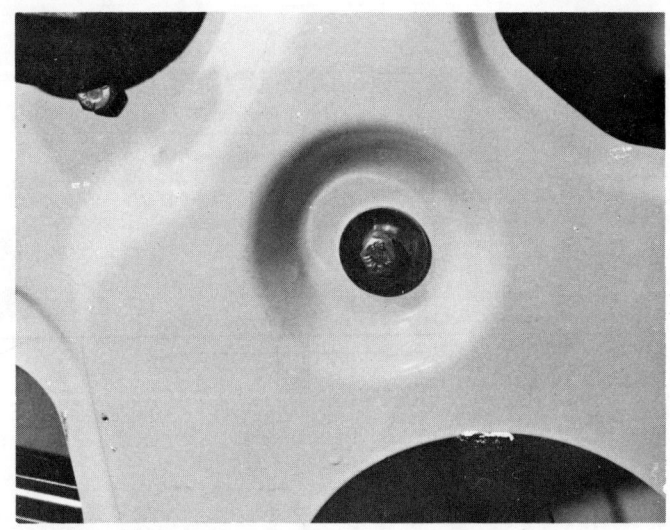

18.3a, b Front glass channel retaining screws

3 Support the glass, then remove the four bolts, spring washers and flat washers securing the regulator.
4 Carefully push the regulator shaft into the door cavity and slide the assembly rearwards and out of the door. Lower the glass, as necessary, to enable the rollers to be disengaged from the glass channels and the channel on the door interior.
5 Manoeuvre the regulator carefully out of the door through the lower aperture.
6 Remove the bolt, spring washer and flat washer securing the rear glass channel. Detach the insert from the channel, then carefully withdraw the channel and the weather curtain.
7 Lower the rear of the glass until it can be withdrawn through the aperture. Take care that it is not scratched on the seal clips.
8 Installation is the reverse of the removal procedures. Take care that the glass channel is positioned so that the glass can be moved freely.

20 Facia and glovebox cowl - removal and installation

1 Detach the battery earth lead then remove the facia instrument cowl, as described in Section 21.
2 Remove the facia switch panel, referring to Chapter 10, as necessary. Note the positions of the harness plugs and the switch identification bulb holders to facilitate refitting.
3 Remove the two screws and take off the A-post trim pads and facia corner finishers.
4 Remove the four screws from the demister vents.
5 Remove the two screws from the facia brackets on the bulkhead.
6 Remove the two bolts and washers from the facia tongues on the bulkhead.
7 Remove the two screws and take off the two halves of the steering column vacelle.
8 Refer to Chapter 9 and remove the two shear-head bolts securing the steering column housing to the body.

Left-hand drive models

9 Remove the two screws securing the facia to the instrument panel support rail.
10 Remove the two screws securing the facia to the support rail below the glovebox.
11 Pull the lid off the component mounting panel inside the glovebox then remove the three component panel mounting screws.
12 Remove the two screws securing the wiring loom to the top of the facia.

Right-hand drive models

13 Remove two screws from each side of the steering column and one from below the bonnet release, which secure the facia to the support rail below the instrument panel.

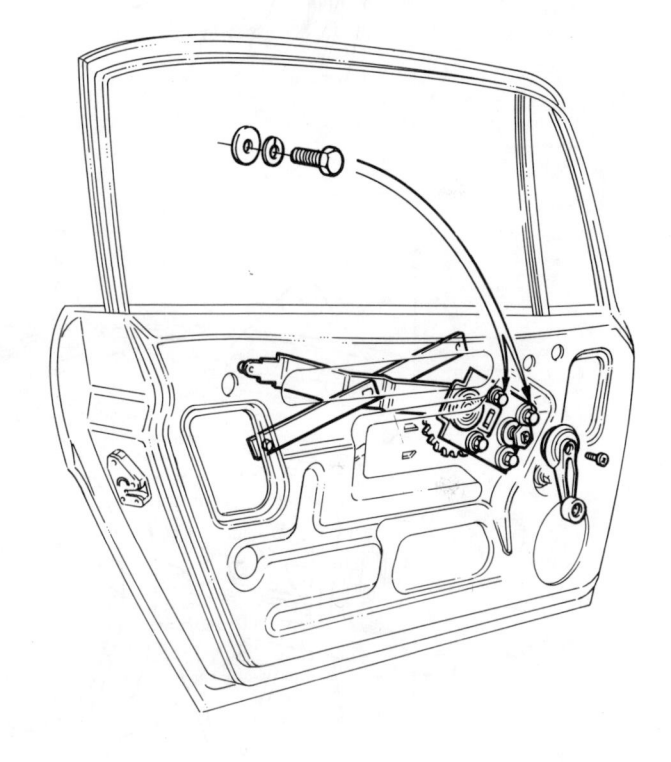

Fig. 12.8. Door glass regulator attachment points

14 Remove the two screws securing the facia to the support rail beneath the glovebox.
15 Remove the two screws securing the control illumination panel.
16 Remove the single screw securing the facia to the bracket on the control cowl.
17 Remove the two screws and take off the lid from the component mounting panel inside the glovebox.
18 Remove the three screws and two brackets securing the component mounting panel.
19 Remove the two screws securing the wiring loom to the top of the facia.

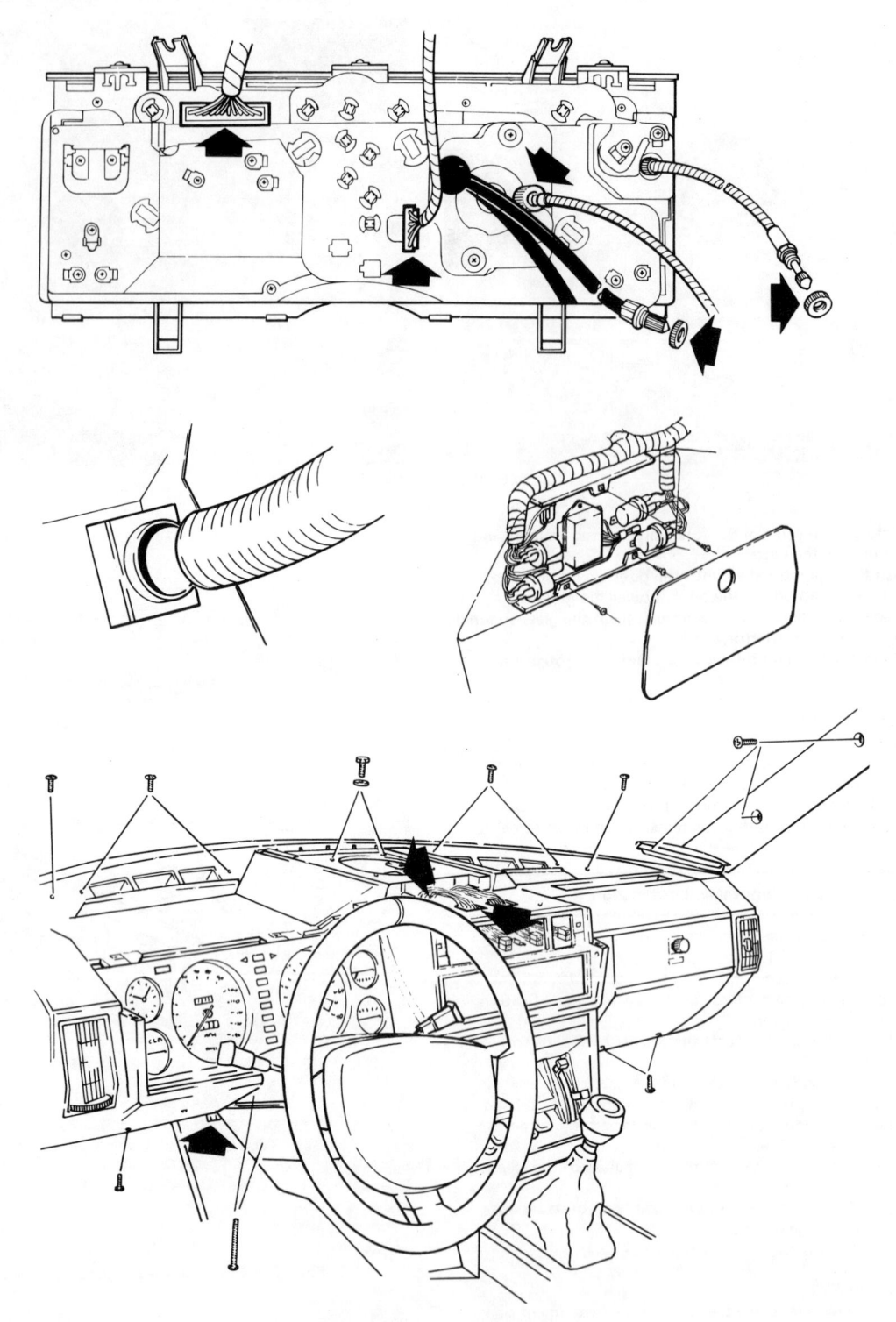

Fig. 12.9. Facia removal - items to be detached

All models
20 Disconnect the speedometer cable (depress the lever to release the catch from the groove in the speedometer boss).
21 Unscrew the knurled nuts, and release the trip and clock resets from their bracket slots.
22 Disconnect the two multi-pin plugs.
23 Where applicable, lower the radio loudspeaker through the facia aperture.
24 Ease the facia rearwards and pull off the hoses from the outer swivelling vents.
25 Manoeuvre the facia carefully out of the car, feeding the three harness plugs and switch identification bulb holders through the four apertures above the switch panel aperture.
26 Installation is the reverse of the removal procedure, but ensure that the demister vents are correctly positioned before the facia is pushed fully into position.

Chapter 12/Bodywork and fittings

21 Centre grille (radio loudspeaker grille) and facia instrument cowl - removal and installation

1 Using a broad-bladed screwdriver prise up the edge furthest from the windscreen and move the grille towards you to disengage it from the five slots.
2 Remove the two screws securing the instrument cowl to the facia above the switch panel.
3 Remove the two screws securing the instrument cowl to the facia above the instrument panel.
4 Where applicable, remove the screw securing the underside of the cowl to the bracket above the switch panel.
5 Remove the two screws securing the outer tongues of the cowl to the facia, then swing the cowl rearwards to disengage the side tongues. The cowl can now be completely removed.
6 Installation is the reverse of the removal procedure.

22 Glovebox lid and lock - removal and installation

Lid
1 Press out the centre piece of the plastic rivets and detach the support straps.
2 Remove the four hinge screws and lift away the lid.
3 Installation is basically the reverse of the removal procedure. Use a suitable tool to install the centre piece of the plastic rivets.

Lock
4 Pull off the knob then remove the two screws securing the lid lock interior.
5 If necessary, remove the lid latch by taking out the two retaining screws.
6 Installation is the reverse of the removal procedure.

23 Console assembly - removal and installation

1 Raise the handbrake lever and pull off the grip; also unscrew the gearlever knob.
2 Remove the handbrake lever surround trim panel (two screws), then lift off the gaiter. (On some models the gaiter has four fasteners).
3 Remove the two screws and washers securing the front console to the transmission tunnel.
4 Remove the two screws and washers securing the consoles and bridge plate to the floor; lift off the bridge plate.
5 Raise the armrest/lid and remove the two catch securing screws. Take off the catch and front console.
6 Remove the two screws and washers securing the rear console to the rear parcel shelf and body, then lift off the rear console.
7 Installation is the reverse of the removal procedure.

24 Control cowl - removal and installation

1 Remove the console assembly, as described in the previous Section.
2 If the choke control is mounted on the control cowl, disconnect the cable at the carburettor then pull out the inner cable from inside the car. Remove the bezel securing the choke outer cable to the cowl.
3 Pull the knobs off the heating and ventilating control levers.
4 Remove the two screws securing the control cowl to the control levers.
5 Remove the two screws securing the illumination panel to the control cowl.
6 Remove the screw securing the control cowl to the bracket below the facia (where applicable).
7 Remove the two screws securing the control cowl to the transmission tunnel bracket.
8 Remove the two screws securing the cowl to the body (where applicable).
9 Engage 4th gear and pull the cowl away from the heater. Disconnect the bulb holder and leads from the cigar lighter. Disconnect the leads from the rheostat. Where applicable detach the leads from the air-conditioning cut-out switch.
10 Withdraw the control cowl completely.
11 Installation is the reverse of the removal procedure.

25 Rear quarter trim and parcel tray - removal and installation

Rear quarter trim
1 Locally pull away the weatherstrip from the door aperture.
2 Carefully prise the trim pad edging away from the body flange.
3 Remove the pad retaining screw cap, the screw and the cap retainer.
4 Carefully prise the pad and clips from the side panel.
5 Installation is the reverse of the removal procedure. Use an adhesive such as Dunlop SP758 on the mating surfaces of the trim pad edging and body flange.

Rear compartment trim pad
6 Move both seats and seat squabs fully forward, then open the rear console lid and remove the two screws and washers securing the console and trim pad to the body.
7 Remove the two screws and cup washers at the lower part of the trim pad, and the four screws and cup washers at the upper edge of the trim pad.
8 Carefully raise the edge of the parcel tray, raise the trim pad clear of the console and remove it from the car.
9 Installation is the reverse of the removal procedure.

Rear parcel tray
10 Move both seats and seat squabs fully forward, then remove the bolts and spring washers securing the seatbelt swivel brackets to the seats.
11 Feed the seatbelts through the cut-outs in the parcel tray, then remove the four screws and cup washers securing the parcel tray and rear compartment trim pad.
12 Remove the four screws and cup washers securing the rear of the parcel tray, then carefully manoeuvre the tray upwards and forwards to remove it from the car.
13 Installation is basically the reverse of the removal procedure, but feed the seatbelts through the cut-outs as the parcel shelf is being positioned in the car.

26 Seats and seat runners - removal and installation

1 With the seats fully rearwards, remove the two long Allen screws and square washers securing the runners to the floor.
2 With the seats fully forward, remove the two short Allen screws, one square washer and one stop plate securing the runners.
3 Remove the bolt and spring washer securing the belt bracket to the seat.
4 Where applicable, the single bolt can be removed to release the two belt warning light switch harness plugs.
5 Lift out the seat, complete with its runners.
6 If the runners are to be detached from the seat, remove the six Allen screws and spring washers, moving the slides as necessary.
7 Installation is the reverse of the removal procedure, but ensure that the packing washers are correctly positioned.

27 Seatbelts - removal and installation

1 Remove the bolt and spring washer securing the swivel bracket to the seat, then feed the seatbelt through the cut-out in the parcel shelf.
2 From inside the luggage compartment, remove the trim panel (two fasteners) from above the wheel arch.
3 Remove the bolt and spring washer securing the reel assembly to the wheel arch, then withdraw the complete assembly from the car.
4 Where applicable, disconnect the belt buckle unit harness plug and integral buckle and switch unit. Do not lose the wave washer, flange bush and distance piece.
5 If necessary, remove the two screws and take out the tongue retainer.
6 Installation is the reverse of the removal procedure.

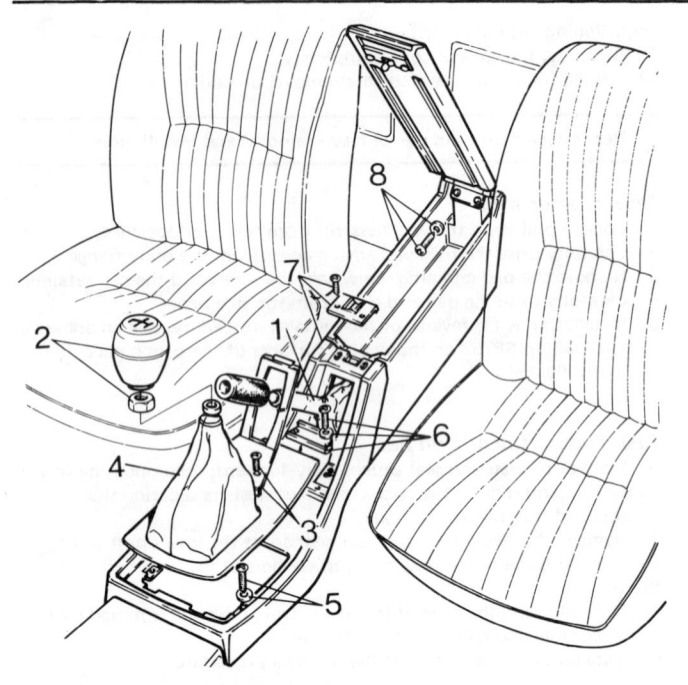

Fig. 12.10. Console assembly removal

1 Handbrake lever
2 Gear lever knob and nut
3 Surround trim panel screws
4 Gaiter
5 Console securing screws
6 Bridge plate and screws
7 Catch and screws
8 Rear attachment screws

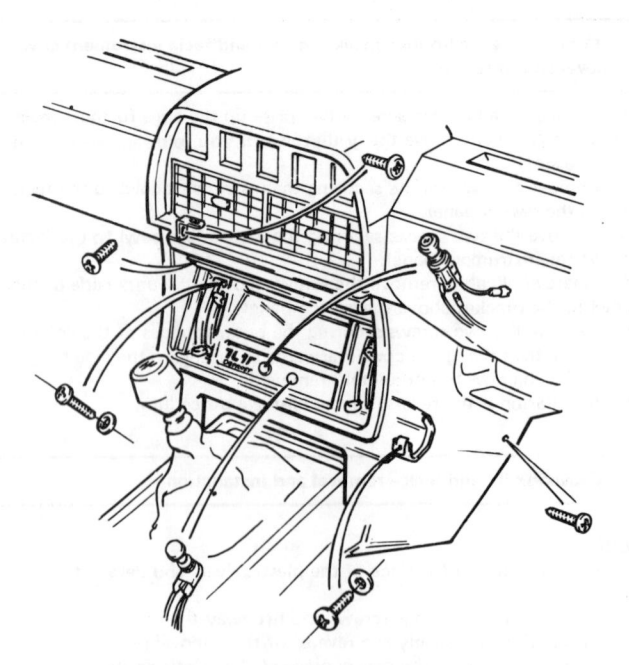

Fig. 12.11. Control cowl removal

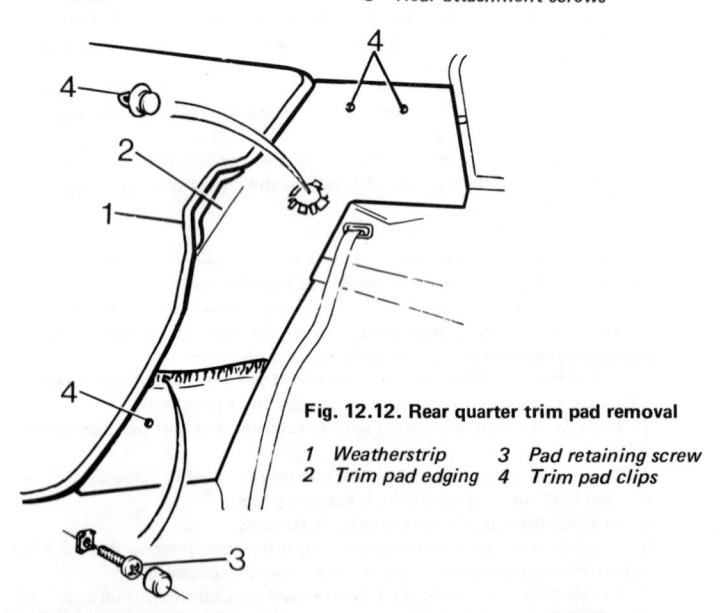

Fig. 12.12. Rear quarter trim pad removal

1 Weatherstrip
2 Trim pad edging
3 Pad retaining screw
4 Trim pad clips

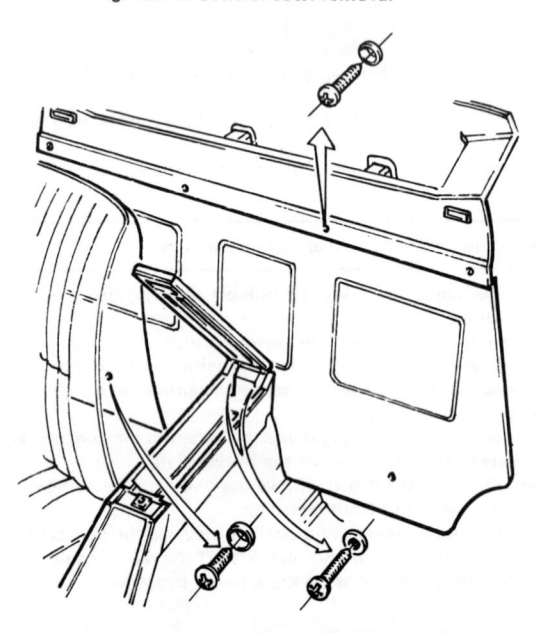

Fig. 12.13. Rear compartment trim pad removal

28.4 Windscreen finisher strip retaining screw and plug

28 Windscreen lower finisher strip - removal and installation

1 Remove the windscreen wiper arms, as described in Chapter 10.
2 Remove the nut, distance piece and rubber washer from the wheelbox spindle on the passenger's side.
3 Pull off the two bulkhead weatherstrips.
4 Remove the plastic plug (where fitted), two screws and washers, and detach the finisher from the body. Take care that the windscreen and body are not scratched (photo).
5 Installation is the reverse of the removal procedure.

29 Windscreen - removal and installation

The windscreen is retained in the aperture by a neoprene based material which is supplied in strip form. It contains a resistance wire core which is heated by electric current to cure the sealing strip after it is installed. Installation of the windscreen requires special techniques and equipment, and should never be attempted by the do-it-yourself owner. This is definitely a job for the Triumph dealer or a windscreen specialist.

30 Heated rear window - removal and installation

1 Using a blunt tool, break the existing seal.
2 Carefully pull away the rear edges of the rear quarter trim pads from the body, and disconnect the window element lead connectors.
3 With help from an assistant, push out the glass taking care that it is not scratched.
4 Remove the weatherstrip from the glass.
5 When installing, ensure that the edges of the glass and the body aperture are clean and dry. Apply a mastic sealing compound such as Seelastik to the glass channel, and place it around the windscreen.
6 Insert a strong draw-cord into the weatherstrip inner channel, allowing the ends to protrude from the lower edge.
7 With an assistant holding the glass centrally in the aperture, apply steady pressure and pull the cord ends so that the weatherstrip is pulled over the body flange.
8 Seal the outer channel to the body flange using the mastic sealer.
9 Reconnect the window element lead connectors.

31 Sliding roof - removal and installation

1 Carefully prise out the four chrome blanking caps, then remove the screws, spring washers and flat washers.
2 Open the roof, carefully pulling it rearwards, on one side at a time, until the four slides are disengaged from the runners. The roof can now be lifted off.
3 Installation is the reverse of the removal procedure. Adjust it, if necessary, to obtain the correct tension before fully tightening the four screws.

32 Demister ducts - removal and installation

Passenger's side
1 Remove the facia, as described in Section 20.
2 Where applicable remove the screw securing the right-hand side of the control cowl to the body.
3 Detach the duct from the heater.
4 Installation is the reverse of the removal procedure.

Driver's side
5 On right-hand drive models, detach the speedometer trip control and the speedometer cable, as described in paragraphs 20 and 21, of Section 20.
6 Where applicable, remove the screw securing the right-hand side of the control cowl to the body.
7 Remove the two screws which secure the demister duct to the facia.
8 Detach the duct from the facia and the heater, separating the two parts if necessary.
9 Installation is the reverse of the removal procedure.

33 Air hoses and swivelling vents - removal and installation

Air hoses
1 If the passenger's side air hose is to be removed, first remove the facia (Section 20).
2 Pull the ends of the hose from the vent duct and heater.
3 Installation is the reverse of the removal procedure. Ensure that components behind the facia are not damaged as the hoses are installed.

Outer swivelling vents
4 Pull the air hose off the vent duct, from behind the facia.
5 Push the vent outwards to release the tongues securing it to the facia (photo).
6 Depress the vent sides to release the tongues from the duct.
7 Remove the duct so that the vent can be removed.
8 Installation is the reverse of the removal procedure.

Centre swivelling vent
9 Depress one side of the vent frame to detach the retaining boss. Pull the vent out (photo).
10 Installation is the reverse of the removal procedure.

34 Heater unit - removal and installation

1 Detach the battery earth lead.
2 Drain the cooling system, referring to Chapter 2, if necessary.
3 Pull the bonnet lock cable clear of the fresh-air duct, then remove the two screws, spring washers and flat washers, and carefully manoeuvre the duct clear of the engine compartment. **Note:** On some models the duct is retained by two 'over-centre' clips (photo).
4 Disconnect both water hoses which connect to the heater.
5 Remove the facia (Section 20).
6 Detach the two air hoses from the heater.
7 Remove the console assembly (Section 23).
8 Remove the control cowl (Section 24).
9 Remove the heater demister ducts.
10 Remove the single bolt securing the air intake to the bulkhead.
11 Remove the two bolts, spring washers and flat washers, and take off the control cowl support bracket.
12 Loosen the two bolts securing the heater to the front of the support bracket on the transmission tunnel.
13 Remove the two bolts, spring washers and flat washers, to release the two heater support brackets.
14 Remove the two bolts, nuts, spring washers and flat washers securing the facia support rails to the transmission tunnel support bracket.
15 Disconnect the fan motor leads.
16 Remove the nut, spring washer and flat washer securing the water pipe bracket.
17 Remove the nut, spring washer and flat washer securing the rear of the heater.
18 Remove the heater, taking care that coolant which is remaining in the heater matrix does not spill on the carpets.
19 Installation is the reverse of the removal procedure. Ensure that the cooling system is filled, as described in Chapter 2.

35 Fan motor - removal and installation

1 Remove the heater, as described in the previous Section.
2 Loosen the trunnion and disconnect the air intake control rod.
3 Remove the six screws and detach the air intake.
4 Drill out the four rivets from the heater casing.
5 Release the clip and detach the fan motor resistor.
6 Detach the supply leads from the motor.
7 Detach the clips (fourteen) and lift off the casing upper half. Lift out the fan motor.
8 Installation is the reverse of the removal procedure. Ensure that the trunnion is positioned to permit full movement of the air intake control lever and flap.

36 Air-conditioning

On certain models, an air-conditioning system is available. For any repair work, including removal and installation, the job should be entrusted to a Triumph dealer or air-conditioning specialist. Where it is necessary to depressurize the system (for example where an engine is to be removed), this also must be entrusted to a Triumph dealer or air-conditioning specialist.

33.5 Removing an outer swivelling vent

33.9 Removing a centre swivelling vent

34.3 Removing the fresh air duct

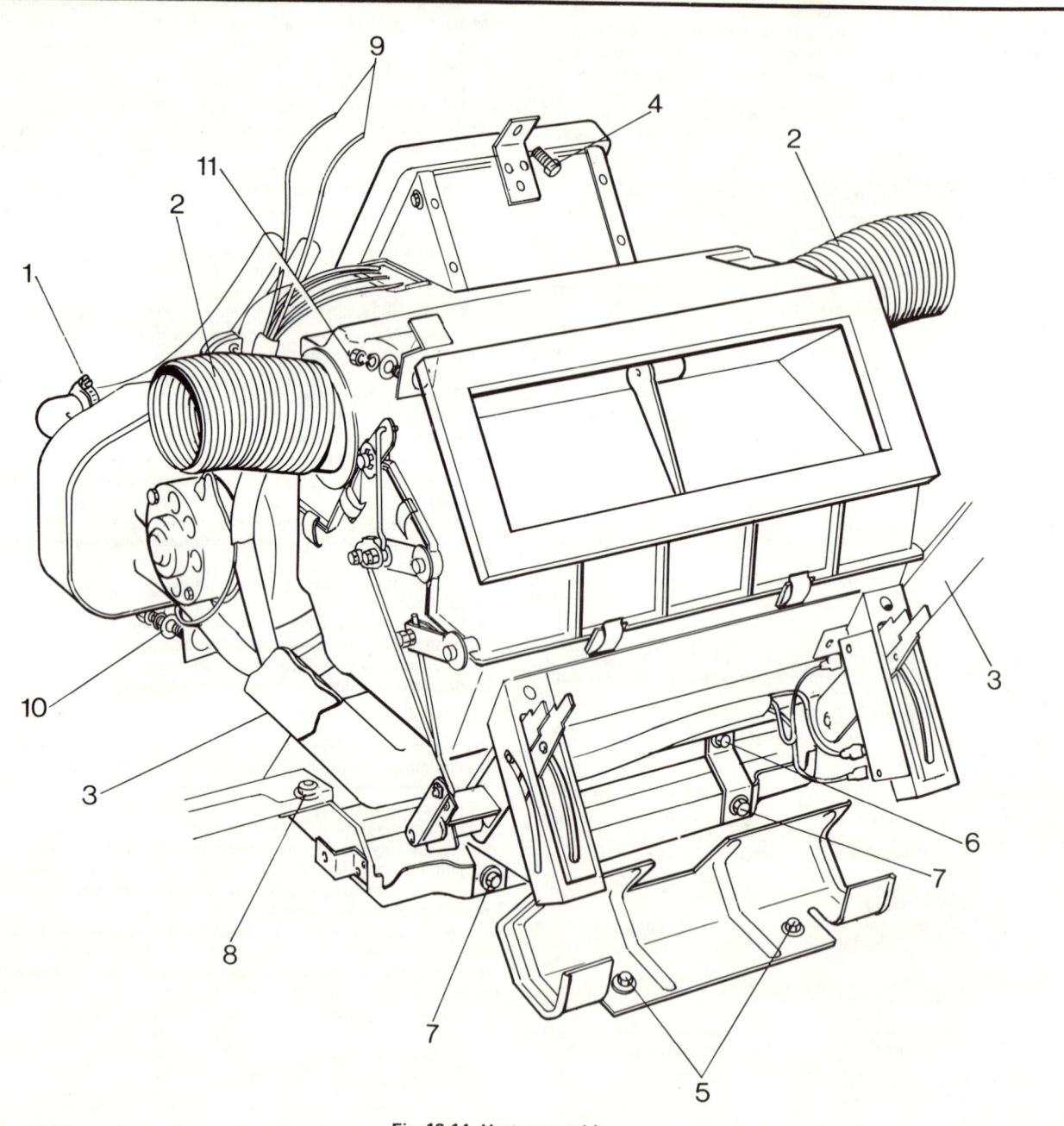

Fig. 12.14. Heater assembly

1 Water hose clips
2 Air hoses
3 Demister ducts
4 Intake-to-bulkhead bolt
5 Control cowl support bracket bolts
6 Heater support bracket-to-transmission tunnel bolts
7 Heater support bracket bolts
8 Facia support rail bolts
9 Motor leads
10 Water pipe bracket bolt
11 Bulkhead attachment bolt at rear of heater

Chapter 13 Supplement

Contents

Introduction ... 1	Absorption canisters, twin canister specification - removal and refitting
Specifications ... 2	Air injection system
Engine ... 3	Exhaust gas recirculation (EGR) system
Timing cover, UK and Europe - refitting	Ignition system ... 6
Timing cover, USA and Canada - refitting	Ignition coil - removal and refitting

Introduction ... 1
Specifications ... 2
Engine ... 3
   Timing cover, UK and Europe - refitting
   Timing cover, USA and Canada - refitting
   Timing cover oil seal (engine in the vehicle) - removal and refitting
   Timing cover oil seal (engine out of the vehicle) - removal and refitting
   Valve timing - checking
   Valve timing - adjusting
   Air conditioning compressor drivebelt - adjustment
   Air conditioning compressor drivebelt - removal and refitting

Cooling system ... 4
   Header tank system - filling
   Header tank, later models - removal and refitting
   Expansion tank, early models - removal and refitting

Fuel system ... 5
   Air cleaner, single carburettor models - removing, refitting, and renewal of filter
   General description - tamperproof carburettors for non-USA markets
   Tamperproof carburettors - dismantling and reassembly (general)
   Twin Zenith-Stromberg carburettors, USA specification excluding California - deceleration by-pass valve checking and adjustment
   Twin Zenith-Stromberg carburettor automatic choke (where fitted) - removal and refitting
   Air intake temperature control system - testing
   Fuel injection system - description
   Air cleaner and element, fuel injection system - removal and refitting
   Throttle cable, fuel injection system - removal and refitting
   Throttle pedal assembly, fuel injection system - removal and refitting
   Fuel injection system - depressurising
   Fuel pump, fuel injection system - removal and refitting
   Fuel tank, fuel injection system - removal and refitting
   Fuel line filters, fuel injection system - removal and refitting
   Fuel pressure regulator, fuel injection system - removal and refitting
   Fuel rail, fuel injection system - removal and refitting
   Plenum chamber, fuel injection system - removal and refitting
   Induction manifold, fuel injection system - removal and refitting
   Injectors, fuel injection system - removal and refitting
   Cold start injector, fuel injection system - removal and refitting
   Emission control and evaporative loss control systems - general

Absorption canisters, twin canister specification - removal and refitting
Air injection system
Exhaust gas recirculation (EGR) system

Ignition system ... 6
   Ignition coil - removal and refitting
   Ignition coil and ballast resistor system - general
   Drive resistor (where fitted) - removal and refitting
   Delco Remy electronic ignition - general description
   Delco Remy electronic distributor, type D302 - general
   Delco Remy ignition coil and electronic module assembly description
   Delco Remy ignition coil and electronic module assembly - removal and refitting
   Ignition system and ballast resistor system - testing

Automatic transmission ... 7
   Downshift cable - initial setting and adjustment
   Downshift cable - removal and refitting
   Selector lever, automatic transmission - removal and refitting
   Oil/fluid filter - removal and refitting

Rear axle ... 8
   Pinion oil seal (semi-floating axle) - removal and refitting
   Rear wheel hub studs - removal and refitting

Electrical system... 9
   Heater control illumination lamps - removal and refitting
   Facia switch panel illumination lamps - removal and refitting
   Hazard warning light lamp - removal and refitting
   Door speaker for radio (where fitted) - removal and refitting
   Rear fog lamp assembly - removal and refitting
   Map/courtesy lamp (where fitted) - removal and refitting
   Timer/buzzer module, for seat belt (where fitted) - removal and refitting
   Alternator, type Lucas 25ACR - brush removal

Suspension and steering ... 10
   Front wheel studs - removal and refitting

Bodywork... 11
   Door hinges - removal and refitting
   Facia glovebox cowl - removal and refitting
   Carpets - removal and refitting
   Ashtray - removal and refitting
   Hood (where fitted) - removal and refitting
   Heater matrix - removal and refitting

Chapter 13/Supplement

1 Introduction

This Chapter has been added to cover the various additions and changes which have occurred to the TR7 since the original manual was published.

Where possible the latest specifications have been included in the existing Chapter. However, this has not always been possible, and the new material will in that case be found included in the Supplement.

The information supplied in the Supplement is normally related to modifications and changes, but there may be certain cases where information has been included which refers to early vehicles, but which was not available at the time of publication of the first edition.

2 Specifications

Fuel injection system
Type ... Bosch, electronic
Fuel pump operating pressure ... 36 lbf/in^2 (2.5 kgf/cm^2)

Torque wrench settings	lbf ft	Nm
Air meter support plate	21	28
Fuel rail to injectors	0.5	0.68
Injector clamp plates to manifold	7	10
Lambda sensor to exhaust manifold	44	60
Pressure regulator to fuel rail	21	28
Temperature sensor to manifold	11	15
Thermo-time switch to manifold	21	28

Carburettor (Canada) - beginning with VIN 402027
Type ... Twin Zenith-Stromberg 175 CD-SEVX
Needle ... B1DH

Carburettor (low compression engines) - 1981
Type ... SU HS6
Needle ... BBF

Ignition (Canada) - beginning with VIN 402027
Dynamic timing (at idle) ... 2° ATDC

Ignition (low compression engines) - 1981
Type ... AC Delco D308

Manual gearbox
Clearances
Baulk plate upper face-to-gear lever bush lower edge ... 0.010 in (0.254 mm)
Gear lever side-to-baulk plate edge ... 0.010 to 0.049 in (0.25 to 1.25 mm)

Ratios
5 speed (beginning with VIN 402027)
- Top (fifth) ... 0.792 : 1
- Fourth ... 1.000 : 1
- Third ... 1.400 : 1
- Second ... 2.09 : 1
- First ... 3.32 : 1
- Reverse ... 3.43 : 1

Overall ratios (with 5 speed gearbox)
UK ... 3.09 : 1
North America ... 2.73 : 1

Automatic transmission (beginning with VIN 402027)
Type ... Borg Warner
Overall ratio ... 3.08 : 1

Rear axle
Capacity ... 1.6 Imp pints (2.0 US pints/0.9 litres)

Braking system
Rear brakes (five-speed gearbox cars)
Type ... Drum, self adjusting
Drum internal diameter ... 8.995 to 9.000 in (228.5 to 228.6 mm)
Drum internal diameter, maximum ... 9.050 in (230 mm)

Chapter 13/Supplement

3 Engine

Timing cover, UK and Europe - refitting
1 If fitted remove the cover from the sump opening.
2 Clean the mating faces. Position new gaskets on the cover. Seal the corners where the crankcase meets the sump.
3 Locate the cover in position. Fit the centre bolt with a fibre washer under the head, and tighten.
4 Refit the fan and viscous coupling (see Chapter 2).
5 Fit the cylinder head to timing cover nuts and bolts, leaving them loose.
6 Fit the alternator adjusting link bracket, leaving it loose.
7 Fit the alternator mounting bracket.
8 Fit the alternator, leaving it loose, and connect the plug.
9 Place the alternator drivebelt in position.
10 Fully tighten the timing cover to cylinder head nuts and bolts.
11 Fit and fully tighten the sump to cover bolts.
12 Fit the crankshaft pulley, and adjust the alternator belt.
13 Refit the radiator (see Chapter 2).
14 Fit the fan guard.
15 Fit the bonnet, and reconnect the battery.

Timing cover, USA and Canada - refitting
16 Proceed as in paragraphs 1 to 3.
17 Secure the fan and Torquatrol unit to the timing cover with the two uncommitted bolts (also see Chapter 2).
18 Secure the air conditioning compressor adjustment bracket with the rear bolt, leaving it loose.
19 Fit the cylinder head to timing cover nuts and bolts, leaving them loose.
20 Fit and tighten the air conditioning compressor steady bracket.
21 Tighten the rear bolt to the air conditioning compressor adjustment bracket.
22 Fit, but leave loose, the air conditioning compressor.
23 Fit the air pump bracket and lifting hook (if applicable).
24 Fit the diverter and relief valve bracket with valve.
25 Proceed as in paragraphs 6 to 8.
26 Loosely fit the air pump (if applicable).
27 Loosely fit all the drivebelts, as applicable.
28 Proceed as described in paragraphs 10 and 11.
29 Fit the pulleys.
30 Tighten the four air conditioning compressor to carrier engine bolts.
31 Adjust all the drivebelts.
32 Refit the hose on the diverter/relief valve.
33 Refit the fan blades, and proceed as in paragraphs 13 to 15.

Timing cover oil seal (engine in the vehicle) - removal and refitting
34 Remove the timing cover as described in Chapter 1, Section 11.
35 Lever out the old seal.
36 Liberally oil the new seal.
37 Tap the new seal in squarely, lip leading, until flush with the outside of the cover.
38 Refit the timing cover (see paragraphs 1 to 33).

Timing cover oil seal (engine out of the vehicle) - removal and refitting
39 Remove the timing cover as described in Chapter 1, Section 10.
40 Proceed as in paragraphs 35 to 38.

Valve timing - checking
41 Disconnect the battery.
42 Rotate the crankshaft to put the pulley timing mark opposite the zero mark on the timing cover scale.
43 Remove the fresh air duct.
44 Check the rotor arm position (Chapter 1, Section 41, paragraph 25).
45 Remove the camshaft cover (Chapter 1, Section 6, paragraphs 1 and 2).
46 Check that the timing marks on the camshaft spigot and bearing cap are in line. If so, the valve timing is correct. If the valve timing is not correct proceed as described in paragraphs 48 to 64.
47 Refit all parts in reverse order to dismantling.

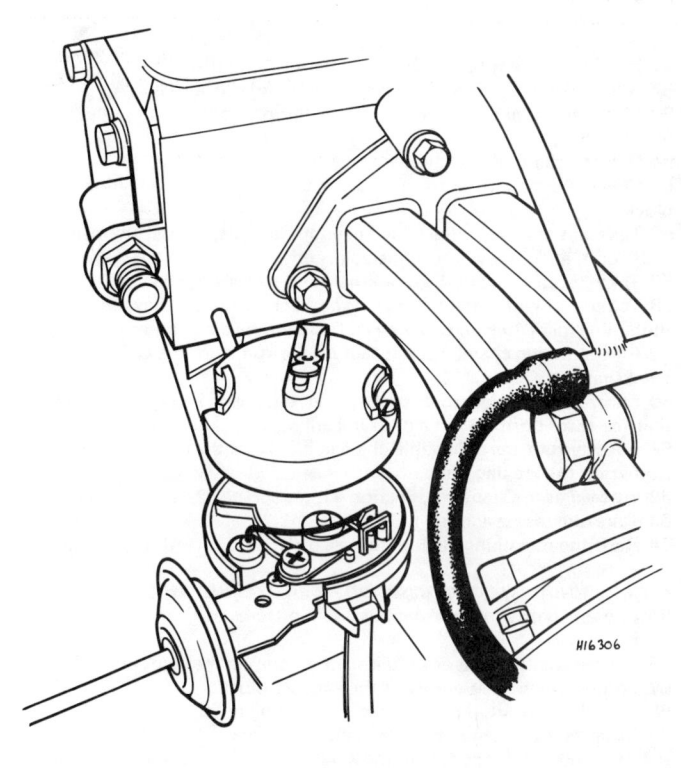

Fig. 13.1. The correct position for the rotor arm with No 1 piston at TDC firing (ie pointing at the head of the rear inlet manifold securing bolt (Sec. 3)

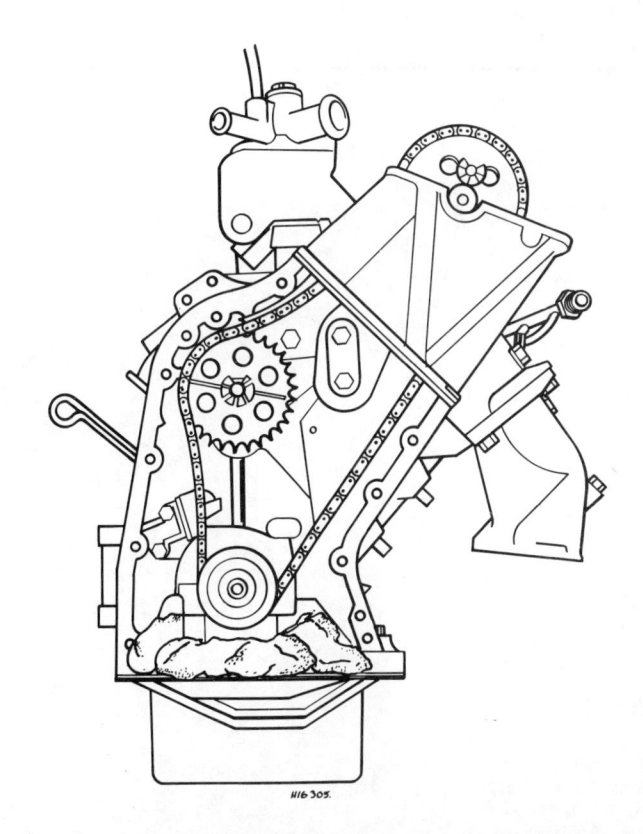

Fig. 13.2. The correct position for the scribed line on the jackshaft sprocket with No 1 piston at TDC firing (Sec. 3)

Valve timing - adjusting

48 Disconnect the battery.
49 Remove the fresh air duct.
50 Remove the timing cover (see Chapter 1, Section 11).
51 Remove the camshaft cover (see Chapter 1, Section 6).
52 Align the marks on the camshaft spigot and bearing cap.
53 Check that the rotor arm points to the rear bolt on the inlet manifold (Fig. 13.1).
54 Remove the timing chain tensioner.
55 Fit a slave bolt to the timing cover centre hole in the cylinder block.
56 Take out the two adjustable chain guide bolts, loosen the other guide bolt, and remove the adjustable guide.
57 Take the chain from the crankshaft and jackshaft pulleys.
58 Temporarily refit the timing cover and crankshaft pulley, and set the timing mark (see paragraph 42). Remove the cover and pulley.
59 Fit the timing chain. Keep it taut on the run from the camshaft to crankshaft sprockets.
60 Ensure that the jackshaft scribed line remains as in Fig. 13.2, and that the rotor position remains as in paragraph 53.
61 Loosely refit the adjustable chain guide. Note the plain and spring washer on the locking bolt.
62 Proceed as in Chapter 1, Section 41, paragraphs 27 to 31.
63 Take out the slave bolt.
64 Refit the remaining parts in reverse order to dismantling procedure.

Air conditioning compressor drivebelt - adjustment

65 Loosen the two bolt assemblies at the transverse support bracket (Fig. 13.3).
66 Loosen the two bolt assemblies at the longitudinal support bracket.
67 Loosen one bolt assembly at the exhaust bracket.
68 Free the locknut, and three main mounting nuts.
69 Rotate the adjustment nut to tighten or slacken the belt, to give 0.75 to 1.00 in (19 to 25 mm) movement at the mid-point.
70 Tighten the three main mounting nuts, and check the adjustment. Readjust as necessary.
71 Reverse the procedure in paragraphs 65 to 68.

Air conditioning compressor drivebelt - removal and refitting

72 Proceed as in paragraphs 65 to 68.
73 Turn the adjustment nut anti-clockwise, to lower the compressor.
74 Free the fan blade bolt assemblies, and slide the blades forward to provide adequate clearance.
75 Run the belt off over the crankshaft pulley (not the clutch pulley).
76 To refit, fit the belt and tighten the fan blade bolt assemblies.
77 Carry out the procedure given in paragraphs 69 to 71.

4 Cooling system

Header tank system - filling

1 Refit the cylinder block drain plug.
2 Refit the bottom hose.
3 Refit the thermostat housing filler plug.
4 Remove the header tank cap.
5 Place the heater controls to maximum heat.
6 Fill the system, until the coolant is 1 in (25 mm) below the header tank neck.
7 Refit the header tank cap.
8 Run the engine at about 1500 rpm until the thermostat opens.
9 Taking the proper precautions (see Chapter 2, Section 2) remove the header tank cap.
10 Top up if necessary to within 1 in (25 mm) of the filler neck.

Header tank, later models - removal and refitting

11 This operation must only be carried out with a cold or cool engine.
12 Remove the lower hose from the tank, draining the coolant.
13 Remove the top and overflow hoses at the tank.
14 Remove the two bolts holding the tank, and lift it away.
15 To refit, reverse the dismantling procedure, and refill the cooling system.

Expansion tank, early models - removal and refitting

16 This operation must only be carried out with a cold or cool engine.
17 Remove the expansion pipe from the thermostat housing.
18 Remove the expansion tank pressure cap, draining the coolant.
19 Loosen the tank clamp nut and bolt, and withdraw the tank.
20 Remove the overflow pipe at the tank.
21 To refit, reverse the removal procedure.
22 Half-fill the tank with coolant, and refit the pressure cap.

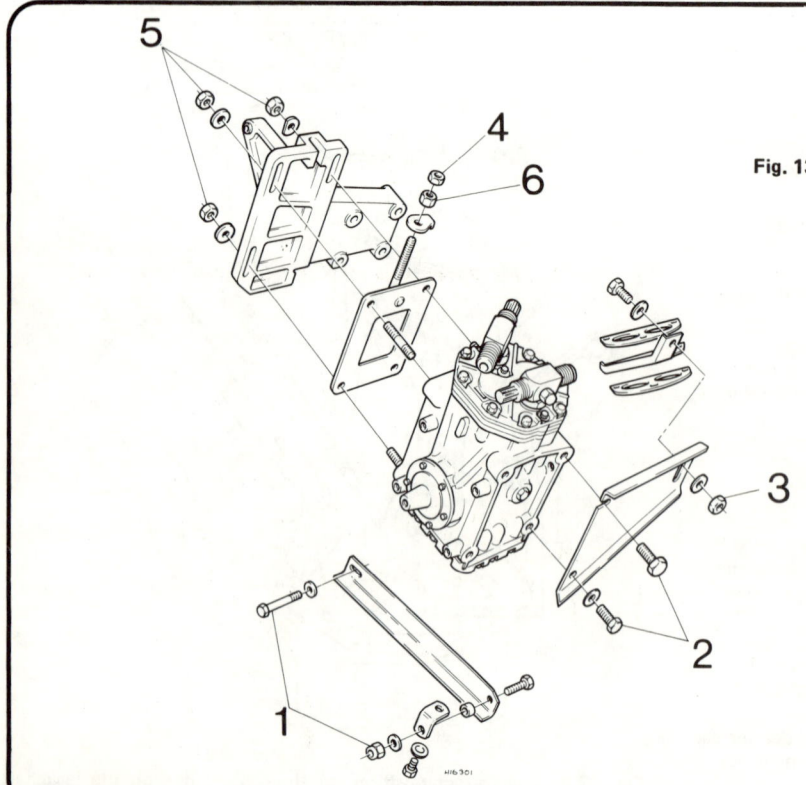

Fig. 13.3. Air conditioning compressor mountings (Sec. 3)

1 Nut and bolt
2 Bolt and bracket
3 Nut
4 Locknut
5 Main mounting nuts
6 Adjustment nut

5 Fuel system

Air cleaner, single carburettor models - removing, refitting, and renewal of filter
1 Disconnect the emission control pipes at the air cleaner backplate, and the hot air hose from the intake.
2 Remove the two bolts, plain washers and rubber washers, and remove the air cleaner complete with the air temperature control unit.
3 To remove the control unit, slacken a hose clip and remove the unit.
4 To renew the filter, it is not necessary to carry out the operation described in paragraph 3.
5 To renew the filter, separate the air cleaner front container from the backplate.
6 Clean the container and backplate thoroughly.
7 Fit a new element and sealing rings into the container, and refit the backplate with the container peg located in the backplate slot.
8 To reassemble, reverse the removal procedure. Ensure that the gasket is in good condition.

General description - tamperproof carburettors for non-USA markets
9 Certain later vehicles can be found fitted with tamperproof carburettors, and local legislation should be checked and complied with where relevant.
10 These carburettors provide a more stringent control of the air/fuel mixture, and consequently of the exhaust gas emissions.
11 The only normal adjustment is to the fast idle speed screw.

Tamperproof carburettor - dismantling and reassembly (general)
12 Attention to the float chamber is permissible, and is as described in Chapter 3, Section 9.
13 Normally the only other dismantling permitted is removal of the piston damper assembly, the suction chamber, and the piston and needle assembly. Cleaning may therefore be carried out to these items, or the needle may be renewed.
14 Further dismantling, and breaking of seals, may render the vehicle user liable to legal action where legislation is in force.
15 If legislation permits, the seals may be removed and attention given to the slow-running and the mixture as described in Chapter 3, Section 8. On completion, check the emission with an approved CO meter, and fit new approved type seals.
16 Needle replacements should always be made as an assembly, i.e. with its related components.

Twin Zenith-Stromberg carburettors (USA specification excluding California) - deceleration by-pass valve checking and adjustment
17 Proceed as described in Chapter 3, Section 10, paragraphs 1 to 17.
18 With the engine at the normal running temperature and idling speed, slowly turn the adjustment screw on the rear carburettor clockwise from the fully loaded position until the engine rpm starts to increase (Fig. 13.12).
19 Turn the screw three complete revolutions anti-clockwise.
20 Repeat the operation described in paragraph 18 on the front carburettor, and then turn the screw two complete revolutions anti-clockwise.
21 Turn the screw on the rear carburettor 1 revolution anti-clockwise.
22 The by-pass valves may be renewed only as complete units.
23 Refit the air cleaner and fresh air duct.

Twin Zenith-Stromberg carburettor automatic choke (where fitted) - removal and refitting
24 This device, which is fitted to the front carburettor only, must be removed as a complete unit.
25 Remove the air cleaner.
26 Loosen the clip on the automatic choke fuel mixture outlet pipe hose, and remove the lead from the automatic choke heater.
27 Remove the carburettors.
28 Remove the three screws holding the automatic choke to the carburettor.
29 Remove the choke and gasket.
30 To refit, reverse the removal procedure.

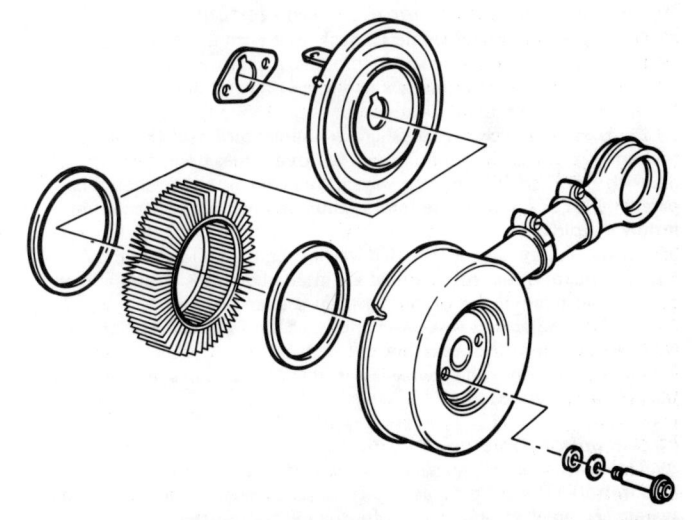

Fig. 13.4. Air cleaner assembly, single carburettor models - exploded view (Sec. 5)

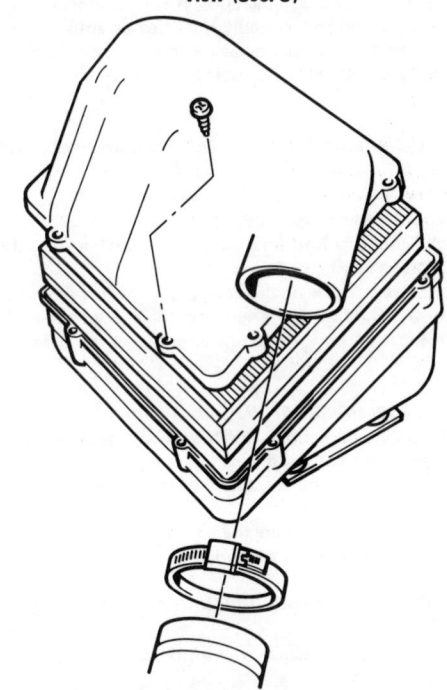

Fig. 13.5. Air cleaner assembly, fuel injection vehicles (Sec. 5)

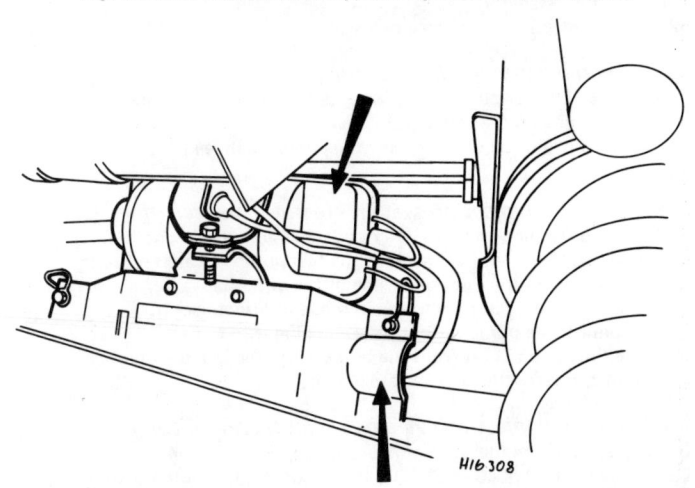

Fig. 13.6. Fuel line filter/s, fuel injection system (arrowed) (Sec. 5)

Air intake temperature control system - testing

31 **On single carburettor models**, check that with a cold engine the sponge rubber flap valve makes an effective seal at the intake.
32 Check that as the engine warms up the valve progressively closes the aperture from the hot air supply.
33 **On twin-carburettor models**, check the condition of the hoses to the temperature sensor unit, and of the hot air inlet hose.
34 With a cold engine, the flap valve in the cold air intake should be parallel with the axis of the intake, thus permitting the maximum inflow of cold air.
35 On starting the engine, the flap valve should immediately move to a point where it cuts off the cold air intake completely, and allows only hot air to enter the air cleaner from the hot air hose.
36 Increase the engine speed for a few seconds and check that the valve remains in the hot position.
37 Ensure that, once the engine is hot, the flap valve has returned to the cold air position.

Fuel injection system - description

38 The fuel injection system may be fitted as an alternative to carburettors. The system has two main parts, namely, the injection system and an electronic control for the injection system.
39 The system is complex in design, and apart from the tasks which are described in the following paragraphs, it is recommended that all work should be entrusted to a suitably qualified establishment.
40 Note that it is essential to depressurise the fuel system as described in paragraphs 53 and 54 before disconnecting *any* part of the fuel system.

Air cleaner and element, fuel injection system - removal and refitting

41 Disconnect the intake pipe from the case.
42 Take out eight screws to split the case.
43 Lift away the case top, and take out the element, noting the correct position for ease of refitting.
44 To remove the remainder of the case, disconnect it from the air-flow meter, and take out the four bolts securing the case to the body.
45 To refit, reverse the removal procedure. Clean the case before refitting.

Throttle cable, fuel injection system - removal and refitting

46 Take off the plenum chamber, and release the outer, then the inner, cables from the throttle body bracket.
47 Remove the clip inside the car, and disconnect the inner cable at the pedal.
48 Withdraw the cable into the engine compartment.
49 To refit, reverse the removal procedure.

Throttle pedal assembly, fuel injection system - removal and refitting

50 Disconnect the battery, remove the spring clip, and take the cable from the pedal.
51 Remove the two bolts holding the pedal assembly to the bulkhead. Withdraw the assembly.
52 To refit, reverse the removal procedure.

Fuel injection system - depressurising

53 Remove the fuel pump earth wire, switch on the ignition, and crank the engine on the starter for a few moments.
54 Switch the ignition off, and refit the pump earth lead.

Fuel pump, fuel injection system - removal and refitting

55 Depressurise the system as described in paragraphs 53 and 54.
56 Disconnect the electrical leads at the fuel pump (located in front of the rear axle on the right-hand side of the underbody).
57 Clamp off the input pipe to the pump, and disconnect it.
58 Disconnect the pump output pipe and plug it.
59 Take the nut and bolt from the bracket and remove the pump.
60 To refit, reverse the removal procedure.

Fuel tank, fuel injection system - removal and refitting

61 Depressurise the fuel system. See paragraphs 53 and 54.
62 Proceed as described in Chapter 3, Section 17, but note the extra fuel return pipe from the pressure regulator to the tank which must be disconnected before the tank can be removed.

Fuel line filter(s), fuel injection system - removal and refitting

63 Depressurise the fuel system. See paragraphs 53 and 54.
64 From the filter(s) (located just below the fuel pump), take out the four screws and lift off the cover.
65 Disconnect and plug the input and output fuel pipe(s).
66 Take out the clamp bolt and nut, and manoeuvre the filter(s) out.
67 To refit, reverse the dismantling procedure.

Fuel pressure regulator, fuel injection system - removal and refitting

68 Depressurise the fuel system. See paragraphs 53 and 54.
69 Disconnect the vacuum pipe from the regulator (Fig.13.7).
70 Disconnect the fuel inlet pipe, and plug it.
71 Remove the nut and screw from the regulator bracket, and release the regulator from the rail using two spanners. Do not strain the rail.
72 To refit, reverse the removal procedure.

Fuel rail, fuel injection system - removal and refitting

73 Proceed as in paragraphs 68 to 71.
74 Take the fuel pipe connections from the rail, and remove the rail securing bolts.
75 Remove the rail (Fig.13.7).
76 To refit, reverse the removal procedure.

Plenum chamber, fuel injection system - removal and refitting

77 If a water heated plenum chamber is fitted, partially drain the cooling system and take off the plenum chamber water pipes.
78 Remove the battery.
79 Free the throttle bracket from the plenum chamber.
80 Free the dipstick tube from the chamber bracket.
81 Take off the plenum chamber to inlet manifold connecting pipe.
82 From the plenum chamber disconnect the extra air valve pipe, pressure regulator vacuum pipe, brake servo vacuum pipe and cold start injector.
83 Take out the five bolts, and lift off the plenum chamber (Fig.13.7).
84 To refit, reverse the removal procedure.

Induction manifold, fuel injection system - removal and refitting

85 Disconnect the battery, and drain the cooling system (see Chapter 2).
86 Remove the plenum chamber (see paragraphs 77 to 83).
87 Take the electrical connections from the temperature switch, temperature sensor and thermo-time switch.
88 Depressurise the fuel system (see paragraphs 53 and 54).
89 Remove the fuel rail and fuel injectors (see paragraphs 73 to 75 and 94 to 97).

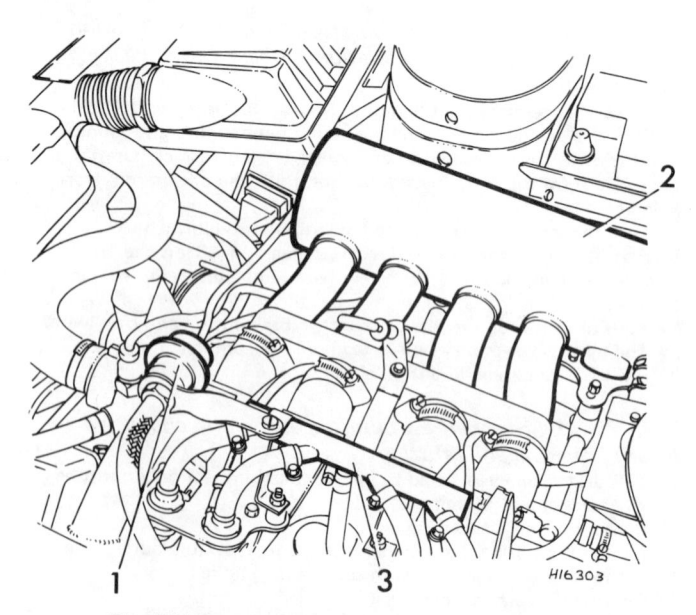

Fig. 13.7. Detail parts, fuel injection system (Sec. 5)

1 Fuel pressure regulator
2 Plenum chamber
3 Fuel rail

Chapter 13/Supplement

90 Disconnect the manifold outlet water pipes and top radiator hose.
91 Remove the thermostat (see Chapter 2).
92 Take out the four bolts, and lift the manifold away.
93 To refit, clean all surfaces and renew the gaskets as necessary. Reverse the removal procedure.

Injectors, fuel injection system - removal and refitting
94 Depressurise the fuel system.
95 Remove the electrical lead and fuel feed pipe from the injector.
96 Remove the screws from the injector clamp bracket, lift off the bracket and withdraw the injector.
97 To refit, renew the sealing ring and reverse the dismantling procedure.

Cold start injector, fuel injection system - removal and refitting
98 The injector is located beneath the plenum chamber. The removal and refitting procedure is similar to that described in paragraphs 94 to 97.

Emission control and evaporative loss control systems - general
99 More recent models have been fitted with varying layouts of emission and evaporative loss control equipment, and space does not permit a diagram to be given of every arrangement. Owners are therefore recommended to make a tidy sketch of hoses, pipework and any other items which might otherwise cause confusion during reassembly. Details of the individual items will be found in either Chapter 3, or in the following paragraphs.

Absorption canisters, twin canister specification - removal and refitting
100 Make a note of the pipe connections by labelling, or by some other suitable method.
101 Remove the pipes from the upper canister, loosen the clamping nut, and release the canister.
102 Remove the pipework from the lower canister, loosen the clamp nut and release the canister.
103 Refitting is a reversal of the removal procedure.

Air injection system
104 Fig.13.9 gives the system layout, and should be worked on in conjunction with the comments in Chapter 3, Section 21, paragraphs 6 to 10.

Exhaust gas recirculation (EGR) system
105 Whilst the earlier layout is given in Fig.3.27, space does not permit the inclusion of diagrams covering subsequent layouts. A careful note of the arrangement should therefore be made before dismantling.
106 Note that on certain layouts an EGR control valve is not used.

6 Ignition system

Ignition coil and ballast resistor system - general
1 This arrangement, designed for easier starting, has a ballast resistor built into the wiring harness in series with the normal supply to the coil. This resistor drops the circuit voltage from 12 volts, to a figure suitable for powering the 6 volt rated coil.
2 When the starter is operated, the ballast resistor is bypassed, allowing the full circuit voltage (reduced, however, as a result of the starter motor load) to be applied to the coil. The slightly overloaded coil thus provides an increased voltage at the spark plugs.

Ignition coil - removal and refitting
3 To remove the coil (located on the right-hand side of the bulkhead), remove the protective cover and pull off the HT lead. Disconnect the two LT connections, remove the two nuts and washers, and take the drive resistor lug (where fitted) from the body stud. Remove the coil.
4 To refit the coil, reverse the removal procedure.

Drive resistor (where fitted) - removal and refitting
5 The resistor is located on the right-hand side of the engine bulkhead, next to the coil, and is removed by pulling off the two connections, removing the nut and washers, and then the drive resistor.
6 To refit, reverse the removal procedure. The connectors may be attached either way round.

Delco Remy electronic ignition - general description
7 The system consists of two basic units, a distributor and an electronic module assembly.
8 The distributor is an angle transducer which supplies pulses to the module assembly. The module assembly controls the collapse of the primary field in the ignition coil, and hence the spark at the plug points.

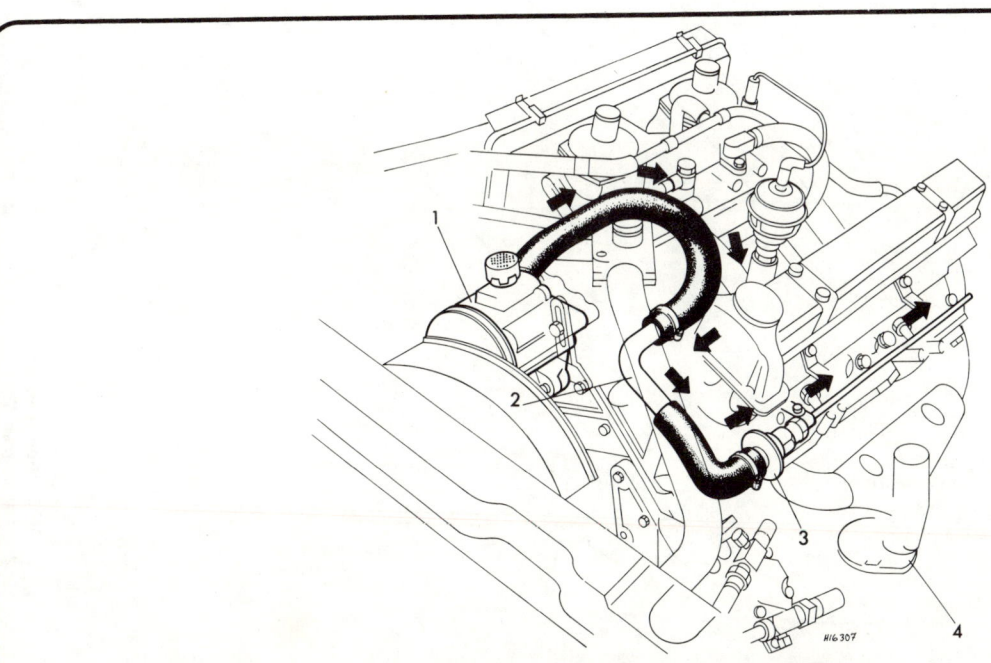

Fig. 13.8. Air injection system - layout (Sec. 5)

| 1 Air pump | 2 Air rail | 3 Check valve | 4 Exhaust manifold |

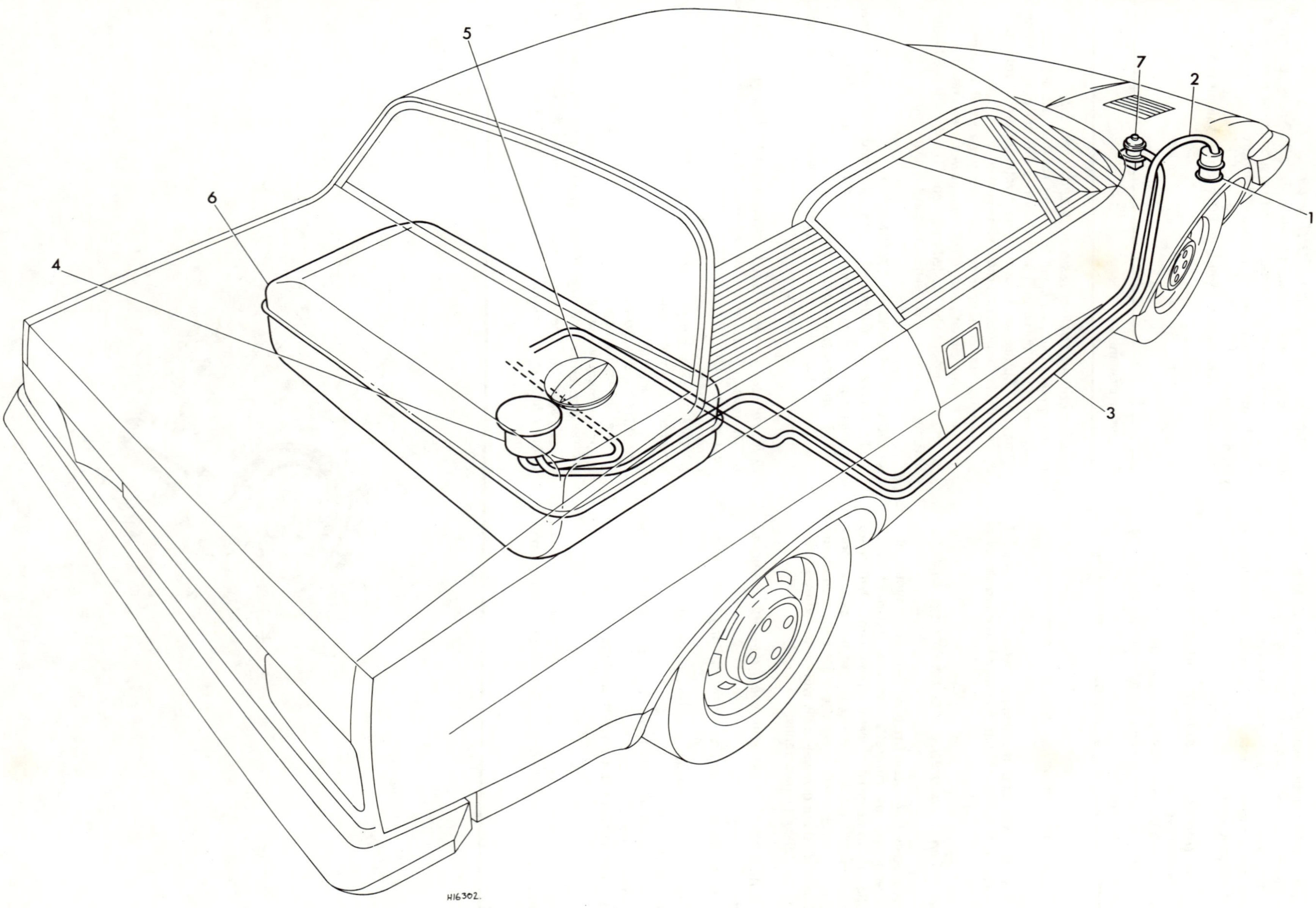

Fig. 13.9. Evaporative loss control system - typical layout (Sec. 5)

1 Absorption canister
2 Vapour feed line
3 Fuel feed line
4 Fuel vapour separator
5 Sealed filler cap
6 Limited fill fuel tank
7 Fuel pump

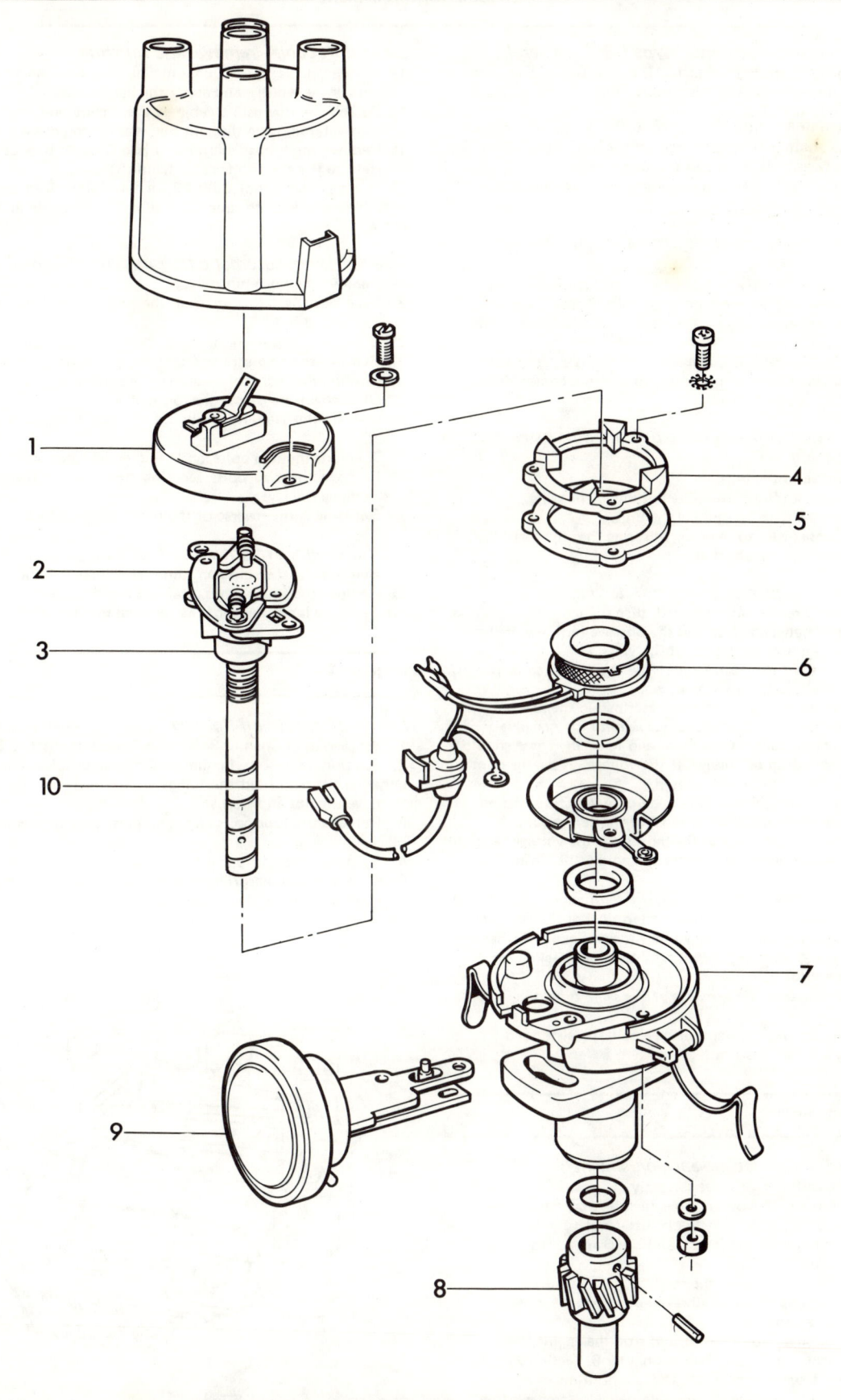

Fig. 13.10. Delco Remy Series D302 electronic ignition distributor (Sec. 6)

1 High tension rotor arm
2 Centrifugal advance mechanism
3 Angle transducer timing rotor
4 Angle transducer static component
5 Angle transducer permanent magnet
6 Angle transducer winding
7 Body
8 Drivegear
9 Retard unit
10 Plug to electronic module assembly

Chapter 13/Supplement

Delco Remy electronic distributor, type D302 - general

9 The distributor is illustrated in Fig.13.10 and is basically an angle transducer in a conventional AC Delco body, with the centrifugal advance and retard unit still retained.
10 A vacuum retard unit is fitted, and permits the permanent magnet and winding to be rotated through a limited angle, thereby altering the relationship between the rotor and static assembly.
11 At the time of compiling this Supplement, no servicing details were available for this unit.

Delco Remy ignition coil and electronic module assembly - description

12 The assembly consists of a light alloy casing, containing an ignition coil winding and an electronic module. Pulses from the ignition distributor are amplified by the module, and switch a power transistor in the on–off modes.
13 The primary coil current is collapsed by the power transistor "off" condition, giving sparks at the plug points in the conventional manner.

Delco Remy ignition coil and electronic module assembly - removal and refitting

14 Remove the wiring to the unit.
15 Remove the front bolt by freeing the nut under the wing.
16 Remove the rear bolt, and withdraw the unit.
17 Refit in the reverse order to removal, ensuring that the earth lead tag is in position under the front bolt.

Ignition coil and ballast resistor system - testing

18 With the contact breaker points closed, turn the ignition switch on. Connect a voltmeter between earth and the positive terminal of the coil. The voltmeter should register about 6.5 volts.
Caution: *Do not connect the positive terminal directly to earth (i.e. with a test wire) while the ignition is switched on.*
19 With the voltmeter still connected as described in the previous step and the ignition switched on, connect a test wire to the negative terminal of the coil and connect the other end to earth. There should not be any noticeable drop of voltage. If the voltage drops, the contact breaker points are dirty or not fully closed.
20 Crank the engine with the test wire and voltmeter still installed. The voltage reading at the positive terminal of the coil should increase to the amount of voltage at the battery. If the voltage reading remains the same or decreases, the starter solenoid or its wiring is at fault.
21 Disconnect the test wire from the negative terminal of the coil and start the engine. The voltage at the positive terminal of the coil should be between 9.5 and 11.5 volts. If the voltage is significantly higher, the ballast resistor is at fault and its wiring should be checked for poor connections and/or damage.
Note: *If the coil is replaced, be sure it is of the same type previously fitted.*
Ignition coils differ between systems using a ballast resistor and systems not using a ballast resistor.

7 Automatic transmission

Downshift cable - initial setting and adjustment

1 Ensure that the carburettor is set correctly.
2 Free the locknut on the outer cable ferrule, and adjust the ferrule in the bracket by means of the adjuster nut to give 1/16 in (1.5 mm) clearance between the ferrule and the crimped stop.
3 Tighten the locknut.
4 Check the gear shift speeds on the road.
5 To adjust further, apply the handbrake and chock the wheels.
6 Start the engine, and select D.
7 Adjust the idling speed to 750 rpm, and stop the engine.
8 Remove the sump pan as described in Chapter 6, Section 27.
9 Ensure that the downshift cam is in the idle position.
10 Have someone fully depress the throttle pedal, and check that the downshift cam is in the kickdown position.
11 Adjust the cable if necessary, until the idling and kickdown positions are correctly obtained on the downshift cam.
12 Tighten the locknut.
13 Refit the sump pan.

Downshift cable - removal and refitting

14 Have available an adequate means of lifting the vehicle. Select 'N', chock the wheels, and apply the handbrake.
15 Detach the downshift cable at the engine end, by loosening the locknut and removing the split pin, washer and clevis pin.
16 Remove the transmission sump (see Chapter 6, Section 27).
17 Release the downshift cable from the cam.
18 Using Leyland tool CBW 62, or a suitable sized tube or socket over the ferrule to contract it, pull out the ferrule and remove the cable.

Selector lever, automatic transmission - removal and refitting

19 Remove the push button cap.
20 Take out the screw securing the push button, and remove the button and spring.
21 Unscrew the gear knob.
22 Remove the two screws from the gear lever surround, disconnect the illumination light, and remove the surround.
23 Disconnect the gear indicator light.
24 Operate the gear lever fully in both directions to release the steel cover.
25 Take out the two bolts, and remove the roller assembly.
26 Remove the two bolts securing the lever to the turret shaft, and remove the lever.
27 Refitting is the reverse of the removal procedure.

Oil/fluid filter - removing and refitting

28 Remove the transmission sump (see Chapter 6, Section 27).
29 Remove the four screws retaining the filter, and remove the filter.
30 Refitting is a reversal of the removal procedure.

8 Rear axle

Pinion oil seal (semi-floating axle) - removal and refitting

1 Proceed as in Chapter 8, Section 7, paragraphs 1 to 5.
2 Position two bolts in the pinion flange holes, and use a bar between them to hold the flange whilst loosening the pinion nut. Remove the nut and washer.
3 Remove the four bolts and spring washers, and remove the pinion oil seal housing.
4 Remove the seal from the housing.
5 Fit a new seal, ensuring that the lip is towards the differential.
6 Place a narrow strip of masking tape over the step on the pinion

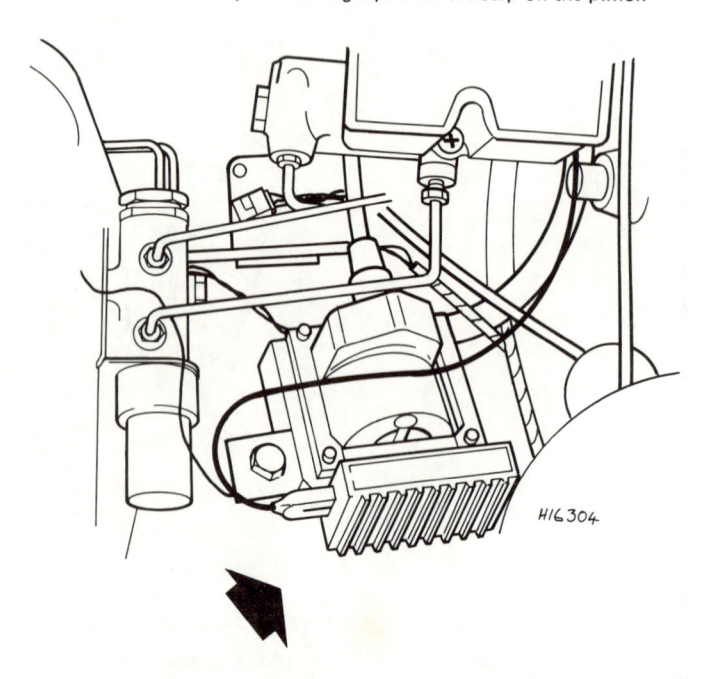

Fig. 13.11. Ignition coil and electronic module assembly (arrowed) (Sec. 6)

shaft, to prevent damage to the seal during reassembly.
7 Oil the masking tape and seal lip.
8 Fit the seal and housing, tapping it gently and evenly into place. Fit and tighten the four bolts and washers.
9 Refit the flange, and secure with the nut and washer. Tighten the nut to the recommended torque.
10 Refit the propeller shaft.
11 Lower the vehicle.
12 Check the oil level.

Rear wheel hub studs - removal and refitting
13 Raise the rear of the vehicle.
14 Remove the hub, as described in Chapter 8, Section 3.
15 Drive out the studs.
16 To fit a stud, ensure that the mating faces of the hub and stud are clean.
17 Enter the stud in the hub, ensuring that the splines engage, and press or drive it home.
18 Refit the hub.

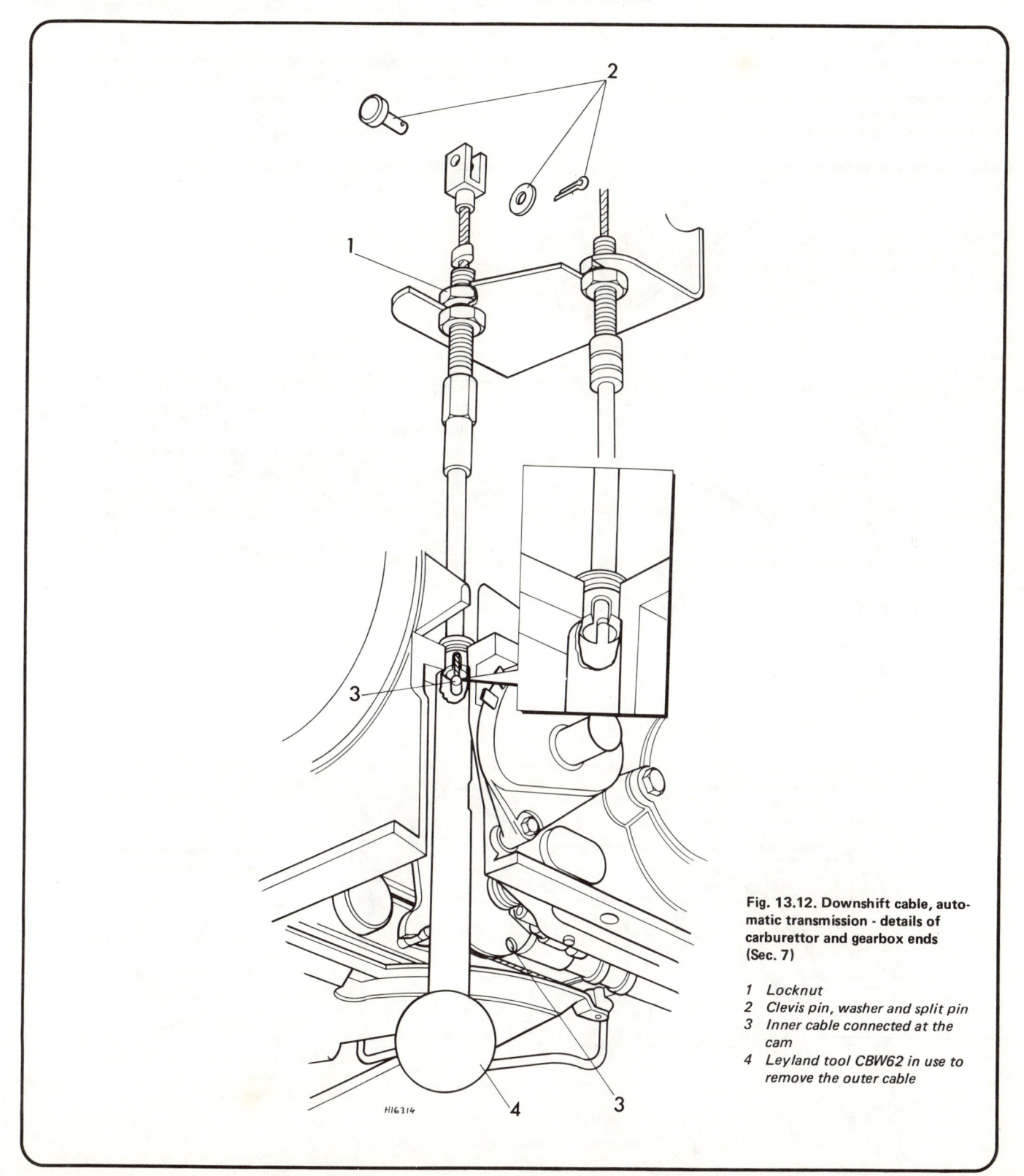

Fig. 13.12. Downshift cable, automatic transmission - details of carburettor and gearbox ends (Sec. 7)

1 Locknut
2 Clevis pin, washer and split pin
3 Inner cable connected at the cam
4 Leyland tool CBW62 in use to remove the outer cable

Chapter 13/Supplement

9 Electrical system

Heater control illumination lamps - removal and refitting
1. Remove the two cross-head screws.
2. Lower the panel.
3. Tilt the panel and remove the bulb holder(s) as required.
4. Withdraw the bayonet fitting bulbs.
5. To refit, reverse the removal procedure.

Facia switch panel illumination lamps - removal and refitting
6. Disconnect the battery.
7. Remove the two cross-head screws and washers, and take out the panel.
8. Carefully pull the bulb holder(s) from the housing.
9. Pull the bulb from the holder.
10. To refit, reverse the removal procedure.

Hazard warning light lamp - removal and refitting
11. Disconnect the battery.
12. Remove the two cross-head screws and washers, and take out the panel.
13. Unplug the hazard switch harness, and pull out the bulb.
14. To refit, reverse the removal procedure.

Door speaker for radio (where fitted) - removal and refitting
15. Remove the door trim (see Chapter 12, Section 14).
16. Remove the four cross-head screws and spring clips, and release the speaker.
17. Disconnect the leads and withdraw the speaker.
18. To refit, reverse the removal procedure.

Rear fog lamp assembly - removal and refitting
19. Disconnect the lamp leads inside the luggage compartment.
20. Pull the leads out through the grommet.
21. Remove the two bolts, nuts and washers, and the lamp.
22. To refit, reverse the removal procedure.

Map/courtesy lamp (where fitted) - removal and refitting
23. Disconnect the battery.

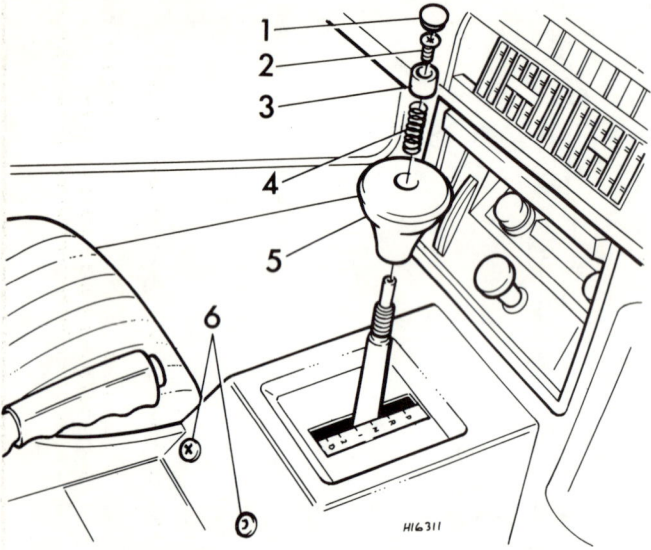

Fig. 13.13. Hand selector lever, automatic transmission (Sec. 7)
1. Push button cap
2. Push button retaining screw
3. Push button
4. Spring
5. Knob
6. Surround retaining screws

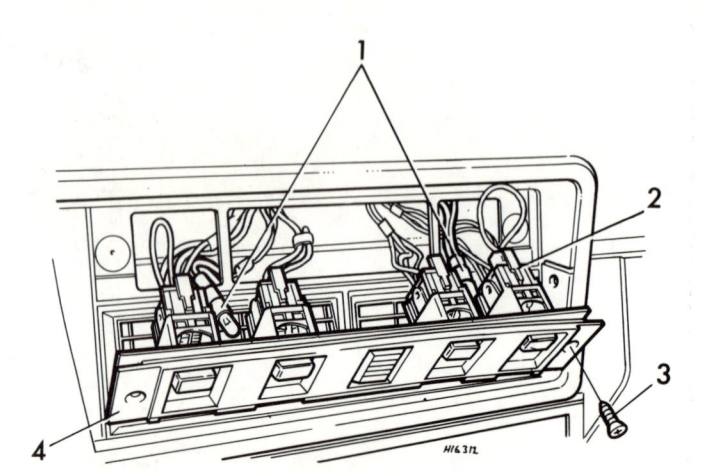

Fig. 13.14. Switch panel details (Sec. 9)
1. Panel illumination bulbs
2. Hazard warning light holder
3. Panel screws
4. Panel

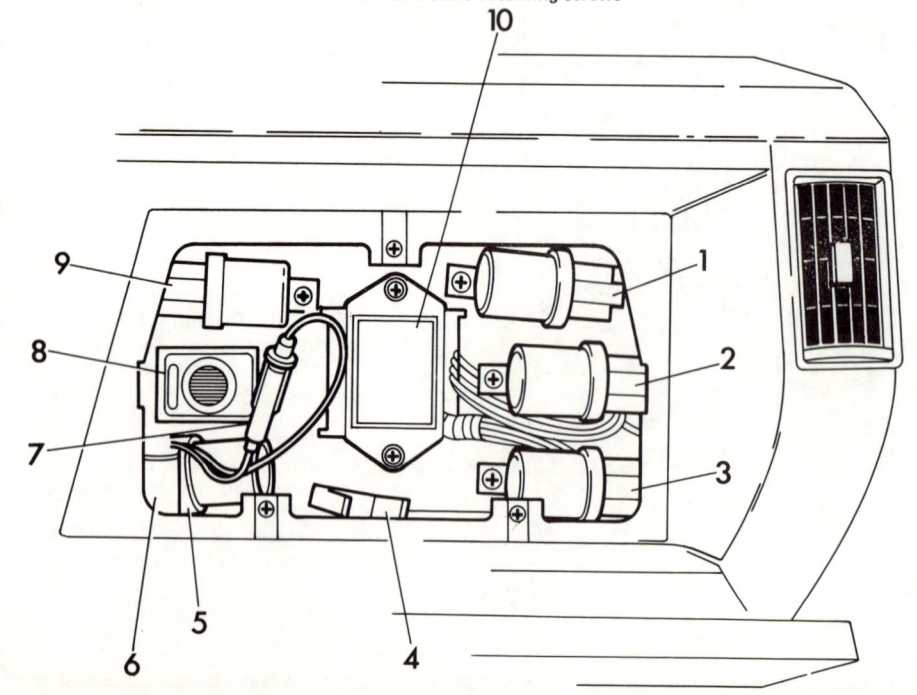

Fig. 13.15. Component mounting plate - certain USA vehicles (Sec. 9)

1. Starter motor relay
2. Air conditioning control relay
3. Headlamp right-hand run/stop relay
4. Headlamp circuit breaker
5. Headlamp left-hand run/stop relay
6. Hazard flasher unit
7. In line fuse
8. Seat belt timer/buzzer module
9. Horn relay
10. Fuse box

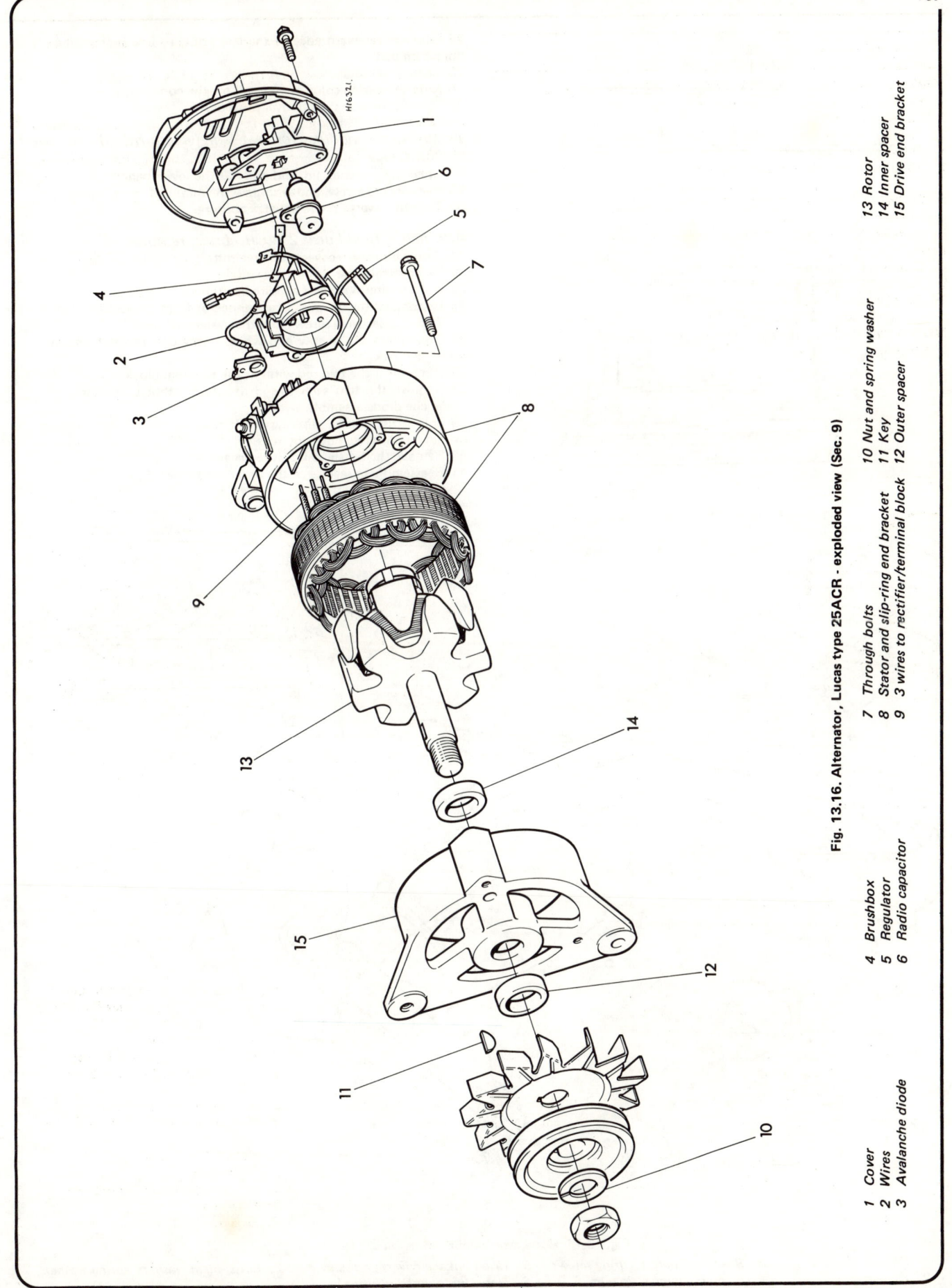

Fig. 13.16. Alternator, Lucas type 25ACR - exploded view (Sec. 9)

1 Cover
2 Wires
3 Avalanche diode
4 Brushbox
5 Regulator
6 Radio capacitor
7 Through bolts
8 Stator and slip-ring end bracket
9 3 wires to rectifier/terminal block
10 Nut and spring washer
11 Key
12 Outer spacer
13 Rotor
14 Inner spacer
15 Drive end bracket

24 Prise the rearward edge of the lamp out, to free it from the door trim pad.
25 Remove the bulb.
26 Note the wiring colours, and pull off the connectors.
27 To refit, reverse the removal procedure.

Timer/buzzer module, for seat belt (where fitted) - removal and refitting
28 Open the glovebox lid, and pull out the lower panel.
29 Remove the screw, and withdraw the module from the plug.
30 To refit, reverse the removal procedure.

Alternator, type Lucas 25ACR - brush removal
31 Remove the two cover retaining bolts.
32 Remove the cover.
33 Make a careful note of all wire positions.
34 Disconnect the yellow Lucar connector from the rectifier pack.
35 Take the securing screw from the avalanche diode.
36 Disconnect the yellow and white Lucar connectors from the terminal block.
37 Disconnect the two red wires at the terminal block.
38 Remove the two screws, and lift the brushbox, regulator and avalanche diode assembly away.
39 Polish discoloured slip-rings with fine glasspaper, crocus paper, or metal polish. Remove any residue.
40 Remove the brushes and fit replacements.
41 Reassemble in reverse of the order used when dismantling.

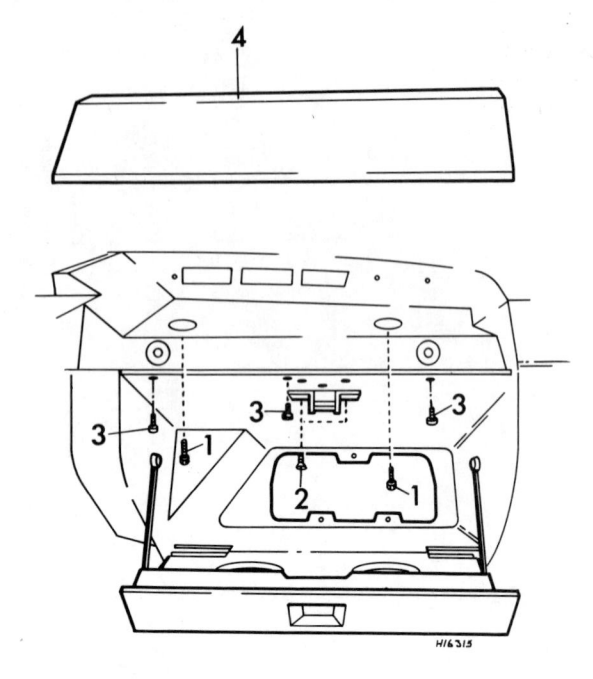

Fig. 13.17. Facia glovebox cowl (Sec. 11)

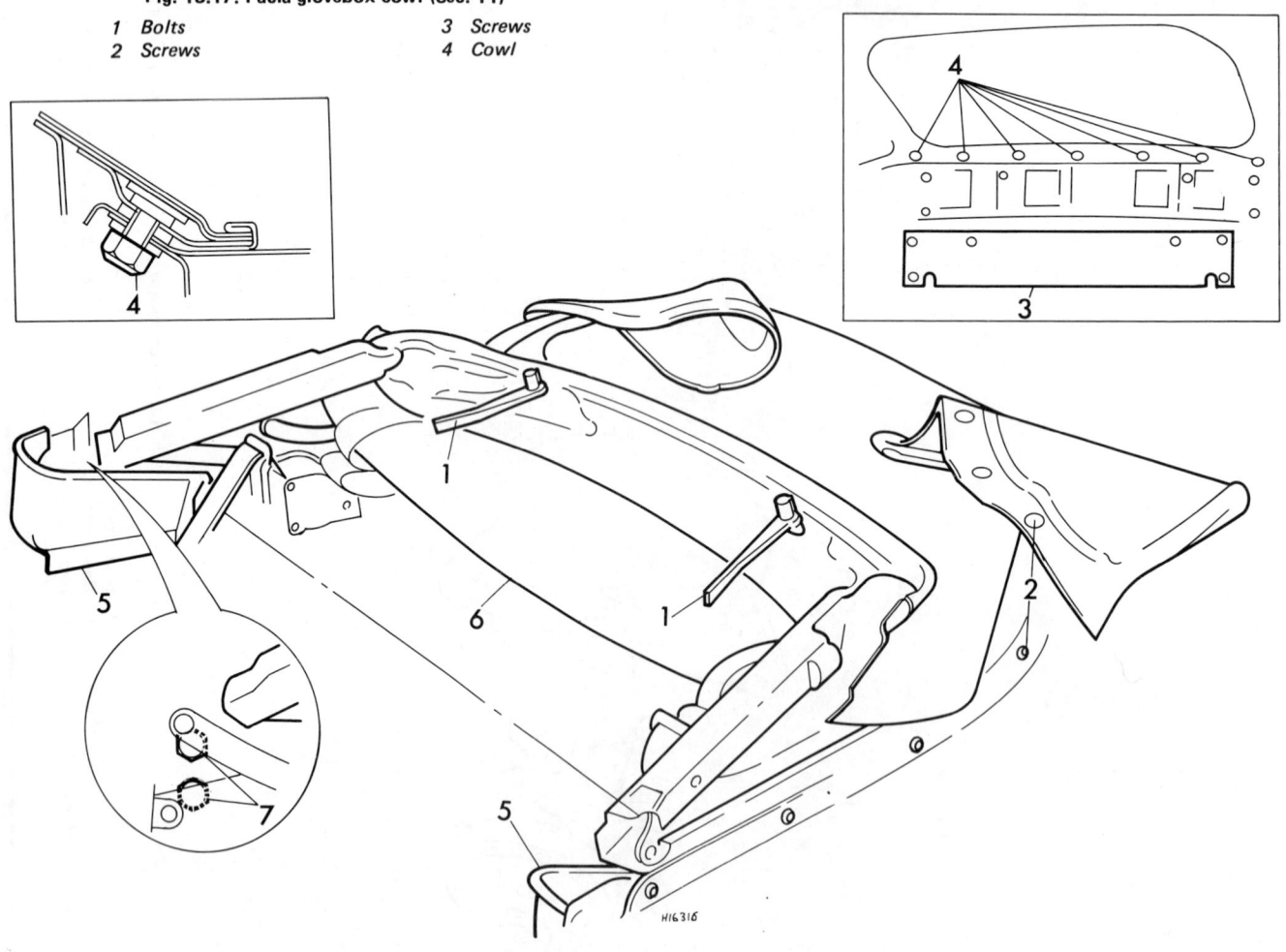

1 Bolts	3 Screws
2 Screws	4 Cowl

Fig. 13.18. Hood attachment points (Sec. 11)

1 Release levers	3 Rear trimboard securing screws	5 Hood linkage covers and screws	7 Bolts, plain washers, spring washers
2 Fasteners	4 Hood retaining strip nuts	6 Hood (in lowered position)	

10 Suspension and steering

Front wheel studs - removal and refitting
1 Remove the front hub as described in Chapter 11, Section 5.
2 Remove the hub from the disc, by taking out the four retaining bolts.
3 Press or carefully drive the studs from the hub.
4 With the mating surfaces clean, enter the stud, align the splines, and press the stud into place.
5 Refit the hub to the disc, and tighten the four bolts to the correct torque figure.
6 Refit the hub.

11 Bodywork

Door hinge - removal and refitting
1 The body half of the hinge is welded to the body and is not normally renewed. The door half, and the hinge pins are, however, renewable.
2 The door half must always be renewed together with the pins.
3 Drive the upper pin upwards with a chisel under the head until the serrations are clear.
4 Drive the lower pin down until the serrations are clear.
5 Remove the door as described in Chapter 12, Section 14.

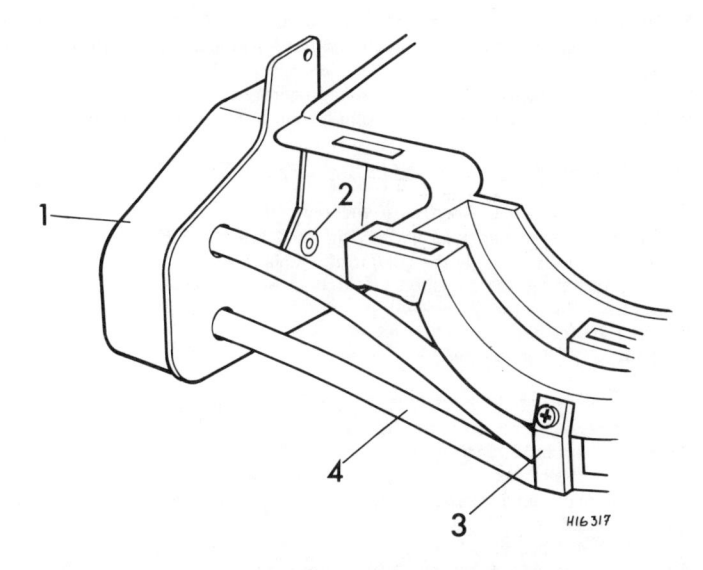

Fig. 13.19. Heater matrix pipe flange plate (Sec. 11)

1 Seal
2 Rivet
3 Screws
4 Pipe assembly

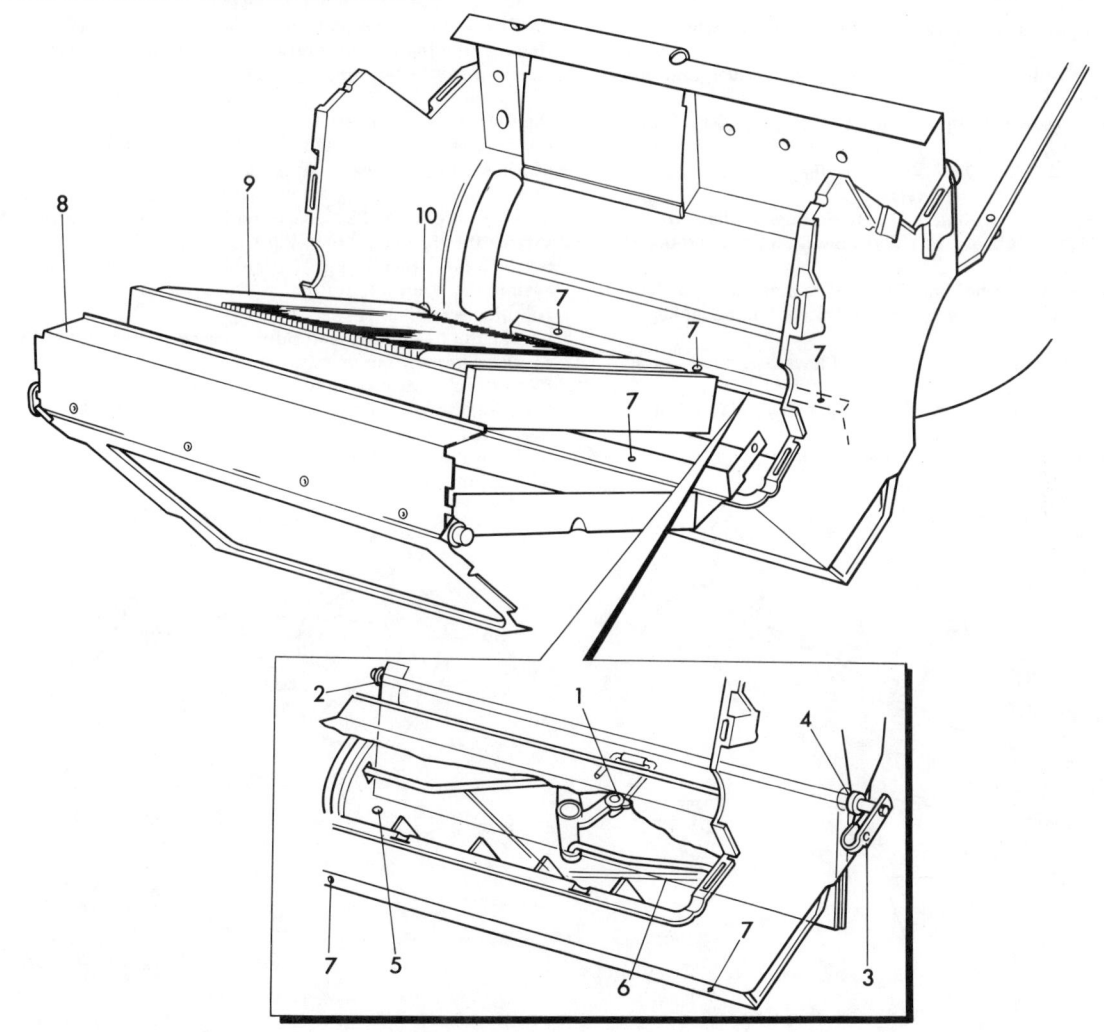

Fig. 13.20. Heater matrix details (Sec. 11)

1 Spire nut
2 Spire nut
3 Operating lever and spring clip
4 Lower flap spacer
5 Rivet
6 Lower side flap
7 Rivets
8 Matrix housing
9 Matrix
10 Matrix pipe seals

6 Remove the pins, the door half of the hinge, and the spring.
7 To refit, lubricate the hole in the body half hinge, and position the spring and door half hinge so that the door is half open.
8 Compress the spring, and fit the pins fingertight.
9 Engage the pin serrations by tapping them home with a hammer and drift.
10 Refit the door.

Facia glovebox cowl - removal and refitting
11 Remove the facia as described in Chapter 12, Section 20, and take out the two bolts securing the cowl to the facia inside.
12 Take out the three glovebox lid latch screws, and also the three screws from inside the glovebox.
13 Take off the cowl.
14 To refit, reverse the removal procedure.

Carpets - removal and refitting
15 To remove the carpet over the transmission tunnel, remove the console (Chapter 12, Section 23).
16 Lift out the carpet.
17 To remove the rear carpet, remove the seat as described in Chapter 12, Section 26.
18 Lift out the carpet.
19 To refit, reverse the removal procedures.

Ashtray - removal and refitting
20 Press down the cigarette stubber, thus freeing the top edge of the bowl.
21 Disengage the bowl bottom retainers from the surround, and take out the bowl.
22 Carefully lift the surround retainers with a screwdriver, and pull the surround clear.
23 To refit, snap the surround into place, and refit the bowl.

Hood (where fitted) - removal and refitting
24 Free the hood front using the levers.
25 Release the eight fasteners holding the hood sides to the body.
26 Remove the eight trimboard to body screws. Pull the trimboard forward.
27 Take off the seven nuts holding the hood retainer strip to the rear deck, and remove the three screws from each hood linkage cover.
28 Take care not to damage the material, and lower the hood to the rearmost position.
29 Obtain help to lift the hood away, after removing the four bolts, plain, and spring washers.
30 To refit, position the hood and loosely fit the four bolts with washers.

31 Check for freedom of movement of the hood, and fully secure the bolts.
32 Complete refitting by reversing the removal procedure.

Heater matrix - removal and refitting
33 Disconnect the battery.
34 Remove the heater unit and fan motor (see Chapter 12, Sections 34 and 35).
35 Take the seal from the pipe flange plate, and drill out the rivet holding the plate to the heater.
36 Remove the two screws securing the pipe mounting bracket to the heater, and lift the bracket and packing piece away.
37 Carefully withdraw the pipe assembly from the matrix bushes.
38 Free the three trunnions and take the flap rods from the trunnions.
39 Take off the face level flap and heater inlet flap.
40 Drill the four rivets from the control level mounting plate, remove the two screws holding the lower flange, and lift off the mounting plate.
41 Remove the spire nut holding the lower side flap operating rod to the lever, and release the rod.
42 Remove the spire nut holding the lower flap spindle to the matrix housing.
43 Pull the operating lever and spring clip from the lower flap spindle, detach the flap from the matrix housing, and collect the spacer.
44 Drill the two rivets from the lower side flap assembly to the heater box, and remove the flap assembly.
45 Drill the seven rivets from the matrix housing to heater box. Remove the matrix and housing from the heater box, and take the matrix from the housing.
46 To refit, first place two new pipe seals in the matrix, and refit the foam packing piece.
47 Reverse the removal sequence given in paragraphs 38 to 46.
48 Close the face level ventilation flap, place the control lever to "OFF", and tighten the trunnion screw (Fig.13.22, items 1, 2).
49 Close the lower flap, place the control lever to SCREEN, and tighten the trunnion screw (Fig.13.21, item 1).
50 With the cam as in Fig.13.21 (item 2), set the air inlet flap against the aperture and the control lever to OFF, and tighten the trunnion screw.
51 Set the heater flap and outlet flap as shown in Fig.13.22 (item 4), tighten the trunnion screw (3).
52 Set the control lever to COLD (Fig.13.22, item 5) and tighten the trunnion screw (item 6).
53 Refit the motor.
54 Refit the heater unit.
55 Reconnect the battery.

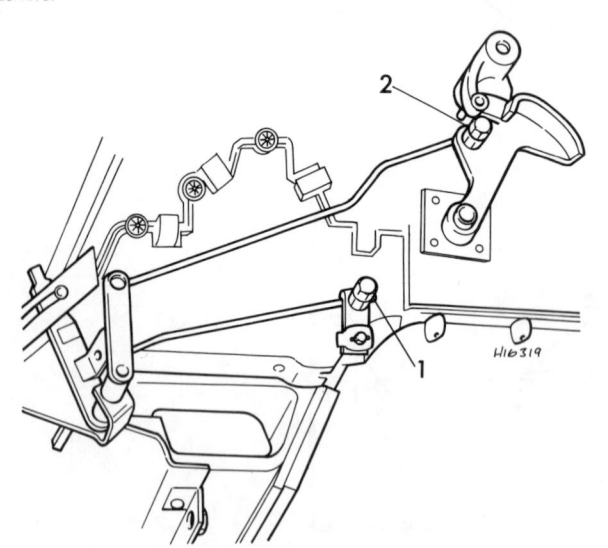

Fig. 13.21. Setting positions for heater control mechanisms (see text) (Sec. 11)

1 Trunnion screw, lower flap control
2 Cam and trunnion

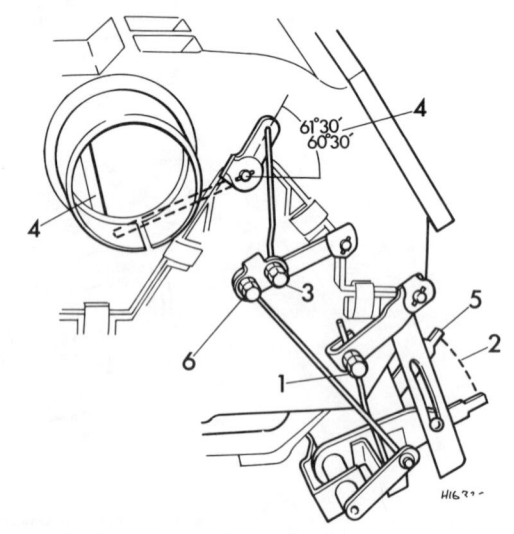

Fig. 13.22. Setting positions for heater control mechanisms (see text) (Sec. 11)

1 Trunnion screw
2 Face ventilation flap position
3 Trunnion screw
4 Inlet and outlet flap position
5 Control lever (cold position)
6 Trunnion screw

Wiring diagrams commence overleaf

Fig. 13.23. Key to wiring diagram - UK and European models (1978 to 1979)

1 Battery
2 Alternator
3 Starter motor
4 Starter motor relay
5 Start inhibit switch (automatic transmission)
6 Ignition switch
7 Headlamp motor circuit breaker
8 Headlamp relay
9 Headlamp relay
10 Headlamp flash control
11 Headlamp flash relay
12 Master light switch
13 Headlamp motor
14 Headlamp motor
15 Loudspeakers
16 Radio
17 Main dip flash switch
18 Hazard switch
19 Flasher unit
20 Main beam warning light
21 Main beam
22 Main beam
23 Dip beam
24 Dip beam
25 Sidelight
26 Sidelight
27 Side indicator
28 Side repeater connection
29 Side indicator
30 Side repeater connection
31 Front fog lamps junction
32 Panel rheostat
33 Heater control illumination
34 Selector panel illumination (automatic transmission only)
35 Cigar lighter illumination
36 Switch panel illumination
37 Panel illumination
38 Fog lamps switch
39 Rear fog lamps warning light
40 Rear fog lamps junction
41 Left-hand indicator warning light
42 Right-hand indicator warning light
43 Direction indicator switch
44 Hazard unit
45 Cigar lighter
46 Horn push
47 Horn relay
48 Horns
49 Windscreen wiper motor
50 Windscreen wipe/wash switch
51 Windscreen washer motor
52 Door switch
53 Door switch
54 Courtesy light
55 Courtesy light
56 Clock
57 Reverse light switch
58 Stop light switch
59 Engine thermostat
60 Fully automatic starter device
61 Boot lamp
62 Boot lamp switch
63 Fog switch
64 Rear fog lamp warning light
65 Boot lamp and switch wiring assembly. Left-hand drive
66 Boot lamp
67 Boot lamp switch
68 Heated rear screen assembly (Coupe)
69 Heated rear screen switch
70 Heated rear screen warning light
71 Heated rear screen
72 Wiring condition for front, side and front and rear indicator lights. Left-hand drive
73 Sidelight
74 Sidelight
75 Side repeater
76 Side indicator
77 Side repeater
78 Side indicator
79 Rear side indicator
80 Rear side indicator
81 Direction indicator switch
82 Side indicator
83 Side indicator
84 Stop light
85 Stop light
86 Reverse light
87 Reverse light
88 Number plate lights
89 Tail light
90 Tail light
91 Seat belt warning light
92 Driver's buckle switch
93 Passenger belt switch
94 Passenger seat switch
95 Distributor
96 Coil (6 volt)
97 Eureka wire
98 Tank unit
99 Low fuel delay unit
100 Brake pressure differential switch
101 Handbrake switch
102 Oil pressure switch
103 Ignition warning light
104 Oil warning light
105 Brake warning light
106 Choke warning light
107 Low fuel warning light
108 Choke warning light switch
109 Fuel gauge
110 Tachometer
111 Temperature gauge
112 Battery condition indicator
113 Temperature transmitter
114 Air conditioning blower unit
115 Heater/air conditioning blower unit
116 Thermostat
117 Right-hand condenser fan
118 Left-hand condenser fan
119 Radiator thermostat
120 Fan relay
121 Full throttle cut-off switch
122 Air conditioning cut-out switch
123 Air conditioning relay
124 Delay unit
125 Throttle jack
126 Ranco valve high pressure cut-out
127 Clutch
128 Delay circuit relay

Colour code

B	Black	N	Brown	S	Slate
G	Green	O	Orange	U	Blue
K	Pink	P	Purple	W	White
LG	Light green	R	Red	Y	Yellow

173

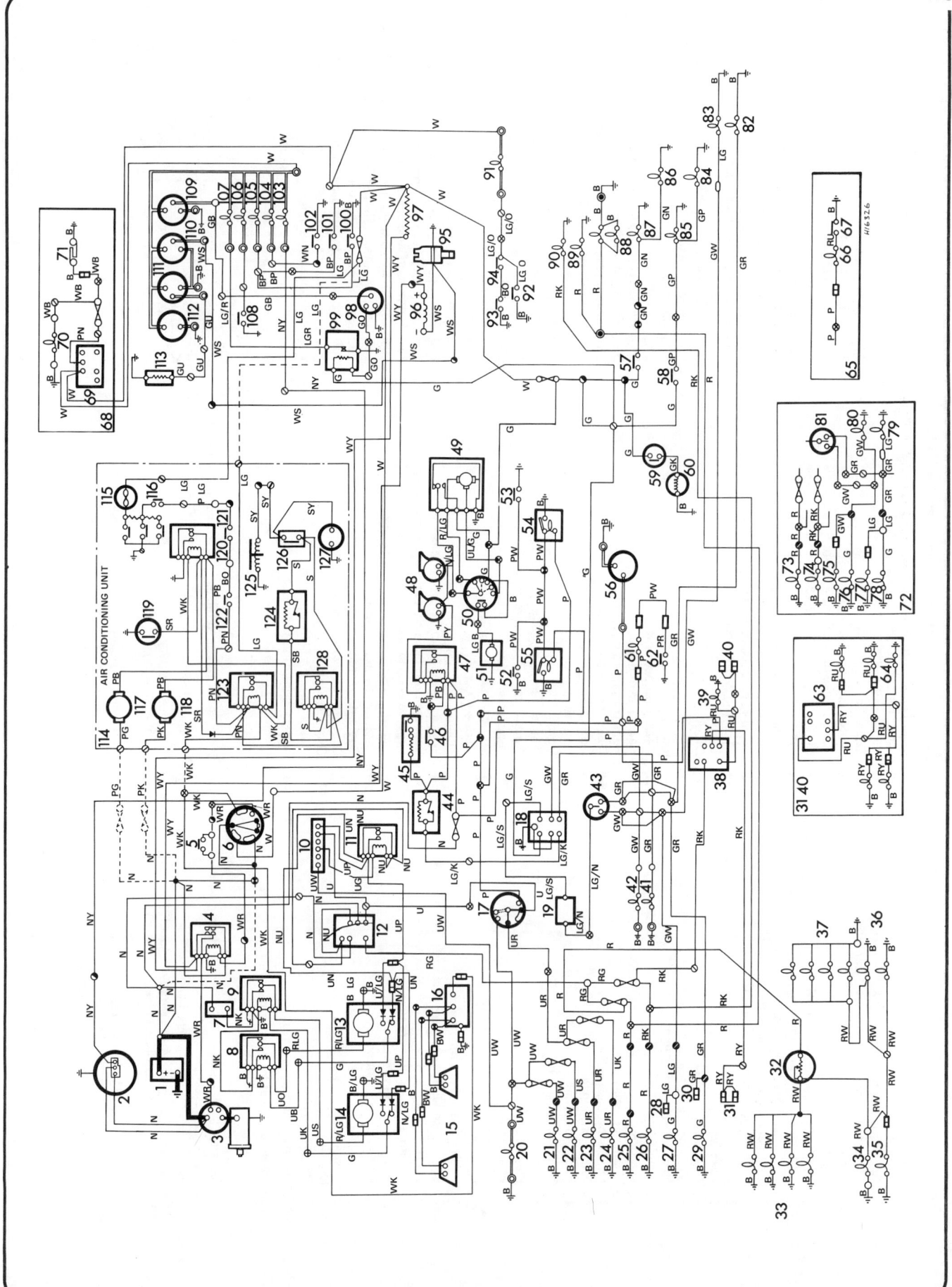

Fig. 13.24. Key to wiring diagram - UK and European models (1980)

1. Battery
2. Alternator
3. Ignition/starter switch
4. Start inhibit switch (automatic transmission only)
5. Starter relay
6. Starter motor
7. Headlamp motor circuit breaker
8. Left-hand headlamp run/stop relay
9. Right-hand headlamp run/stop relay
10. Left-hand headlamp actuator
11. Right-hand headlamp actuator
12. Radio
13. Speaker
14. Master light switch
15. Main/dip/flash switch
16. Main beam warning light
17. Main beam
18. Dip beam
19. Headlamp delay unit
20. Headlamp flash relay
21. Front parking lamp
22. Front fog lamp
23. Rear fog lamp
24. Tail lamp
25. Plate illumination lamp
26. Fog switch
27. Rear fog lamp warning light
28. Panel rheostat
29. Heater control illumination
30. Selector panel illumination (automatic transmission only)
31. Cigarette lighter illumination
32. Instrument illumination
33. Switch panel illumination

Air conditioning
34. Master relay
35. Manual cut-out switch
36. Full throttle cut-out switch

37. Cold thermostat
38. Blower motor switch
39. Blower motor
40. Delay unit
41. Clutch delay relay
42. Ranco valve high pressure cut-out
43. Clutch
44. Throttle jack
45. Condenser fan relay
46. Left-hand condenser fan
47. Right-hand condenser fan
48. Radiator thermostat
49. Diode
50. Diode

All models
51. Hazard flasher unit
52. Hazard switch
53. Hazard warning light
54. Cigarette lighter
55. Horn push
56. Horn relay
57. Horn
58. Windscreen wash/wipe switch
59. Windscreen wiper motor
60. Windscreen washer pump
61. Map/courtesy lamp
62. Door switch
63. Clock
64. Luggage boot lamp. Right-hand drive
65. Luggage boot lamp switch. Right-hand drive
66. Luggage boot lamp switch. Left-hand drive
67. Luggage boot lamp switch. Left-hand drive
68. Coolant low indicator unit
69. Coolant low sensor
70. Coolant low warning light
71. Choke switch
72. Choke warning light
73. Battery condition indicator

74. Temperature indicator
75. Temperature transmitter
76. Tachometer
77. Fuel indicator
78. Fuel tank unit
79. Low fuel delay unit
80. Low fuel warning light
81. Brake warning light
82. Brake pressure differential switch
83. Handbrake warning light
84. Handbrake switch
85. Oil pressure warning light
86. Oil pressure switch
87. Ignition warning light
88. Ballast resistor wire
89. Coil (6 volt)
90. Distributor
91. Heated rear windscreen switch
92. Heated rear windscreen
93. Heated rear windscreen warning light
94. Seat belt warning light
95. Passenger seat switch
96. Passenger belt switch
97. Driver belt switch
98. Reverse lamp switch
99. Reverse lamp
100. Stop lamp switch
101. Stop lamp
102. Turn signal flasher unit
103. Turn signal switch
104. Left-hand turn signal warning light
105. Right-hand turn signal warning light
106. Left-hand front flasher lamp
107. Left-hand front flasher repeater lamp
108. Left-hand rear flasher lamp
109. Right-hand front flasher lamp
110. Right-hand front flasher repeater lamp
111. Right-hand rear flasher lamp

Colour code
B black
G Green
K Pink
LG Light green
N Brown
O Orange
P Purple
R Red
S Slate
U Blue
W White
Y Yellow

Wire colour codes of flasher lamp circuit

Wire identity	Item	Right-hand drive		Left-hand drive	
		Heater vehicles	Air conditioning vehicles	Heater vehicles	Air conditioning vehicles
*1	Left-hand supply	G/LG	GR	LG/G	GR
*2	Left-hand front lamp	G/LG	GR	LG/G	GR
*3	Left-hand rear lamp	GR	GR and LG	GR and LG	GR and LG
*4	Right-hand supply	LG/G	GW	G/LG	GW
*5	Right-hand front lamp	LG/G	GW	G/LG	GW
*6	Right-hand rear lamp	GW and LG	GW	GW	GW

175

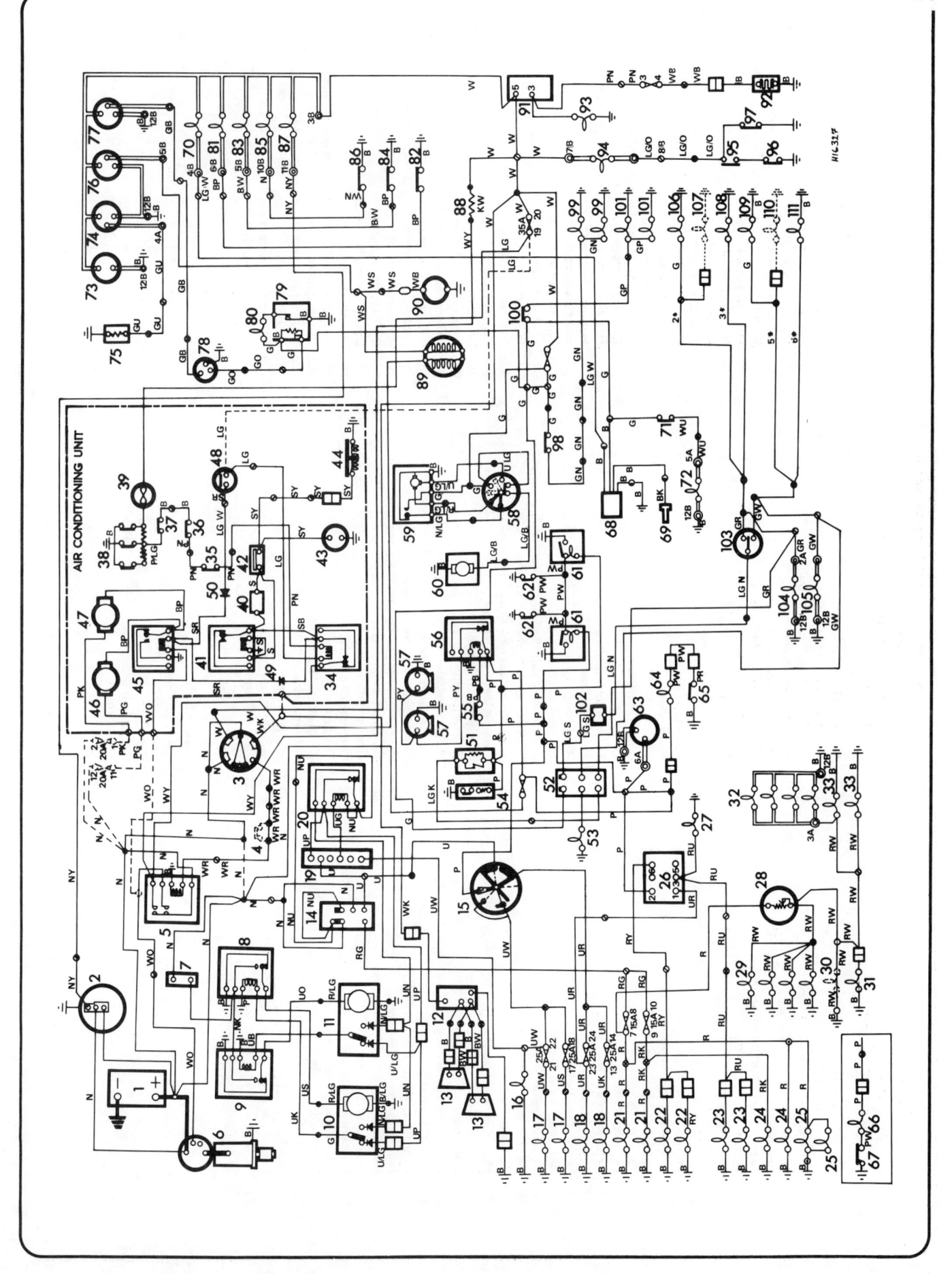

Fig. 13.25. Key to wiring diagram - USA models (1976)

1 Alternator
2 Ignition warning light
3 Battery
4 Battery condition indicator
5 Ignition/starter switch
6 Radio supply
7 Radio
8 Speaker
9 Starter motor relay
10 Starter motor
11 Ballast resistor wire
12 Coil
13 Distributor
14 Drive resistor
15 Battery lead connector
16 Master light switch
17 Actuator - limit switch
18 Circuit breaker
19 Headlamp - run/stop relay
20 Actuator motor
21 Main/dip/flash switch
22 Dip beam
23 Main beam
24 Main beam warning light
25 Fuse (35 amp)
26 Rear fog lamp switch
27 Rear fog lamp warning light
28 Rear fog lamp
29 Front marker lamp
30 Front parking lamp
31 Plate illumination lamp
32 Rear marker lamp
33 Tail lamp
34 Panel rheostat
35 Heater control illumination
36 Facia panel switch illumination
37 Selector panel illumination
38 Cigarette lighter illumination
39 Instrument illumination
40 Fuse (50 amp)
41 Buzzer/timer module
42 Driver's belt switch
43 Fasten belt warning light
44 Left-hand door switch
45 Key switch
46 Horn push
47 Horn relay
48 Horn
49 Cigarette lighter
50 Clock
51 Front fog lamp switch
52 Front fog lamp
53 Roof lamp
54 Right-hand door switch
55 Service indicator diodes (catalyst vehicles only)
56 EGR service indicator
57 EGR service warning light
58 Catalyst service indicator
59 Catalyst warning light
60 Fuse (50 amp)
61 Radiator switch
62 Condenser fan motor
63 Air conditioning control relay
64 Air conditioning delay circuit flasher unit
65 Air conditioning delay circuit relay
66 High pressure cut-out
67 Compressor clutch
68 Throttle jack
69 Air conditioning cut-out switch
70 Cold thermostat
71 Blower motor
72 Blower motor switch
73 Fuse (35 amp)
74 Windscreen washer/wipe switch
75 Windscreen wiper motor
76 Windscreen washer pump
77 Fuse (50 amp)
78 Reverse lamp switch
79 Reverse lamp
80 Stop lamp switch
81 Stop lamp
82 Turn signal flasher unit
83 Turn signal switch
84 Left-hand flasher lamp
85 Left-hand turn signal warning light
86 Right-hand flasher lamp
87 Right-hand turn signal warning light
88 Hazard flasher unit
89 Hazard switch
90 Hazard warning light
91 Fuel indicator
92 Fuel tank unit
93 Fuel warning light
94 Fuel delay unit
95 Tachometer
96 Temperature indicator
97 Temperature transmitter
98 Oil pressure warning light
99 Oil pressure switch
100 Anti-run-on valve
101 Brake warning light
102 Brake line failure switch
103 Handbrake switch
104 Choke warning light (non-catalyst vehicles only)
105 Choke switch
106 Heated rear windscreen switch
107 Fuse (15 amp)
108 Heated rear windscreen
109 Heated rear windscreen warning light

Colour code
B   Black
G   Green
K   Pink
LG Light green
N   Brown
O   Orange
P   Purple
R   Red
S   Slate
U   Blue
W   White
Y   Yellow

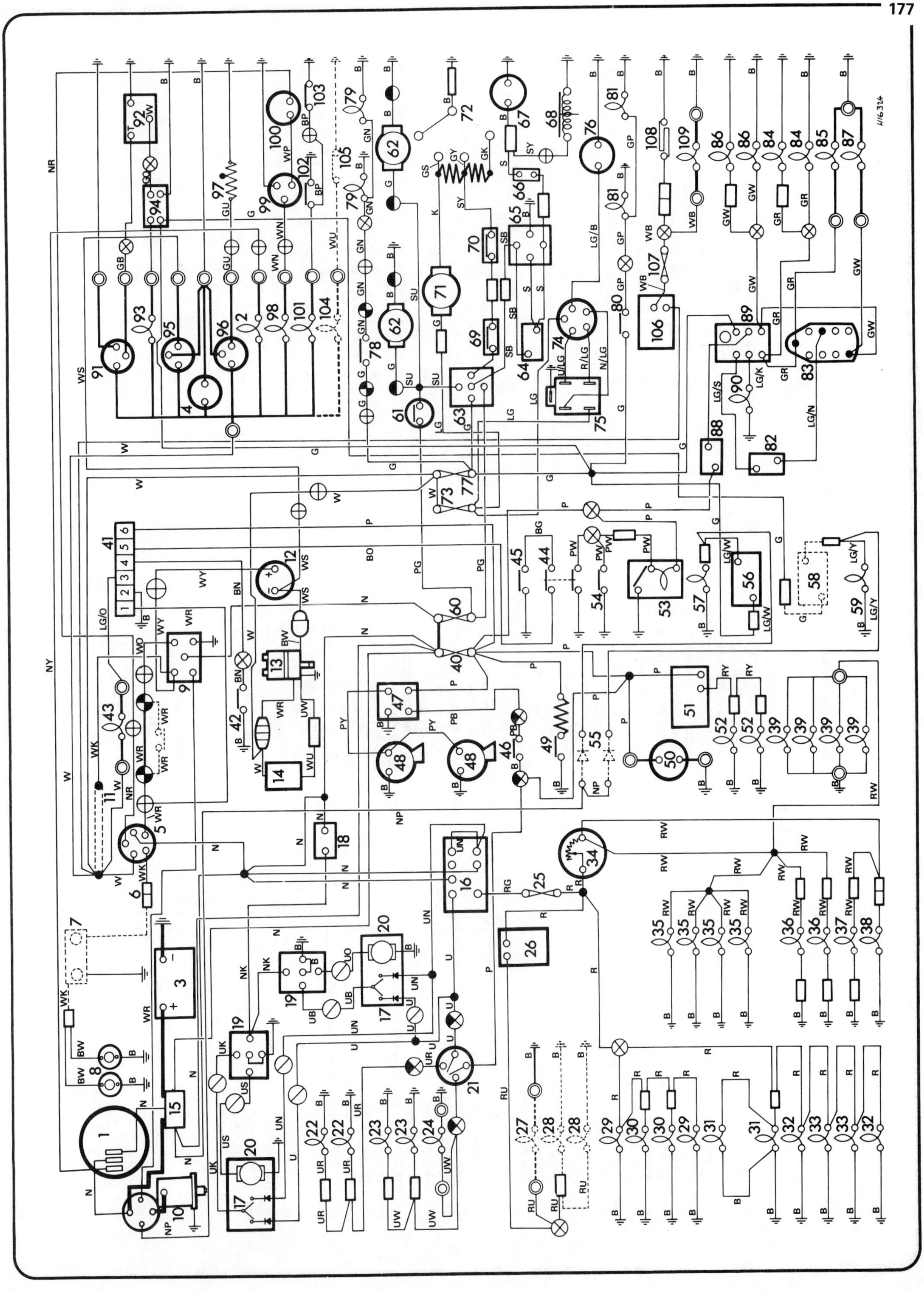

Fig. 13.26. Key to wiring diagram - USA models (1977)

1 Alternator
2 Ignition warning light
3 Battery
4 Battery condition indicator
5 Ignition starter switch
6 Radio supply
7 Radio
8 Speaker
9 Starter motor relay
10 Starter motor
11 Ballast resistor wire
12 Coil
13 Distributor
14 Drive resistor
15 Battery lead connector
16 Master light switch
17 Actuator - limit switch
18 Circuit breaker
19 Headlamp - run/stop relay
20 Actuator motor
21 Main/dip/flash switch
22 Dip beam
23 Main beam
24 Main beam warning light
25 Fuse (35 amp)
29 Front marker lamp
30 Front parking lamp
31 Number plate illumination lamp
32 Rear marker lamp
33 Tail lamp
34 Panel rheostat
35 Heater control illumination
36 Facia switch panel illumination
37 Selector panel illumination (automatic transmission)
38 Cigarette lighter illumination
39 Instrument illumination
40 Fuse (50 amp)
41 Buzzer/timer module
42 Driver's belt switch
43 Fasten belt warning light
44 Left-hand door switch
45 Key switch
46 Horn push
47 Horn relay
48 Horn
49 Cigarette lighter
50 Clock
53 Roof lamp
54 Right-hand door switch
60 Fuse (50 amp)
61 Radiator switch
62 Condenser fan motor
63 Air conditioning control relay
66 High pressure cut-out
68 Throttle jack
69 Air conditioning cut-out switch
71 Blower motor
72 Blower motor switch
73 Fuse (35 amp)
74 Windscreen washer
75 Windscreen wiper motor
76 Windscreen washer pump
77 Fuse (50 amp)
78 Reverse lamp switch
79 Reverse lamp
80 Stop lamp switch
81 Stop lamp
82 Turn signal flasher unit
83 Turn signal switch
84 Left-hand flasher lamp
85 Left-hand turn signal warning light
86 Right-hand flasher lamp
87 Right-hand turn signal warning light
88 Hazard flasher unit
89 Hazard switch and warning light
91 Fuel indicator
92 Fuel tank unit
93 Fuel warning light
94 Fuel delay unit
95 Tachometer
96 Temperature indicator
97 Temperature transmitter
98 Oil pressure warning light
99 Oil pressure switch
100 Anti-run-on valve
101 Brake warning light
102 Brake line failure switch
103 Handbrake switch
104 Choke warning light (non-catalyst vehicles only)
105 Choke switch
106 Heated rear windscreen switch
107 Fuse (15 amp)
108 Heated rear windscreen
109 Heated rear windscreen warning light
110 Map light
111 Boot light
112 Full throttle cut-out switch
113 Ignition switch

Colour code

B	Black		P	Purple
G	Green		R	Red
K	Pink		S	Slate
LG	Light green		U	Blue
N	Brown		W	White
O	Orange		Y	Yellow

179

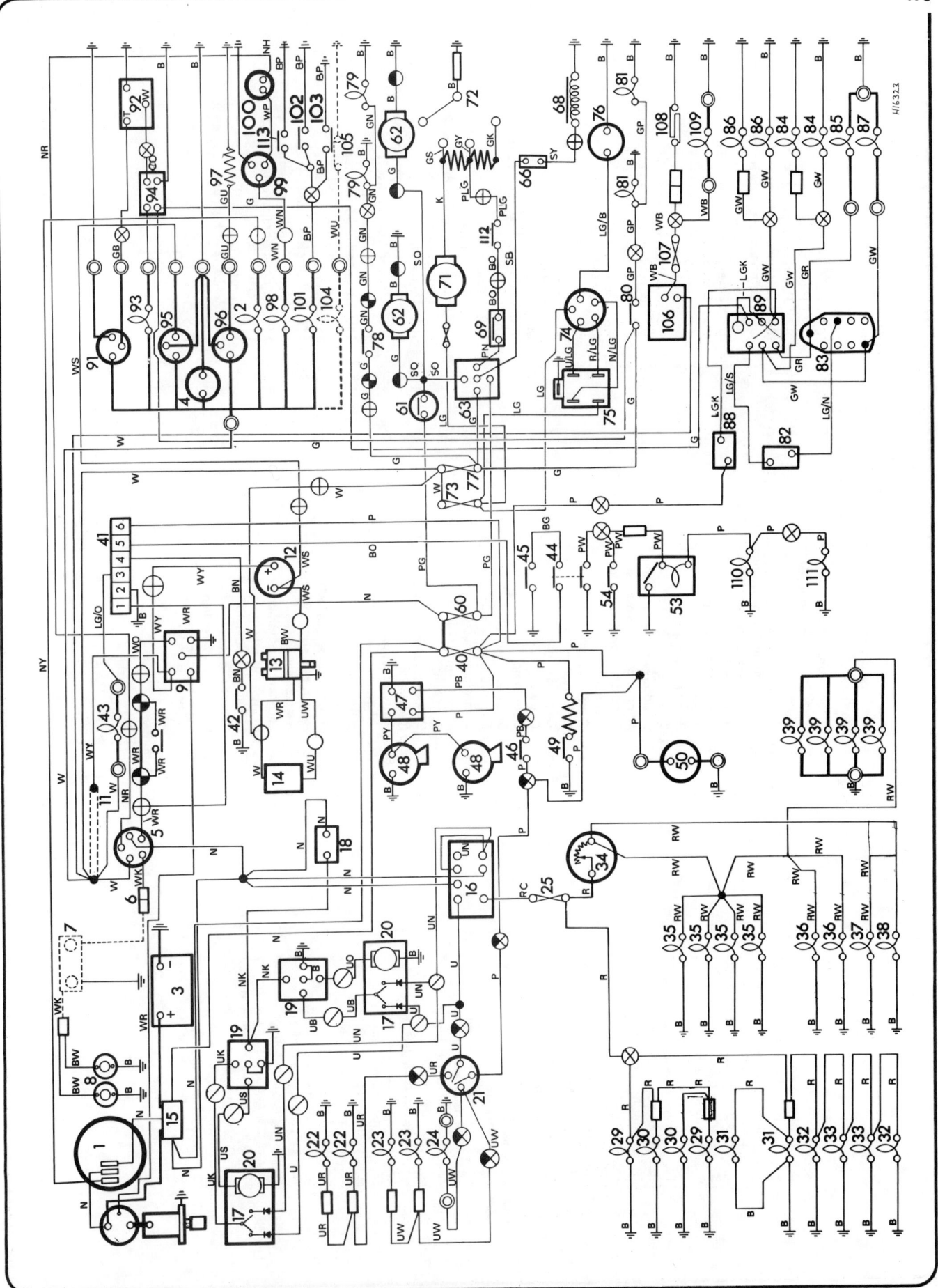

Fig. 13.27. Key to wiring diagram - USA models (1978 to 1979)

1. Battery
2. Alternator
3. Starter motor
4. Starter motor relay
5. Right-hand headlamp relay
6. Left-hand headlamp relay
7. Headlamp motor circuit breaker
8. Start inhibit switch (automatic transmission)
9. Ignition switch
10. Buzzer timer unit
11. Left-hand headlamp motor
12. Right-hand headlamp motor
13. Master light switch
14. Seat belt warning light
15. Buckle switch
16. Door switch
17. Audible warning
18. Cigar lighter
19. Loudspeakers
20. Radio
21. Hazard unit
22. Hazard switch
23. Main/dip/flash switch
24. Main beam warning light
25. Right-hand main beam
26. Left-hand main beam
27. Right-hand dip beam
28. Left-hand dip beam
29. Right-hand sidelight
30. Left-hand sidelight
31. Right-hand side marker
32. Left-hand side marker
33. Right-hand side indicator
34. Left-hand side indicator
35. Front fog lamps junction
36. Heater control illumination
37. Selector panel illumination (automatic transmission)
38. Cigar lighter illumination
39. Panel rheostats
40. Switch panel illumination
41. Panel illumination
42. Rear fog lamps illumination
43. Rear fog lamps warning light
44. Fog lamps switch
45. Left-hand indicator warning light
46. Right-hand indicator warning light
47. Flasher unit
48. Direction indicator switch
49. Horn push
50. Horn relay
51. Horns
52. Windscreen wiper motor
53. Windscreen wipe/wash switch
54. Windscreen washer motor
55. Door switch
56. Courtesy light
57. Door switch
58. Courtesy light
59. Clock
60. Boot lamp
61. Boot lamp switch
62. Fully automatic starter device
63. Engine thermostat
64. Reverse light switch
65. Stop light switch
66. Oil warning light
67. Ignition switch
68. Anti-run-on valve
69. Oil pressure switch
70. Anti-run-on valve (California market only)
71. Heated rear screen (Coupé)
72. Heated rear screen switch
73. Heated rear screen warning light
74. Heated rear screen
75. Left-hand side indicator
76. Right-hand side indicator
77. Left-hand stop light
78. Right-hand stop light
79. Left-hand reverse light
80. Right-hand reverse light
81. Number plate lights
82. Right-hand side marker
83. Right-hand tail lights
84. Left-hand tail lights
85. Left-hand side marker
86. Drive resistor
87. Eureka wire (1.3 to 1.5 ohms)
88. Distributor
89. Coil
90. Fuel tank unit
91. Low fuel relay unit
92. Brake pressure differential switch
93. Handbrake switch
94. Oil pressure switch
95. Ignition warning light
96. Oil warning light
97. Brake warning light
98. Choke warning light
99. Low fuel warning light
100. Choke warning light switch
101. Fuel gauge
102. Tachometer
103. Temperature gauge
104. Battery condition indicator
105. Temperature transmitter
106. Thermostat
107. Heater/air conditioning blower unit
108. Air conditioning circuit
109. Right-hand condenser fan
110. Left-hand condenser fan
111. Radiator thermostat
112. Fan relay
113. Full throttle cut-off switch
114. Air conditioning cut-out switch
115. Air conditioning relay
116. Delay unit
117. Throttle jack
118. Ranco valve high pressure cut-out
119. Clutch
120. Delay circuit relay

Colour code

B	Black		P	Purple
G	Green		R	Red
K	Pink		S	Slate
LG	Light Green		U	Blue
N	Brown		W	White
O	Orange		Y	Yellow

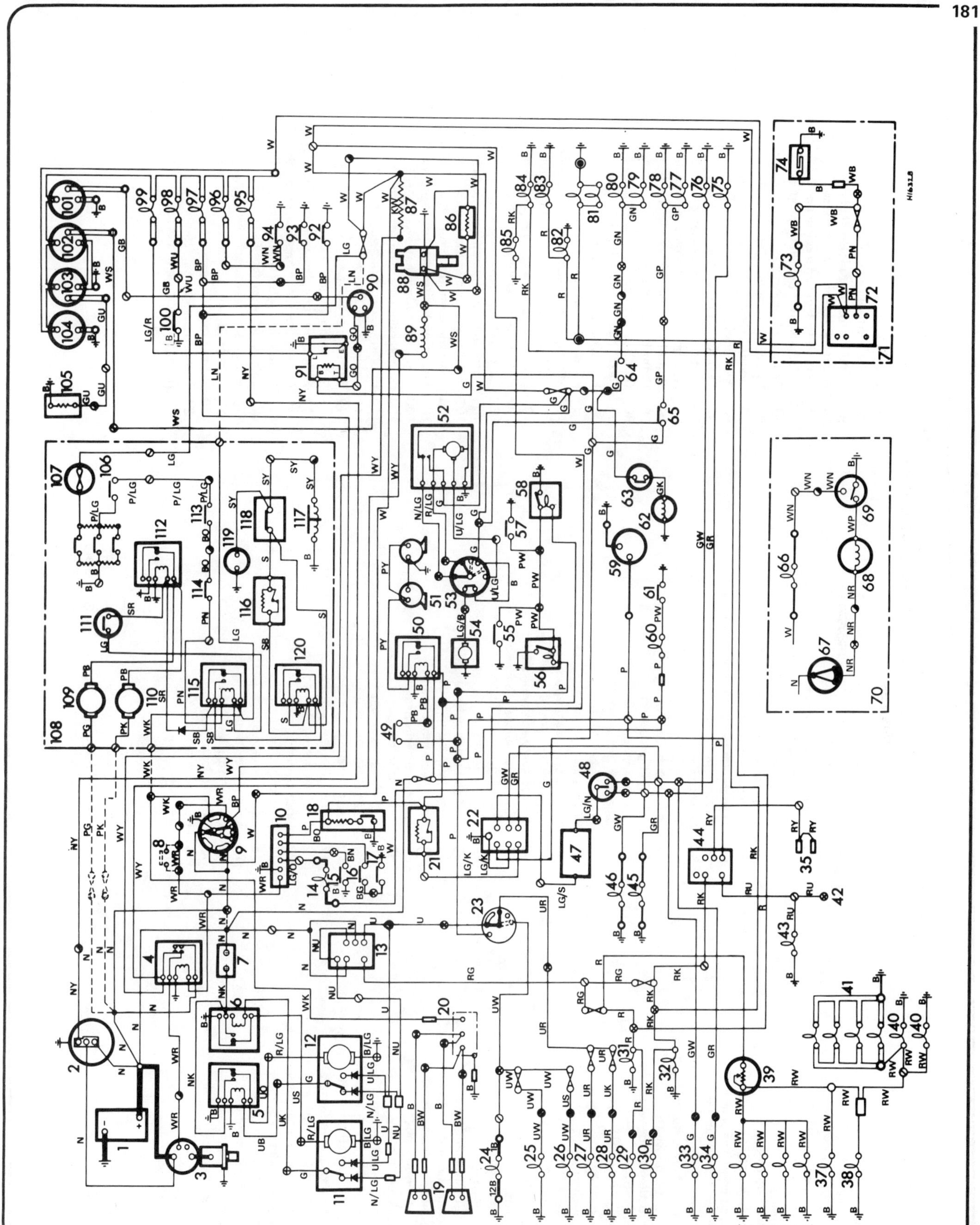

Fig. 13.28. Key to wiring diagram - USA models with carburettors (1980)

1 Battery
2 Alternator
3 Starter relay
4 Starter inhibitor switch (automatic transmission only)
5 Ignition/starter switch

Air conditioning
6 Condenser fan relay
7 Clutch delay relay
8 Left-hand condenser fan
9 Right-hand condenser fan
10 Delay unit
11 Ranco valve high pressure cut-out
12 Manual cut-out switch
13 Full throttle cut-out switch
14 Cold thermostat
15 Blower motor switch
16 Blower motor
17 Radiator thermostat

All models
18 Temperature indicator
19 Tachometer
20 Fuel indicator
21 Coolant low warning light
22 Oil pressure warning light
23 Brake warning light
24 Handbrake warning light
25 Ignition warning light
26 Handbrake switch
27 Brake pressure differential switch
28 Starter motor
29 Right-hand headlamp run/stop relay
30 Headlamp motor circuit breaker
31 Left-hand headlamp run/stop relay
32 Left-hand headlamp actuator
33 Right-hand headlamp actuator
34 Master light switch
35 Seat belt timer/buzzer unit
36 Seat belt warning light
37 Belt switch
38 Door switch
39 Key switch

40 Master relay
41 Clutch
42 Throttle jack
43 Low fuel warning light
44 Low fuel delay unit
45 Tank unit
46 Oil pressure switch
47 Anti-run-on valve
48 Heated rear windscreen warning light
49 Heated rear windscreen switch
50 Speakers
51 Radio
52 Main beam warning light
53 Column light switch
54 Hazard flasher unit
55 Cigarette lighter
56 Horn push
57 Horn relay
58 Windscreen washer pump
59 Horn
60 Windscreen wash/wipe switch
61 Windscreen wiper motor
62 and
63 Coil and electronic module assembly
64 Distributor
65 Right-hand main beam
66 Left-hand main beam
67 Right-hand dip beam
68 Left-hand dip beam
69 Hazard warning light
70 Hazard switch
71 Turn signal flasher unit
72 Door switch
73 Door switch
74 Map/courtesy lamp
75 Map/courtesy lamp
76 Engine thermostat
77 Carburettor fuel heater
78 Reverse lamp switch
79 Reverse lamp
80 Reverse lamp

81 Stop lamp
82 Stop lamp
83 Right-hand front marker lamp
84 Right-hand front parking lamp
85 Left-hand front marker lamp
86 Left-hand front parking lamp
87 Front fog lamp
88 Front fog lamp
89 Rear fog lamp
90 Rear fog lamp
91 Left-hand tail lamp
92 Left-hand rear marker lamp
93 Number plate illumination lamp
94 Right-hand rear marker lamp
95 Right-hand tail lamp
96 Clock
97 Rear fog lamp warning light
98 Fog switch
99 Panel rheostat
100 Selector panel illumination (automatic transmission only)
101 Cigarette lighter illumination
102 Heater control illumination
103 Luggage boot lamp
104 Luggage boot lamp switch
105 Instrument illumination
106 Instrument illumination
107 Instrument illumination
108 Instrument illumination
109 Switch panel illumination
110 Coolant low indicator unit
111 Coolant low sensor
112 Turn signal switch
113 Stop lamp switch
114 Left-hand turn signal warning light
115 Right-hand turn signal warning light
116 Right-hand front flasher lamp
117 Left-hand front flasher lamp
118 Right-hand rear flasher lamp
119 Left-hand rear flasher lamp
120 Heated rear windscreen
121 Temperature transmitter
122 Battery condition indicator

Colour code
B Black
G Green
K Pink
LG Light green
N Brown
O Orange
P Purple
R Red
S Slate
U Blue
W White
Y Yellow

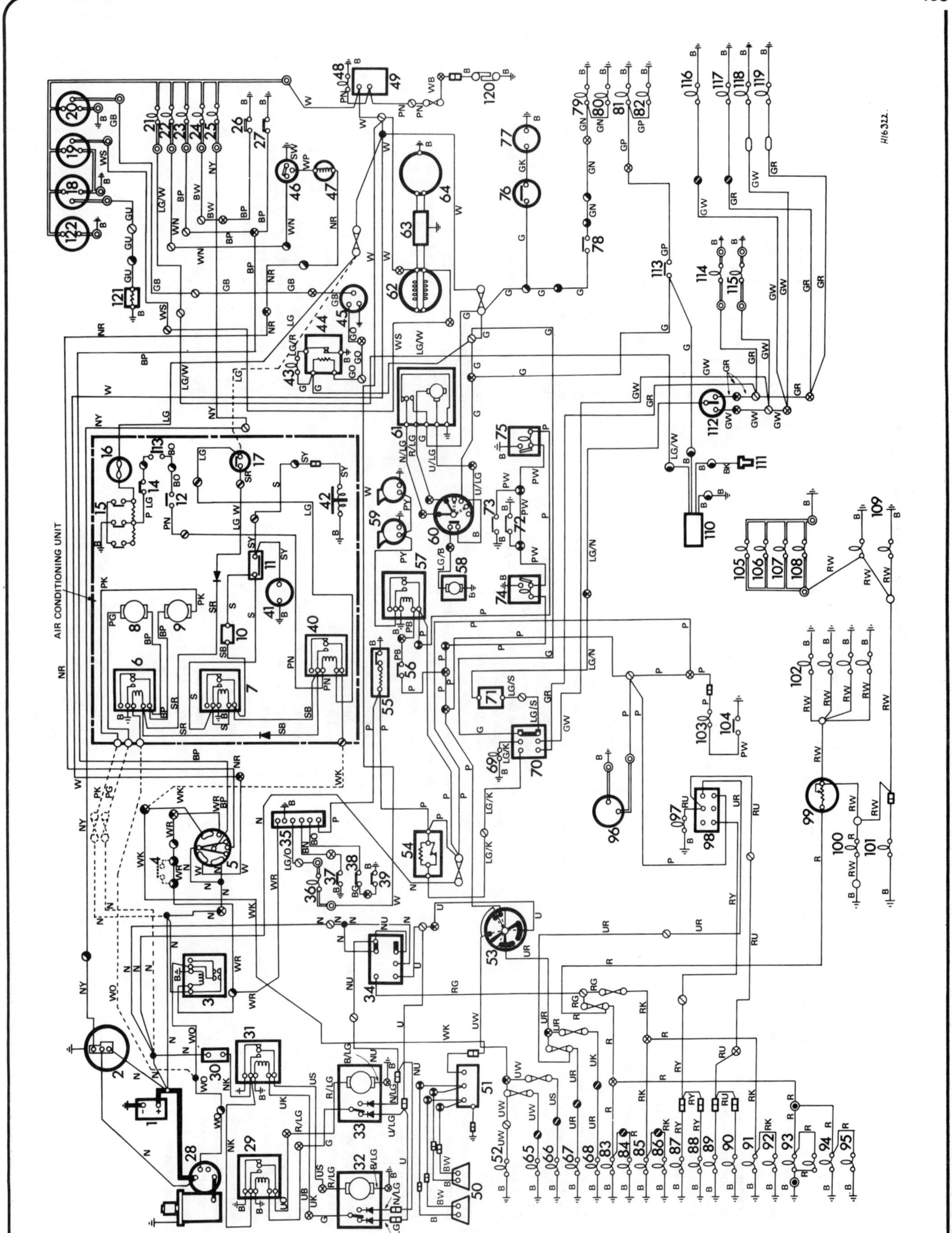
AIR CONDITIONING UNIT
H16322

Fig. 13.29. Key to wiring diagram - USA models with fuel injection (1980)

1 Battery
2 Alternator
3 Starter motor
4 Right-hand headlamp run/stop relay
5 Left-hand headlamp run/stop relay
6 Headlamp motor circuit breaker
7 Starter relay
8 Starter inhibitor switch (automatic transmission only)
9 Ignition/starter switch

Air conditioning
10 Condenser fan relay
11 Clutch delay relay
12 Left-hand condenser fan
13 Right-hand condenser fan
14 Delay unit
15 Diode
16 Ranco valve high pressure cut-out
17 Manual cut-out switch
18 Full throttle cut-out switch
19 Cold thermostat
20 Radiator thermostat

All models
21 Temperature transmitter
22 Battery condition indicator
23 Temperature indicator
24 Tachometer
25 Fuel indicator
26 Coolant low warning light
27 Brake warning light
28 Handbrake warning light
29 Oil pressure warning light
30 Ignition warning light
31 Left-hand headlamp actuator
32 Right-hand headlamp actuator
33 Master light switch
34 Speakers
35 Radio
36 Seat belt timer/buzzer unit
37 Seat belt warning light
38 Belt switch
39 Door switch

40 Key switch
41 Hazard flasher unit
42 Cigarette lighter
43 Horn push
44 Horn relay

Air conditioning
45 Master relay
46 Clutch
47 Throttle jack

All models
48 Windscreen washer pump
49 Horn
50 Windscreen wash/wipe switch
51 Windscreen wiper motor
52 Low fuel warning light
53 Low fuel delay unit
54 Fuel tank unit
55 Coil and electronic module assembly
56 Distributor
57 Oil pressure switch
58 Handbrake switch
59 Brake pressure differential switch
60 Heated rear windscreen warning light
61 Heated rear windscreen switch
62 Heated rear windscreen
63 Main beam warning light
64 Right-hand main beam
65 Left-hand main beam
66 Right-hand dip beam
67 Left-hand dip beam
68 Right-hand front marker lamp
69 Right-hand front parking lamp
70 Left-hand front marker lamp
71 Left-hand front parking lamp
72 Front fog lamp
73 Front fog lamp
74 Rear fog lamp
75 Rear fog lamp
76 Left-hand tail lamp
77 Left-hand rear marker lamp

78 Number plate illumination lamp
79 Right-hand rear marker lamp
80 Right-hand tail lamp
81 Selector panel illumination (automatic transmission only)
82 Cigarette lighter illumination
83 Column light switch
84 Hazard warning light
85 Hazard switch
86 Turn signal flasher unit
87 Clock
88 Rear fog lamp warning light
89 Fog switch
90 Panel rheostat
91 Luggage boot lamp
92 Luggage boot lamp switch
93 Heater control illumination
94 Map/courtesy lamp
95 Door switch
96 Service interval counter
97 Coolant low indicator unit
98 Coolant low sensor
99 Oxygen sensor warning light
100 Turn signal switch
101 Reverse lamp switch
102 Stop lamp switch
103 Left-hand turn signal warning light
104 Right-hand turn signal warning light
105 Reverse lamp
106 Reverse lamp
107 Stop lamp
108 Stop lamp
109 Right-hand front flasher lamp
110 Left-hand front flasher lamp
111 Right-hand rear flasher lamp
112 Left-hand rear flasher lamp
113 Instrument illumination
114 Switch panel illumination

Air conditioning
115 Blower motor switch
116 Blower motor

Colour code
B Black
G Green
K Pink
LG Light green
N Brown
O Orange
P Purple
R Red
S Slate
U Blue
W White
Y Yellow

185

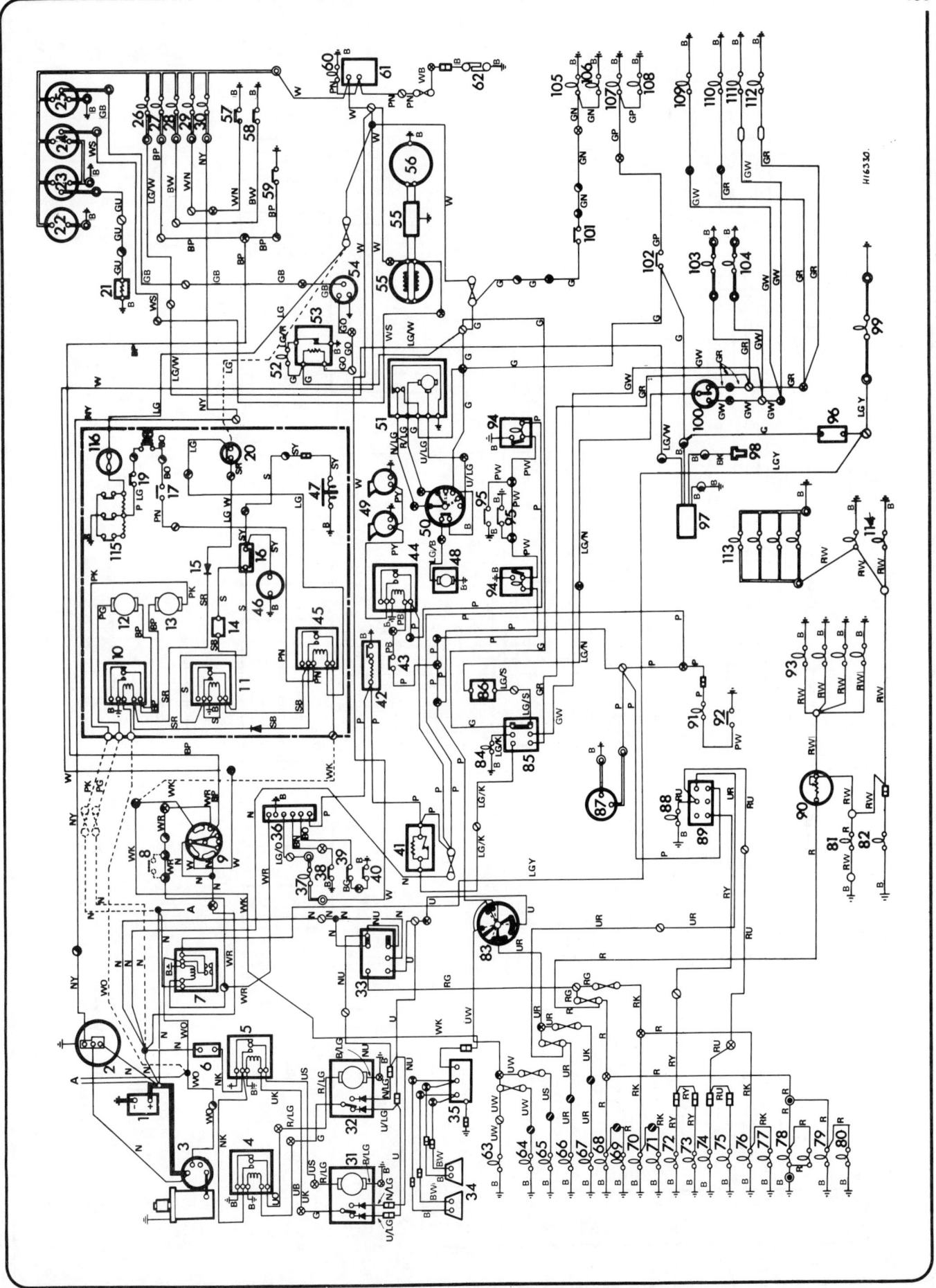

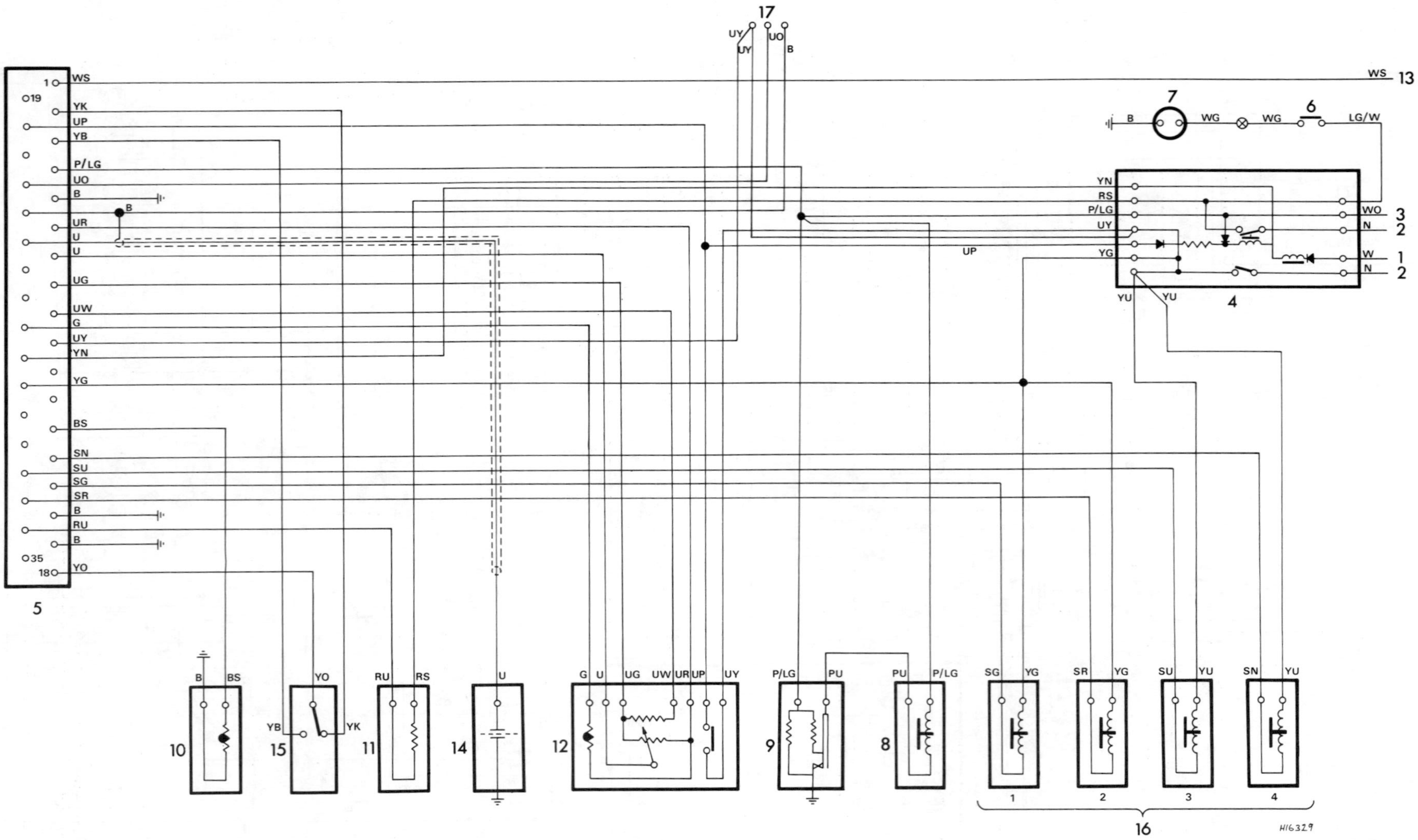

Fig. 13.30. Wiring diagram - fuel injection system

1 Ignition control input
2 Battery input
3 Start input
4 Relay
5 Electronic control unit - ECU
6 Inertia switch
7 Fuel pump
8 Cold start injector
9 Thermotime switch
10 Water temperature sensor
11 Extra air valve
12 Air flow meter
13 Engine speed - input from ignition coil negative
14 Lambda sensor
15 Throttle switch
16 Injector - 4 off
17 Test facility - feedback monitor plug

Colour code
B Black
G Green
K Pink
LG Light green
N Brown
O Orange
P Purple
R Red
S Slate
U Blue
W White
Y Yellow

187

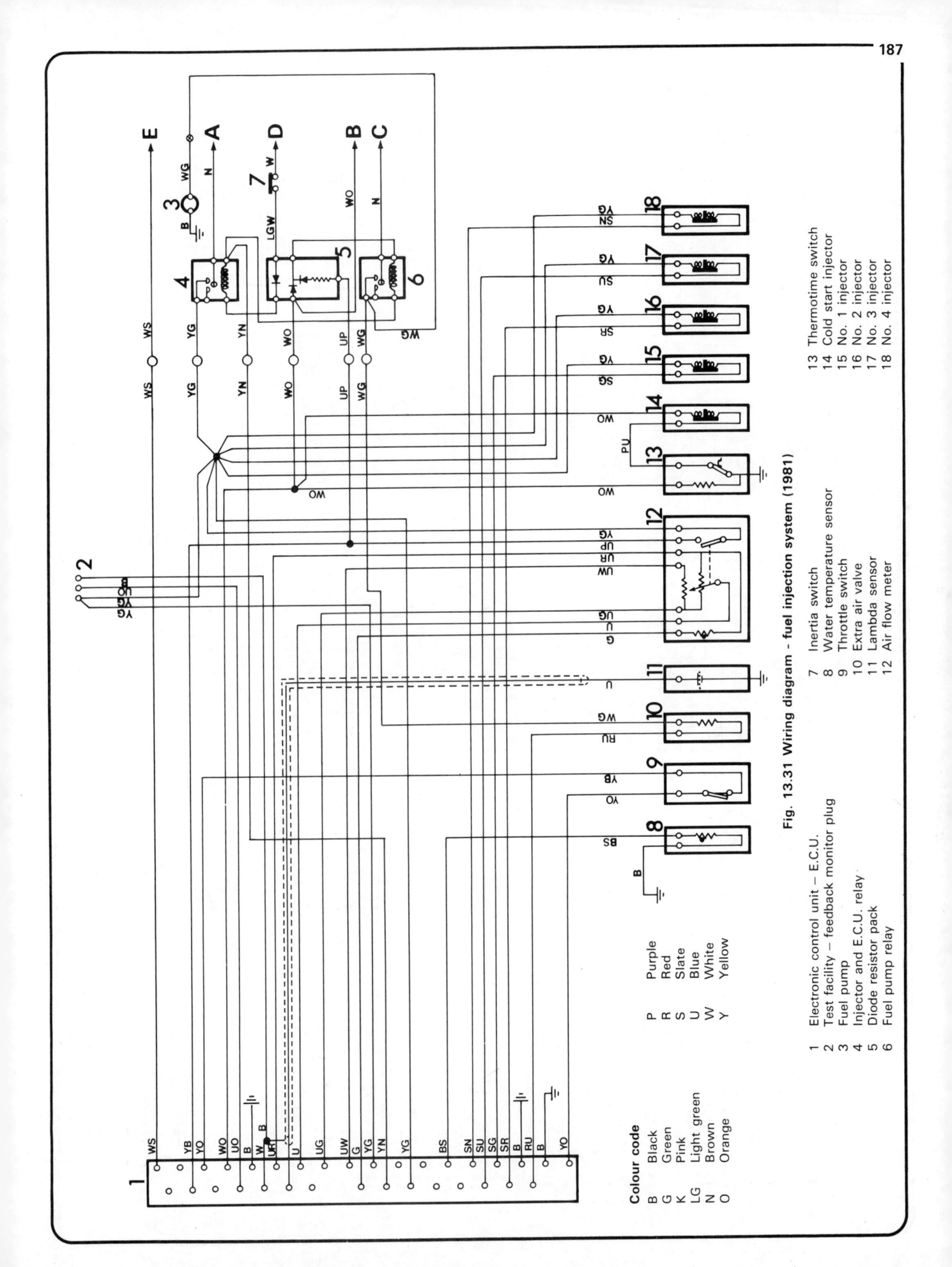

Fig. 13.31 Wiring diagram - fuel injection system (1981)

1 Electronic control unit – E.C.U.
2 Test facility – feedback monitor plug
3 Fuel pump
4 Injector and E.C.U. relay
5 Diode resistor pack
6 Fuel pump relay
7 Inertia switch
8 Water temperature sensor
9 Throttle switch
10 Extra air valve
11 Lambda sensor
12 Air flow meter
13 Thermotime switch
14 Cold start injector
15 No. 1 injector
16 No. 2 injector
17 No. 3 injector
18 No. 4 injector

Colour code

B Black
G Green
K Pink
LG Light green
N Brown
O Orange
P Purple
R Red
S Slate
U Blue
W White
Y Yellow

Fig. 13.32 Key to wiring diagram - North American models (1981)

1 Battery
2 Alternator
3 Starter motor
4 Headlamp relay – R.H.
5 Headlamp relay – L.H.
6 Headlamp motor circuit breaker
7 Starter relay
8 Starter inhibitor switch (automatic transmission only)
9 Ignition/starter switch
10 Condenser fan relay – L.H.
11 Condenser fan relay – R.H.
12 Condenser fan – L.H.
13 Condenser fan – R.H.
14 Diode
15 Pressure cut-in switch
16 Air conditioning cut-out switch
17 Full throttle cut-out switch
18 Thermostat
19 Radiator thermostat
20 Temperature transmitter
21 Voltmeter
22 Temperature gauge
23 Tachometer
24 Fuel gauge
25 Coolant low warning light
26 Brake warning light
27 Handbrake warning light
28 Oil pressure warning light
29 Ignition warning light
30 Headlamp motor – L.H.
31 Headlamp motor – R.H.
32 Master light switch
33 Speaker
34 Radio
35 Seat belt timer/buzzer unit
36 Seat belt warning light
37 Belt switch
38 Door switch
39 Audible warning switch
40 Hazard flasher unit
41 Cigarette lighter
43 Horn-push
44 Clutch relay
45 Clutch
46 Engine air solenoid valve
47 Windscreen washer motor
48 Horn
49 Windscreen wash/wipe switch
50 Wiper delay unit
51 Windscreen wiper motor
52 Low fuel warning light
53 Low fuel delay unit
54 Tank unit
55 Ignition coil
56 Amplifier
57 Ignition distributor
58 Oil pressure switch
59 Handbrake switch
60 Brake pressure differential switch
61 Heated back-light warning light
62 Heated back-light switch
63 Heated back-light
64 Main beam warning light
65 Main beam – R.H.
66 Main beam – L.H.
67 Dip beam – R.H.
68 Dip beam – L.H.
69 Front marker lamp – R.H.
70 Front parking lamp – R.H.
71 Front marker lamp – L.H.
72 Front parking lamp – L.H.
73 Front fog lamp
74 Rear fog guard lamp
75 Tail lamp – L.H.
76 Rear marker lamp – L.H.
77 Number plate illumination lamp
78 Rear marker lamp – R.H.
79 Tail lamp – R.H.
80 Selector panel illumination (automatic transmission only)
81 Cigarette lighter illumination
82 Main/dip/flash switch
83 Hazard warning light
84 Hazard switch
85 Turn signal flasher unit
86 Clock
87 Rear fog guard warning light
88 Fog switch
89 Panel rheostat
90 Luggage boot lamp
91 Luggage boot lamp switch
92 Heater control illumination
93 Map/courtesy lamp
94 Door switch
95 Service interval counter
96 Coolant low indicator unit
97 Coolant low switch
98 Oxygen sensor warning light
99 Turn signal switch
100 Reverse lamp switch
101 Stop lamp switch
102 Turn signal warning light – L.H.
103 Turn signal warning light – R.H.
104 Reverse lamp – R.H.
105 Reverse lamp – L.H.
106 Stop lamp – R.H.
107 Stop lamp – L.H.
108 Front flasher lamp – R.H.
109 Front flasher lamp – L.H.
110 Rear flasher lamp – R.H.
111 Rear flasher lamp – L.H.
112 Instrument illumination
113 Switch panel illumination
114 Blower motor switch
115 Blower motor

A To injector and ECU relay
B To diode resistor pack
C To fuel pump relay
D To diode resistor pack
E Fuel injection timing signal

see also
Fuel Injection
Wiring Diagram

Colour code

B	Black		P	Purple
G	Green		R	Red
K	Pink		S	Slate
LG	Light green		U	Blue
N	Brown		W	White
O	Orange		Y	Yellow

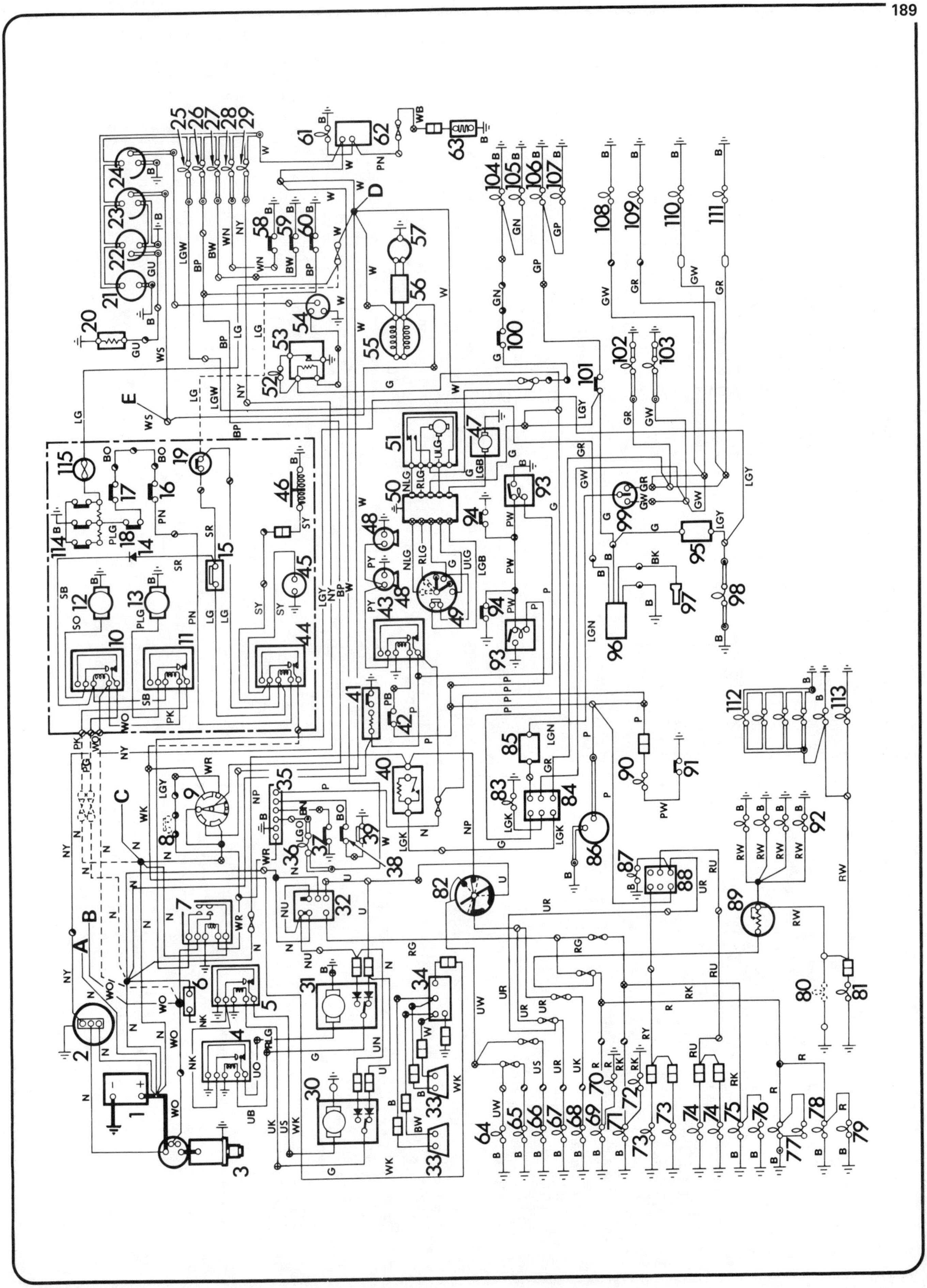

Fig.13.33 Key to wiring diagram - UK and European models (1981)

1 Battery
2 Alternator
3 Ignition/starter switch
4 Start inhibit switch (automatic transmission only)
5 Starter relay
6 Starter motor
7 Headlamp motor circuit breaker
8 Headlamp relay – L.H.
9 Headlamp relay – R.H.
10 Headlamp motor – L.H.
11 Headlamp motor – R.H.
12 Radio
13 Speakers
14 Master light switch
15 Main/dip/flash switch
16 Main beam warning light
17 Main beam
18 Dip beam
19 Headlamp flash unit
20 Headlamp flash relay
21 Front parking lamp
22 Front fog lamp
23 Rear fog guard lamp
24 Tail lamp
25 Number plate illumination lamp
26 Fog switch
27 Rear fog guard lamp warning light
28 Panel rheostat
29 Heater control illumination
30 Selector panel illumination (automatic transmission only)
31 Cigarette lighter illumination
32 Panel illumination
33 Switch panel illumination
34 Clutch relay
35 Air conditioning cut-out switch
36 Full throttle cut-out switch
37 Thermostat
38 Blower motor switch
39 Blower motor
40 Pressure cut-in switch
41 Clutch
42 Throttle jack
43 Condenser fan relay – L.H.
44 Condenser fan relay – R.H.
45 Condenser fan – L.H.
46 Condenser fan – R.H.
47 Radiator thermostat
48 Diode
49 Hazard flasher unit
50 Hazard switch
51 Hazard warning light
52 Cigarette lighter
53 Horn-push
54 Horn relay
55 Horn
56 Windscreen wash/wipe switch
57 Wiper delay unit
58 Windscreen wiper motor
59 Windscreen washer motor
60 Map/courtesy lamp
61 Door switch
62 Clock
63 Luggage boot lamp
64 Luggage boot lamp switch
65 Luggage boot lamp
66 Luggage boot lamp switch
67 Coolant low indicator unit
68 Coolant low sensor
69 Coolant low warning light
70 Choke switch
71 Choke warning light
72 Voltmeter
73 Temperature gauge
74 Temperature transmitter
75 Tachometer
76 Fuel gauge
77 Tank unit
78 Low fuel delay unit
79 Low fuel warning light
80 Brake warning light
81 Brake pressure differential switch
82 Handbrake warning light
83 Handbrake switch
84 Oil pressure warning light
85 Oil pressure switch
86 Ignition warning light
87 Ballast resistor wire
88 Ignition coil – 6 volt
89 Ignition distributor
90 Heated rear screen switch
91 Heated rear screen
92 Heated rear screen warning light
93 Seat belt warning light
94 Passenger seat switch
95 Passenger belt switch
96 Driver belt switch
97 Reverse lamp switch
98 Reverse lamp
99 Stop lamp switch
100 Stop lamp
101 Direction indicator flasher unit
102 Direction indicator switch
103 Indicator warning light – L.H.
104 Indicator warning light – R.H.
105 Front flasher lamp – L.H.
106 Front flasher repeater lamp connection – L.H.
107 Rear flasher lamp – L.H.
108 Front flasher lamp – R.H.
109 Front flasher repeater lamp connection – R.H.
110 Rear flasher lamp – R.H.

Colour code

B	Black	P	Purple
G	Green	R	Red
K	Pink	S	Slate
LG	Light green	U	Blue
N	Brown	W	White
O	Orange	Y	Yellow

Wire colour codes of flasher lamp circuit

Wire identity	Item	Right-hand steering		Left-hand steering		
		Heater vehicles	Air conditioning vehicles	Heater vehicles	Left-hand steering Air conditioning vehicles	
*1	L.H. supply	G/LG	GR	LG/G	GR	
*2	L.H. front lamp	G/LG	GR	LG/G	GR	
*3	L.H. rear lamp	GW	GR and LG	GW	GR and LG	
*4	R.H. supply	LG/G	GW	G/LG	GW	
*5	R.H. front lamp	LG/G	GW	G/LG	GW	
*6	R.H. rear lamp	GR and LG	GW and LG	GW	GW	

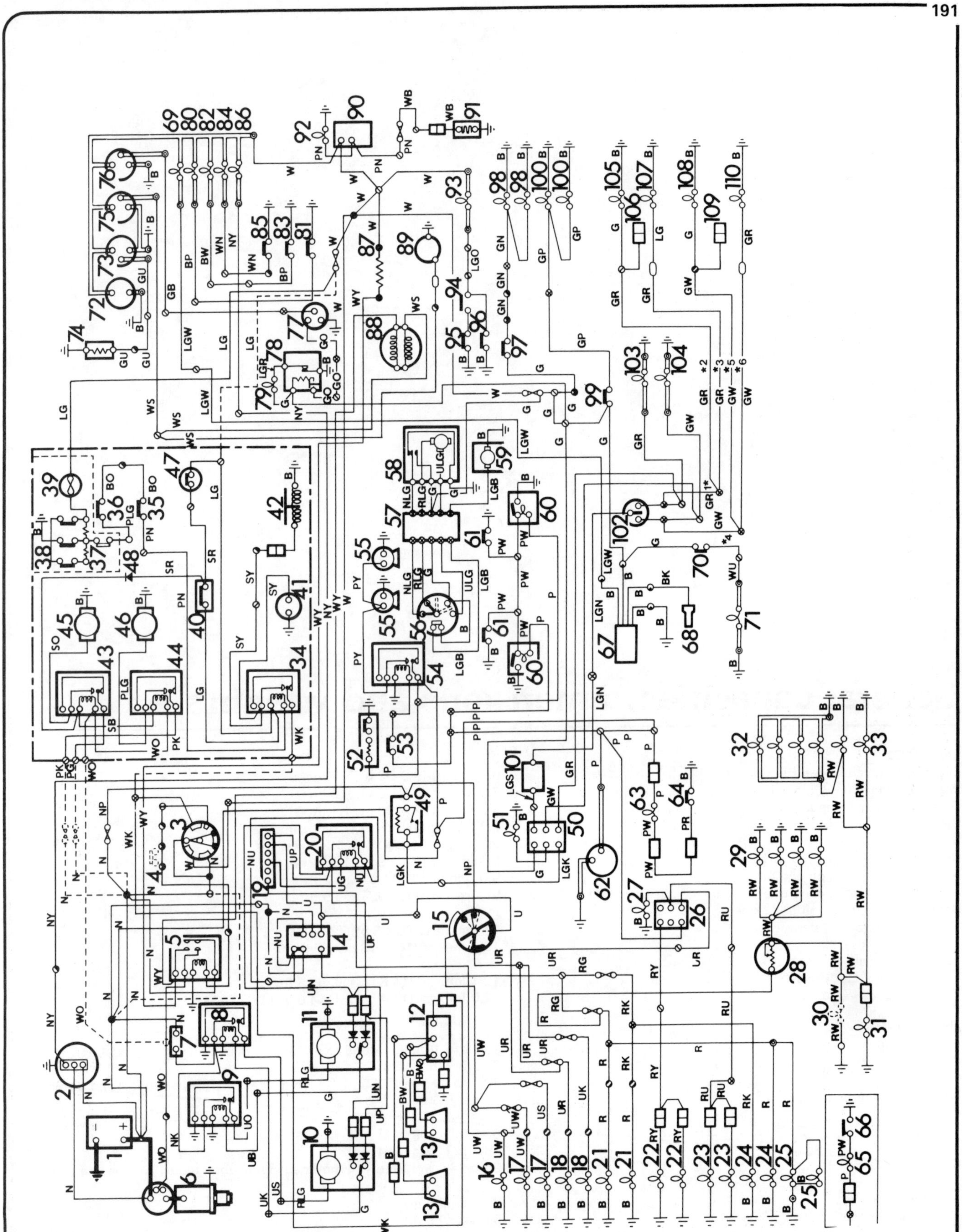

General capacities, dimensions and weights

Capacities
Engine oil capacity, with filter	8.0 Imp pt (9.5 US pt, 4.5 litres)
Engine oil capacity, without filter	7.0 Imp pt (8.5 US pt, 4.0 litres)
Gearbox, 4-speed manual, from dry	2.0 Imp pt (2.4 US pt, 1.1 litres)
Gearbox, 5-speed manual, from dry	2.7 Imp pt (3.3 US pt, 1.5 litres)
Rear axle (4-speed gearbox fitted), from dry	2.35 Imp pt (2.7 US pt, 1.3 litres)
Rear axle (5-speed gearbox fitted) from dry	2.0 Imp pt (2.4 US pt, 1.1 litres)
Automatic transmission without oil cooler	9.0 Imp pt (11.0 US pt, 5.2 litres)
Automatic transmission with oil cooler	9.5 Imp pt (11.6 US pt, 5.5 litres)
Cooling system, reservoir and heater:	
Expansion tank system	13 Imp pt (15.6 US pt, 7.4 litres)
Header tank system	13.4 Imp pt (16.1 US pt, 7.6 litres)
Fuel tank capacity:	
Federal and California	12.1 Imp gal (14.6 US gal, 55.3 litres)
Others	12.0 Imp gal (14.4 US gal, 54.5 litres)

Dimensions
Overall length	160.0 to 164.3 in (4065 to 4173 mm) depending upon model
Width	66.2 in (1681 mm)
Height	49.4 to 49.9 in (1255 to 1268 mm) depending upon model and loading
Wheelbase	85 in (2160 mm)
Track:	
Front	55.5 in (1409 mm)
Rear	55.3 in (1404 mm)
Ground clearance	3.5 to 4.5 in (90 to 114 mm) depending upon model and loading
Turning circle:	
Between kerbs	29.0 ft (8.8 m)
Between walls	31.5 ft (9.5 m)

Weights
Unladen weight:	
Minimum	2341 lb (1062 kg)
Maximum	2509 lb (1138 kg)
Roof rack load, maximum	110 lb (50 kg)
Towing capacity	1680 lb (762 kg)

Conversion factors

Length (distance)
Inches (in)	X	25.4 = Millimetres (mm)	X	0.039	= Inches (in)
Feet (ft)	X	0.305 = Metres (m)	X	3.281	= Feet (ft)
Miles	X	1.609 = Kilometres (km)	X	0.621	= Miles

Volume (capacity)
Cubic inches (cu in; in³)	X	16.387 = Cubic centimetres (cc; cm³)	X	0.061	= Cubic inches (cu in; in³)
Imperial pints (Imp pt)	X	0.568 = Litres (l)	X	1.76	= Imperial pints (Imp pt)
Imperial quarts (Imp qt)	X	1.137 = Litres (l)	X	0.88	= Imperial quarts (Imp qt)
Imperial quarts (Imp qt)	X	1.201 = US quarts (US qt)	X	0.833	= Imperial quarts (Imp qt)
US quarts (US qt)	X	0.946 = Litres (l)	X	1.057	= US quarts (US qt)
Imperial gallons (Imp gal)	X	4.546 = Litres (l)	X	0.22	= Imperial gallons (Imp gal)
Imperial gallons (Imp gal)	X	1.201 = US gallons (US gal)	X	0.833	= Imperial gallons (Imp gal)
US gallons (US gal)	X	3.785 = Litres (l)	X	0.264	= US gallons (US gal)

Mass (weight)
Ounces (oz)	X	28.35 = Grams (g)	X	0.035	= Ounces (oz)
Pounds (lb)	X	0.454 = Kilograms (kg)	X	2.205	= Pounds (lb)

Force
Ounces-force (ozf; oz)	X	0.278 = Newtons (N)	X	3.6	= Ounces-force (ozf; oz)
Pounds-force (lbf; lb)	X	4.448 = Newtons (N)	X	0.225	= Pounds-force (lbf; lb)
Newtons (N)	X	0.1 = Kilograms-force (kgf; kg)	X	9.81	= Newtons (N)

Pressure
Pounds-force per square inch (psi; lbf/in²; lb/in²)	X	0.070 = Kilograms-force per square centimetre (kgf/cm²; kg/cm²)	X	14.223	= Pounds-force per square inch (psi; lbf/in²; lb/in²)
Pounds-force per square inch (psi; lbf/in²; lb/in²)	X	0.068 = Atmospheres (atm)	X	14.696	= Pounds-force per square inch (psi; lbf/in²; lb/in²)
Pounds-force per square inch (psi; lbf/in²; lb/in²)	X	0.069 = Bars	X	14.5	= Pounds-force per square inch (psi; lbf/in²; lb/in²)
Pounds-force per square inch (psi; lbf/in²; lb/in²)	X	6.895 = Kilopascals (kPa)	X	0.145	= Pounds-force per square inch (psi; lbf/in²; lb/in²)
Kilopascals (kPa)	X	0.01 = Kilograms-force per square centimetre (kgf/cm²; kg/cm²)	X	98.1	= Kilopascals (kPa)

Torque (moment of force)
Pounds-force inches (lbf in; lb in)	X	1.152 = Kilograms-force centimetre (kgf cm; kg cm)	X	0.868	= Pounds-force inches (lbf in; lb in)
Pounds-force inches (lbf in; lb in)	X	0.113 = Newton metres (Nm)	X	8.85	= Pounds-force inches (lbf in; lb in)
Pounds-force inches (lbf in; lb in)	X	0.083 = Pounds-force feet (lbf ft; lb ft)	X	12	= Pounds-force inches (lbf in; lb in)
Pounds-force feet (lbf ft; lb ft)	X	0.138 = Kilograms-force metres (kgf m; kg m)	X	7.233	= Pounds-force feet (lbf ft; lb ft)
Pounds-force feet (lbf ft; lb ft)	X	1.356 = Newton metres (Nm)	X	0.738	= Pounds-force feet (lbf ft; lb ft)
Newton metres (Nm)	X	0.102 = Kilograms-force metres (kgf m; kg m)	X	9.804	= Newton metres (Nm)

Power
Horsepower (hp)	X	745.7 = Watts (W)	X	0.0013	= Horsepower (hp)

Velocity (speed)
Miles per hour (miles/hr; mph)	X	1.609 = Kilometres per hour (km/hr; kph)	X	0.621	= Miles per hour (miles/hr; mph)

*Fuel consumption**
Miles per gallon, Imperial (mpg)	X	0.354 = Kilometres per litre (km/l)	X	2.825	= Miles per gallon, Imperial (mpg)
Miles per gallon, US (mpg)	X	0.425 = Kilometres per litre (km/l)	X	2.352	= Miles per gallon, US (mpg)

Temperature

Degrees Fahrenheit (°F) = (°C x $\frac{9}{5}$) + 32

Degrees Celsius (Degrees Centigrade; °C) = (°F − 32) x $\frac{5}{9}$

*It is common practice to convert from miles per gallon (mpg) to litres/100 kilometres (l/100km), where mpg (Imperial) x l/100 km = 282 and mpg (US) x l/100 km = 235

Index

A

Air conditioning
    removal and refitting — 158
    adjustment — 158
Air cleaner
    removal and replacement — 41
Alternator
    description — 110
    drive belt adjustment — 110
    functional check — 110
    overhaul — 110
    refitment — 110
    removal — 110, 168
Antifreeze solution — 59
Automatic transmission
    description — 88
    downshift cable — 88, 90, 164
    extension rear oil seal — 89
    fault diagnosis — 92
    front brake band — 88
    front servo — 88
    governor — 90
    rear brake band — 88
    rear extension — 89
    rear servo — 89
    removal and replacement — 88
    restrictor valves and by-pass pipe — 90
    road test — 92
    selector rod — 90, 164
    specifications — 73
    stall test — 90
    starter inhibitor/reverse lamp switch — 90
    torque wrench settings — 73
    transmission sump — 90

B

Big-end bearings
    removal — 20
    renovation — 23
Bodywork
    description — 141
    doors — 169
    facia panel — 149
    gearbox tunnel cover — 151
    hood — 170
    maintenance — 141
    major repairs — 145
    minor repairs — 144
    seats — 151
Body repair sequence (colour) — 142 and 143
Bonnet
    lock — 145
    lock control cable — 145
    removal and replacement — 145
Boot
    lid — 149
    lid lock and striker — 149
Braking system
    bleeding — 100
    description — 99
    fault diagnosis — 106
    front calipers — 101
    front disc shields — 100
    front discs — 100
    front pads — 100
    handbrake — 104
    hydraulic pipes — 103
    master cylinder — 103
    pedal — 104
    rear backplates — 102
    rear load sensing valve — 103
    rear shoes — 101
    rear wheel cylinders — 102
    servo unit — 105
    specifications — 99
    stop light switch — 121
    torque wrench settings — 99
Bumpers — 145

C

Camshaft
    refitment — 26
    removal — 19
    renovation — 24
Carburation
    balancing — 43, 46
    choke cable renewal — 50
    description — 41
    fault diagnosis — 56
    overhaul — 44, 46
    removal and replacement — 41, 159
    specifications — 41, 159
    throttle cable renewal — 50
    tuning — 43, 46, 49
Carpets
    removal and refitting — 170
Clutch
    bleeding — 67
    description — 67
    fault diagnosis — 71
    judder — 71
    master cylinder — 68
    pedal and support bracket — 104
    refitment — 68
    release mechanism — 70
    removal — 68
    slave cylinder — 68
    slip — 71
    specifications — 67
    spin — 71
    squeal — 71
    torque wrench settings — 67
Condenser
    removal and replacement — 60
    testing — 60
Connecting rods
    dismantling — 20
    reassembly — 28
    removal — 20
Connecting rods to crankshaft reassembly — 28
Contact breaker points
    adjustment — 59
    cleaning — 59
    removal and replacement — 59
Cooling system
    description — 35
    draining — 35
    fault diagnosis — 39
    filling — 36
    flushing — 36
    header tank — 158
    specifications — 35
    torque wrench settings — 36
Crankshaft
    removal — 20
    renovation — 23
    replacement — 26
Cylinder bores renovation — 23

Index

Cylinder head
    decarbonisation — 25
    reassembly — 26
    removal — 18
    replacement — 31

D

Decarbonisation — 25
Distributor
    removal — 61
    repair — 61, 62
    replacement — 61
Door
    dismantling — 148
    hinges — 148
    rattles — 147
    removal and replacement — 147
    striker plate — 148

E

Electrical system
    description — 109
    fault diagnosis — 127
    flasher circuit — 116
    lamp bulbs — 116
    specifications — 107
    steering column mounted switches — 119
Emission control — 161
Engine
    components examination — 23
    components renovation — 23
    crankshaft pulley removal — 19
    crankshaft rear oil seal replacement — 20
    description — 15
    fault diagnosis — 34
    final assembly — 33
    flywheel removal — 19
    flywheel renovation — 25
    flywheel replacement — 28
    initial start-up after overhaul — 33
    inlet manifold replacement — 33
    jackshaft removal — 20
    jackshaft renovation — 24
    jackshaft replacement — 28
    major operations
        engine in the car — 16
        engine removed from car — 17
    oil strainer replacement — 31
    oil transfer adaptor
        removal — 20
        replacement — 31
    reassembly — 26
    removal with gearbox — 18
    replacement — 33
    separating from gearbox — 18
    small end bearings renovation — 24
    specifications — 13
    torque wrench settings — 15
Exhaust system
    general — 51
    removal — 51

F

Fan
    belt adjustment — 36
    removal and replacement — 36
Fault diagnosis
    automatic transmission — 92
    braking system — 106
    carburation — 56
    clutch — 71
    cooling system — 39
    electrical system — 126
    engine — 34
    fuel system — 56
    gearbox (manual) — 88
    ignition system — 66
    rear axle — 98
    steering — 140
    suspension — 140
Fuel injection system
    absorption canisters — 161
    air cleaner — 159
    cold start injector — 161
    depressurising — 160
    fuel line filters — 160
    induction manifold — 160
    injectors — 161
    plenum chamber — 160
    pressure regulator — 160
Fuel pump
    description — 41
    overhaul — 41
    removal — 41
    replacement — 41
    testing — 41
Fuel system
    description — 41, 159
    fault diagnosis — 56
    specifications — 40
Fuel tank
    removal and replacement — 51
Fuses — 119

G

Gearbox (manual 4-speed)
    description — 73
    dismantling — 74
    examination — 76
    fault diagnosis — 88
    input shaft — 73
    mainshaft — 78
    reassembly — 78
    removal and replacement — 73
    renovation — 76
    specifications — 72
    torque wrench settings — 72
Gearbox (manual 5-speed)
    description — 73
    dismantling — 80
    examination — 76, 82
    fault diagnosis — 88
    mainshaft — 82
    reassembly — 84
    removal and replacement — 73
    renovation — 76, 82
    specifications — 72
    top cover and remote control assembly — 86
    torque wrench settings — 72
Gudgeon pin renovation — 24

H

Headlamps — 114
Heater
    airflow control cable — 153
    fan motor switch — 153
    removal and replacement — 153, 170
    water valve — 153
Hubs (front)
    bearing adjustment — 135
    bearing removal and replacement — 135
Hubs (rear) — 96

Index

I

Ignition system
- description — 58, 161
- fault diagnosis — 66
- specifications — 57
- timing — 64

M

Main bearings
- removal — 20
- renovation — 23

O

Oil filter
- removal — 20
- replacement — 31

Oil pump
- dismantling — 20
- removal — 20
- renovation — 25
- replacement — 31

P

Pedal box
- removal and refitting — 170

Pistons
- dismantling — 20
- reassembly — 28
- removal — 20
- renovation — 24
- replacement — 28

Piston rings
- renovation — 24
- replacement — 28

Propeller shaft
- description — 93
- torque wrench settings — 93

R

Radiator
- removal and replacement — 36

Rear axle
- description — 95
- differential — 98
- halfshafts — 95
- pinion oil seal — 98, 164
- removal and replacement — 98
- specifications — 95
- torque wrench settings — 98

Rear wheel
- hub studs
  - removal and refitting — 164

S

Spark plugs — 64
Spark plug chart (colour) — 65
Starter motor
- description — 111
- removal and replacement — 111
- repair — 112
- testing — 112

Steering
- column — 139
- column lock and ignition switch — 140
- description — 133
- fault diagnosis — 140
- front wheel alignment — 138
- rack pinion gear — 137
- specifications — 132
- torque wrench settings — 133
- wheel — 138

Sump
- removal — 20
- replacement — 25

Suspension
- description — 133
- fault diagnosis — 140
- specifications — 132
- torque wrench settings — 133

Suspension (front)
- anti-roll bar — 133
- lower wishbone arm and radius rod — 136
- spring and damper — 133
- vertical link balljoints
  - lower — 136
  - upper — 136

Suspension (rear)
- anti-roll bar — 133
- arm — 136
- damper — 137
- radius rod — 137

T

Tappets renovation — 24
Thermostat
- removal — 36
- replacement — 36
- testing — 36

Timing chain and sprockets
- removal — 20
- renovation — 24
- replacement — 28

Tyre pressures — 132

U

Universal joints
- fitting new bearings — 93
- repair — 93

V

Valves
- clearance checking — 26
- grinding — 24
- removal — 20, 157
- timing — 31, 157

W

Water pump
- overhaul — 38
- removal — 38
- replacement — 38

Water pump/thermostat housing connecting tube — 25
Wheels — 132
Windscreen glass replacement — 152
Windscreen washer — 124
Windscreen wipers
- arms — 124
- blades — 124
- description — 124
- mechanism — 125
- motor — 125

Wiring diagrams — 128 to 130, 171 to 191

Printed by
Haynes Publishing Group
Sparkford Yeovil Somerset
England